LEONARDO

TO THE

INTERNET

JOHNS HOPKINS

STUDIES IN THE HISTORY

OF TECHNOLOGY

Merritt Roe Smith,

Series Editor

THOMAS J. MISA

Leonardo

TECHNOLOGY & CULTURE

FROM THE RENAISSANCE TO THE PRESENT

to the
Internet

THE JOHNS HOPKINS UNIVERSITY PRESS
Baltimore & London

© 2004 The Johns Hopkins University Press
All rights reserved. Published 2004
Printed in the United States of America on acid-free paper
9 8 7 6 5 4 3 2 1

The Johns Hopkins University Press
2715 North Charles Street
Baltimore, Maryland 21218-4363
www.press.jhu.edu

Library of Congress Cataloging-in-Publication Data
Misa, Thomas J.
    Leonardo to the Internet : technology and culture from the
Renaissance to the present / Thomas J. Misa.
        p. cm. — (Johns Hopkins studies in the history of technology)
        ISBN 0-8018-7808-X (hardcover : alk. paper) —
ISBN 0-8018-7809-8 (pbk. : alk. paper)
        1. Technology—History.   2. Technology and civilization.
I. Title. II. Series.
T15.M575 2004
609'.03—dc21   2003006813

A catalog record for this book is available from the British Library.

CONTENTS

Figures & Tables

TABLES

PREFACE

IN THIS BOOK I explore the varied character of technologies in the West over the half-millennium from the Renaissance to the present. The investigation spans the preindustrial past; the age of scientific, political, and industrial revolutions; and on to more recent topics, such as imperialism, modernism, war, and global culture. That particular span of time seems well suited to provide a solid empirical base for exploring the wide-ranging notions about technology that I consider in the final chapter.

This study began more than a decade ago when I was trying to understand the work of Leonardo da Vinci. A little background reading convinced me that the time-honored image of Leonardo as an artist and anatomist, who did some nice-looking technical drawings on the side, just did not capture his life. Leonardo spent his most active years working as a technologist and engineer. It is scarcely an exaggeration to say that his famous paintings and strikingly realistic anatomical drawings were done in the periodic lulls between his technology projects. I puzzled further over the character of Leonardo's technical work. Was he really, as some enthusiasts claimed, the "prophet of automation," the inventor of labor-saving machines (and such wonders as helicopters, airplanes, and automobiles) that catapulted Europe from the "dark ages" directly into the modern era? Thinking about these questions, I began to see a distinctive focus in Leonardo, and in the numerous engineers with whom he shared notebook drawings and technical treatises. The technological activities of these Renaissance engineers related closely to the concerns of the Renaissance courts and city-states that commissioned their work. I failed to find Leonardo much concerned with labor-saving or "industrial" technologies, and for that matter few of his technological projects generated wealth at all. Quite the opposite. Leonardo's technologies were typically wealth-*consuming* ones: the technologies of city building, courtly entertainments and dynastic display, and war making.

The Renaissance court system was the conceptual key. While it is common knowledge that Johann Gutenberg invented the Western system of moveable type printing, it is not well known that he himself was a court pensioner. Printing shops throughout the late Renaissance depended, to a surprising extent, on court-generated demand. Even printed books on technical subjects were objects of courtly patronage. I began to appreciate that Renaissance courts across Western Europe, in addition to their well-known support of famous artists, scientists, and philosophers, were at the time the dominant patrons of the most prominent technologists. These courts included royal courts in Spain and France, ambitious regional rulers in Italy, the papal court in Rome, and court-like city-states such as Florence and Milan. The technical projects they commissioned, from the Florence cathedral to the mechanical robots for courtly entertainment, as well as the printed works on science, history, philosophy, religion, and technology, created and themselves constituted Renaissance culture. This heady mix of technology, politics, and culture is the subject of chapter 1.

There are good reasons to see the industrial revolution as a watershed in world history, but our time-worn inclination to seize on industrial technologies as the only ones that really matter has confounded a proper understanding of the great commercial expansion that followed the Renaissance. Chapter 2 discusses these developments. Economic historians persistently ask the same questions about the great Dutch Golden Age in the seventeenth century: How fast did the Dutch economy grow, and why didn't it industrialize? The Dutch not only had rising per capita incomes and a healthy and diverse national economy; they were also the architects and chief beneficiaries of the first multicentered global economy. Again, on close inspection I found a distinct character in Dutch technological activities. Commerce, like the courts, fostered distinctive but nonindustrial technologies. Dutch merchants and engineers were highly attuned to generating wealth and took active steps to sharpen their focus on making high-quality items with relatively high-paid labor. The typical Dutch cloth was not cheap cotton—the prototypical industrial product—but high-priced woolens, linens, and mohairs. Dutch technologies formed the sinews for that country's unprecedented international trading system, which included shipbuilding, sugar refining, instrument making, and innovations in finance, like joint-stock companies and stockmarkets. I began not only to think of technologies as located historically and spatially in a particular society and shaped by that society's ideas of what was possible or desirable, but also to see how these technologies evolved to shape the society's social and cultural developments. To capture this two-way influence, I took up the notion of distinct "eras" of technology and culture as a way of organizing the material for this book.

This notion of distinct eras provides a kernel for new practical insight into our own social and cultural prospects. My view on technology argues against the common billiard-ball model propounded by many popular writers and journalists who see technologies coming from "outside" society and culture and "impacting" them—for good or ill. Indeed, the question whether technologies are outside a society or within it is a far from trivial matter. If technologies come from outside, the only critical agency open to us is slowing down their inevitable triumph—a rearguard action at best. By contrast, if technologies come from *within* society and are products of on-going social processes, we can, in principle alter them—at least modestly—even as they change us. This book presents an extended empirical evaluation of this question. It shows, in many distinct eras, historical actors actively choosing and changing technologies in an effort to create or sustain their ideas of a good life or desirable social and cultural developments (whatever these happened to be).

With these issues in mind, I came to the industrial revolution in Britain with a new interest in comprehending "industrial society," which I explore in chapter 3. An older view—that the industrial revolution radically transformed Britain in the few decades around 1800 and soon thereafter forcibly propelled the world into the modern age—has just not stood the test of time. This version of the billiard-ball model of how technology changes society is, in its simple one-way formulation, empirically false. In the last two decades or so, historians have learned too much about the vast range of historical experiences during industrialization and the dynamics of industrial change and economic growth to sustain the older view of the industrial revolution. For example, we know now that surprisingly few industrial workers in industrial Britain actually labored in large-scale factories (only about one worker in ten) and that there were simply too few steam engines to do anything like "transform" the entire economy—until well into the nineteenth century.

In a convulsion of skepticism and debunking, historians have all but thrown out the concept of an industrial revolution. I would like to revive it, for I think we can see a distinct and historically specific logic that shaped technologies during the early industrial revolution. In industrial era Britain there were precious few Dutch-style technologists focusing on high-quality materials and high-paid labor. Instead, the predominant focus of British technologists was, let's say, industrial: cutting costs, boosting output, and saving labor. Inventions of the era embodied these socioeconomic goals. Cheap cotton cloth, and lots of it, made by ill-paid factory "hands," was a representative product of industrial-era Britain. If mechanizing industry was not the highest calling in life, as Victorian moralists repeatedly warned, it was nevertheless a central and defining purpose for inventors, engineers, and industrialists of the time. Beyond Britain, commentators and technologists sometimes looked to copy British models of industry but more frequently adapted industrial technologies to their own economic and social contexts. The result was a variety of paths through the industrial revolution.

Given these ideas about court, commerce, and industry as defining purposes for technology, I began thinking more broadly about what helped define technologies in the next two centuries—closer to home, as it were. The task became more difficult. It was impossible to isolate a single distinct "type" of society with a corresponding set of technologies. The legacy of the industrial revolution, it seemed, was not a single "industrial society" with a fixed relationship to technology but rather a multidimensional society with a variety of purposes for technology. Between the middle decades of the nineteenth century and the early decades of the twentieth, at least three varied purposes of technology can be identified— the themes of chapters 4, 5, and 6.

The first of these technology-intensive activities to fully flower was empire building, the effort by Europeans and North Americans to extend economic and political control over wide stretches of land abroad or at home. This is the subject of chapter 4. Imperialists faced unprecedented problems in penetrating unfamiliar lands, often in the face of determined resistance by native peoples, and in consolidating military and administrative control over these lands and their peoples. Steamships, telegraphs, and transcontinental railroads were among the technologies that made imperialism effective and affordable. The "gunboat diplomacy" deployed by the British with great success in China, India, and Africa against poorly armed native peoples depended on the construction of iron-hulled, steam-driven, shallow-draft vessels that were heavily armed. (Also in the mid-nineteenth century, and in parallel with steamboats, the use of quinine gave Europeans for the first time reasonable odds against endemic malaria in Africa and Asia.) It would be foolish not to recognize the economic dimensions of the imperialist venture, since Britain's factory-made cotton textiles were shipped off to captive markets in India to be exchanged for tea and raw cotton (with a side trade in opium).

All the same, more was at play during the imperial era than just the disposing of surplus factory goods or the importing of cheap raw materials, important though these were. No one at the time tried to justify or defend empire in strictly economic terms. Feelings of national and imperial pride, the presumed imperatives of linking colonies to the homeland, the often-bizarre economics of empire (where for instance tremendously expensive steamboats or railroads "saved" money in transporting military forces or colonial officials)—these were the ways Britain's imperialists coaxed from taxpayers money for such extravagant ventures as the round-the-world telegraph system, easily the most demanding high-technology effort of the nineteenth century. Long-distance repeating telegraphs, efficient coal-burning steamships, and undersea telegraph cables tempted imperial officials to exert oversight and control over far-flung possessions. Many imperialists credited the telegraph network with "saving" British rule in India during the Mutiny of 1857–58. The same reasoning—national pride, imperial imperatives, and the economics of empire—helps explain the urgency behind the transcontinental railroads in India, North America, and South Africa. Economics as traditionally understood had little to do with empire or imperial-era technologies.

A second impulse in technology gathering force from the 1870s onward lay in the application of science to industry and the building of large systems of technology, the subject of chapter 5. For the first time, in the rise of the science-based chemical and electrical industries, scientific knowledge became as important as land, labor, and capital as a factor of production. The new importance of science led to the rise of fundamentally new social institutions: research-based universities (universities *per se* were centuries old), government research institutes, and industrial research-and-development laboratories, all of which appeared on a large scale in Germany before 1900 and in the United States a little later. The rise of the chemical, electrical, steel, and petroleum industries, and the associated large corporations that funded and managed them, constituted a second industrial revolution. Britain, the first industrial nation, played a surprisingly small role in the movement.

In chapter 6 I examine the intense interactions between modern technology and modern aesthetics during the first half of the twentieth century. This inquiry clarifies the ways technology in the twentieth century formed not only the backdrop of our economy but also the foreground of our daily lives. The twentieth-century modern movement was built on technological capabilities developed during the science-and-systems era, and yet it led to new and distinctive cultural results. The achievement of mass-produced steel, glass, and other "modern materials" around 1900 reshaped the aesthetic experience of working or walking in our cities and living in our homes. These modern materials were the precondition and artistic inspiration for the modern movement in art and architecture that flourished between 1900 and 1950. Modernism led to avant-garde designs for public-housing blocks and office buildings as well as museums, hospitals, and schools. The movement, through its association with German household reformers and with New York's Museum of Modern Art and other self-appointed arbiters of "good taste," shaped public fascination with new modernist designs for domestic appliances.

The book's middle chapters, besides setting down some engaging stories, advance my argument concerning the "question of technology." Highly articulate figures—the artists, architects, and household reformers of the modern movement—self-consciously embraced technology to achieve their visions of housing the poor, embracing modern urban life, and what they saw as enhancing modern society's cultural and spiritual development. Even if you find the modernists' enthusiasm for technology a bit naïve today, you must allow that they both shaped specific technologies and conditioned cultural developments. Mine is not a relentlessly optimistic worldview. In these middle chapters you will also find industrial technologies implicated in the conditions of filthy, disease-ridden cities and imperial technologies implicated in the slaughtering of natives in India and North America. Technology has been and can be a potent agent in disciplining and dominating. I also discuss the modernists' troubling embrace of a fixed "method" of creativity.

My thinking on this book began during the waning days of the Cold War. Since then, it has become easier to see clearly how important the superpowers' military services were in finding and funding technologies of real or imagined military utility. The "command economies" that took shape during World War II fanned the development of countless technological innovations, of which atomic power, radar, and computers are only the best known examples, as chapter 7 recounts. In the Cold War decades, scientists and engineers learned that the military services had the deepest pockets of all potential technology patrons. For dreamers and schemers of the most massive technology projects, the military was the main chance. The story is told of the Nazi rocket scientist Wernher von Braun preparing his laboratory for surrender in the chaotic closing days of World War II. "Most of the scientists were frightened of the Russians, they felt the French would treat them like slaves, and the British did not have enough money to afford a rocket program. That left the Americans."[1]

It is worth recalling the social and political changes brought about during the military-dominated era of technology. In all the great powers during the Cold War era, state-imposed secrecy pervaded the weapons, aerospace, nuclear, and intelligence sectors. (With the opening of archives we have learned how shockingly similar were the Soviet, American, and French nuclear technocrats, sharing kindred visions of limitless energy, technology-induced social change, and at times contempt of safety concerns, even as they were divided by great-power rivalries.) In the United States the fortunes of research universities like MIT and Stanford and industrial contractors like Bell Labs, Boeing, RCA, IBM, and General Electric depended to a surprising degree on the Pentagon. With the end of the Cold War, the sharp contraction of military research and development budgets traumatized technology-based companies, universities, and government institutes.[2] In the West we are comparatively lucky. In the former Soviet Union, frighteningly enough, high-level nuclear technicians with ready access to dangerous materials have been told in no uncertain terms to find work elsewhere.

The hardest history to write is that of our own time, and yet I believe that "globalization," or "global culture," is a force that oriented technology and society in the final three decades of the twentieth century. This is the topic of chapter 8. Think about the world as recently as 1970, without pervasive fax machines, automatic teller machines, and cellular phones. Take away ready access to email and the Internet and bring back the library's card catalogue. Hike the charge fifty-fold for an overseas telephone call. Do away with NASDAQ, Microsoft, Dell Computer, and Amazon.com. Make it impossible for a middle-class Western person to invest retirement savings in anything but domestic government bonds. For that matter, keep the faith that a foreign takeover of the company you work for is impossible. Now ease your way back into the current world. Where an after-hours phone call to a Texas bank is likely to be answered by a call-center worker in India. Where anyone with an Internet connection can risk a stock investment in the Finnish cellular-phone giant Nokia. Where Daimler owns Chrysler and Bertelsmann owns Random House; where Disney

owns a choice piece of Paris, and where McDonald's owns a small slice
of everywhere. No wonder that the coming of a global culture has brought
both exhilaration and fear.

The intellectual framework set out in this book is necessarily tenta-
tive, really a way of thinking about our relationships with technology and
the perennially puzzling "question of technology," to which I return in the
final chapter. Yet, by adopting this way of thinking, several perplexing observations about technology can be brought into sharper focus. At least
since Alvin Toffler's best-selling *Future Shock* (1970) pundits perennially
declare that the pace of technology is somehow quickening and that technology is forcing cultural changes in its wake, that our plunge into the future is driven by technology gone out of control. The evidence cited for
these conclusions is invariably Moore's Law, an observation of electronics pioneer Gordon Moore that every eighteen months the speed of computer chips doubles. I appreciate the obvious difference between my first,
stand-alone Macintosh from 1984 and my present networked computer
that can process words, images, sounds, and movies—and in its off hours
aid the search for extraterrestrial intelligence.[3] And it is mind-boggling
that the number of transistors manufactured each year now exceeds the
number of individual *letters* printed each year in books, newspapers, and
magazines.

All the same, I'd like to propose Misa's Corollary to Moore's Law. My
corollary states that the size of computer operating systems and software
applications has doubled *at the same pace* as the operational speed of
computer chips, soaking up the presumed power of the hardware and
blunting its impact. Here is some evidence for my proposition: my present vintage-2000 word-processing program takes up fully 1,000 times the
disk-storage space as did my vintage-1985 one, even though both feature
pull-down menus and a graphics-friendly display, fancy fonts and for-
matting, and many other features. For fifteen years, amazingly enough,
the program has doubled in size each and every eighteen months. My
computer operating systems have also doubled in size each eighteen
months; it would take no less than 1,000 old-style floppy disks to store the
latest monstrous-sized one. Because of its so-called bloatware, my present computer—despite its many megahertz central-processing unit—
takes much longer to start up than does my vintage-1984 machine. The
same for launching the respective word processing programs. (This is a
genuine puzzle to me. Except for the surreptitious "Easter eggs," or unau-
thorized computer code that programmers embed as inside jokes [see

www.eeggs.com], I have been unable to locate any sensible explanation for the excesses of this bloatware.) Overall, I am doubtful that computing chips alone can help us comprehend the perception of an increasing pace of change, which reaches back at least to the "Science Holiday" movement of the 1930s that sought to ban science-induced technological changes precisely because of their disruptive social and economic consequences.

Instead of a crude technological determinism derived from Moore's Law, I would trace our perception of quickening to a split between our normative expectations for technology and what we observe in the world around us. It is not so much that our technologies are changing especially quickly but that our sense of what is "normal," about technology and society, cannot keep pace. We know that regulatory laws cannot keep up with the practices of cyberspace, that ethical norms are challenged by cloning and biotechnologies, that our very identities are created and constituted by surveillance technologies as well as cell phones, and that scarcity-based economics may be entirely overthrown by nanotechnologies.

In this regard, the longer-duration "eras" of technology discussed in this book can be interpreted as a deep foundation for our cultural norms. These eras appear to be shortening: the Renaissance spanned nearly two centuries, while the twentieth century alone saw the eras of science and systems, modernism, war, and global culture. It is worth mentioning a quickening also in the *self-awareness* of societies—our capacities to recognize and comprehend change are themselves changing. While the Renaissance was not named until two centuries after the fact (in the 1830s) and the term *industrial revolution* became part of the English language a half-century after the Manchester region had dozens of huge factories, the telephone, invented in 1876, took exactly one decade to register culturally. Just four years after Moscow's first telephone station opened, Anton Chekhov published a short story "On the Telephone" (1886), about a hapless early user trying to make a restaurant reservation with the new machine. By comparison, the notion of "cyberspace" arrived in the 1980s with William Gibson's science fiction novel *Neuromancer* almost before the fact of cyberspace. This self-awareness of major historical change is clearly an instance of "reflexive" modernization in sociologist Ulrich Beck's sense. In this way, then, these eras do capture something real in our historical experience.

Finally, while this book traces a long history from Leonardo to the Internet, it in no way argues "the triumph of the present." Indeed, I am

pretty sure that our next generation's relationships with technology will
not be identical to our recent one, although in what ways they may be
different is risky to prophesy. Indeed, this work has found its mark if it
prompts readers to open up a mental space for thinking both more widely
and more deeply about technology. I would hazard the guess that the perceived quickening of the pace of change will continue, and as a citizen I hope
that we will be able to channel technology toward socially productive purposes. We face immense challenges in climate change, international development, and economic disparities, as well as in many other areas, and
these challenges have not engaged the potential of the world's technologists. This is a pity. I hope that readers, by the end of this book, will understand the complex reasons why this is so. In a nutshell, it goes like this.
Societies, pursuing distinct goals and aspirations, have chosen and sustained certain technologies; and these technologies have powerfully
molded their economic, social, and cultural capabilities. What type of society we wish for the future is at once an open question and a constrained
choice. While our societal debate must be an open one, accessible by all
members, practically speaking our choices about our society's tomorrow
will be framed by the choices we make about our technologies today.

ACKNOWLEDGMENTS

I AM INDEBTED to many people who shared their knowledge and perspectives with me. For comments on specific chapters, I would like to thank Henk van den Belt, Mikael Hård, Kevin Harrington, Cheryl Ganz, Richard John, Thomas Hughes, Donna Mehos, Joel Mokyr, Margaret Power, Merritt Roe Smith, and Ed Todd. Arne Kaijser shared with me his knowledge and enthusiasm about Dutch technology. The chapters also benefited from comments at seminars at University of Pennsylvania, Newberry Library, Technical University of Eindhoven, Science Museum (London), Institute for Architecture and Humanities (Chicago), and Massachusetts Institute of Technology. I tried out ideas about writing large-scale history at the 1999 summer school in history of technology, at Bjerringbro, Denmark, and at a plenary session called "How Do We Write the History of Technology?" at the 1999 annual meeting of the Society for the History of Technology, in Detroit.

For assistance with finding and gathering illustrations, I have several persons to thank. Hayashi Takeo for permission to use his father's haunting photograph of Nagasaki, Siân Cooksey for help with the Leonardo drawing held at Windsor Castle, Jenny O'Neill at the MIT Museum, and Johan Schot, Lidwien Kuipers, and Frank Veraart for the images of globalization. Finally, thanks to Sohair Wastawy for arranging access to and duplication of the images held in Illinois Institute of Technology's special collections.

For the staff at the Johns Hopkins University Press I have the following heartfelt thanks: to Henry Tom for his early enthusiasm for this project (and for his persistence and patience!); to Anne Whitmore for her robust, not to say relentless, editing of the manuscript; and to Glen Burris for the design of the book itself.

Finally, I would like to dedicate this book to the memory of my father, Frank Misa, who was born in the last years of the science-and-systems era, grew up in the era of modernism, lived through the era of war, and glimpsed the possibilities of global culture. He would be pleased that I now have a bit more time for backyard games of basketball with Henry and Christopher and leisurely walks with Ruth.

LEONARDO
TO THE
INTERNET

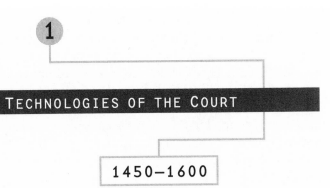

1

TECHNOLOGIES OF THE COURT

1450–1600

EVEN RENAISSANCE MEN struggled with their careers. Niccolò Machiavelli, acclaimed as a political philosopher, and Leonardo da Vinci, hailed as a universal genius, saw tough times during these turbulent years in Italy. Machiavelli (1469–1527) grew up in Florence when it was under the control of Lorenzo de' Medici, who ruled the city-state by what we might now call cultural machine politics. Florence was nominally a republic governed by committees elected from its six thousand guild members. But during the long decades of Medici rule, uncooperative committees were disbanded, while the family created and wielded effective power through an elaborate network of cultural patronage, financial preferences, and selective exile. A Medici-inspired vogue for showy urban palaces, lavish rural villas, and prominent churches, convents, and monasteries kept Florence's artists and builders busy—and the Medici coat of arms embedded in the family's numerous buildings kept its citizens ever mindful of the benefits of Medici rule. Machiavelli's two-decade-long public career, serving his beloved but rickety republic, lasted only so long as the Medici dynasty was out of power.[1]

The Medici dynasty in power was more important for Leonardo da Vinci (1452–1519). Leonardo did his apprenticeship in Medici-ruled Florence, worked then and later on projects sponsored by the Medici, and launched his career with the blessing of the Medici. Whether from the Medici family or from his numerous other courtly patrons, Leonardo's career-building commissions were not as a painter, anatomist, or visionary inventor, as he is typically remembered today, but as a military engineer and architect. Courtly patrons, while offering unsurpassed fame and unequaled resources to the artists, philosophers, architects, and engineers they deemed worthy, did not guarantee them lifetime employment. The Medici's return to power brutally ended Machiavelli's own political career. He was jailed and tortured, then in forced retirement he wrote his famous discourse on political power, *The Prince,* which was dedicated to the new Medici ruler in an unproductive hope of landing a job.

Political turmoil in the court system shaped Leonardo's career as a technologist just as decisively. In 1452, the year of his birth, the city-states of Italy were gaining economic and cultural prominence. Venice had long dominated medieval Europe's trading with the Near East, Florence was flourishing as a seat of banking and finance, and Rome doggedly asserted the primacy of the pope. The princely dominions of Milan and Urbino were ruled outright by noble families, who dispensed artistic and cultural patronage even more lavishly. Leonardo's career was so wide-ranging (he worked at each of these five locations) because the era was politically unstable. And the era was unstable because military conflict was never far away. Rulers up and down the Italian peninsula found themselves at every turn confronting hostile forces: to the east it was the Ottoman Turks, to the north loomed assertive France, and across the peninsula they battled among themselves for valuable land or strategic ports or dynastic advantage. Military engineers such as Leonardo were much in demand. During these decades, secular and religious powers melded uneasily. What else can we say of the cardinals of Rome electing as pope a worldly member of the Borgia family who, at his court, openly showed off his mistress and their four children?

Whether secular or religious, Renaissance courts and city-states had expansive ambitions, and technical activities figured prominently among their objects of patronage. The history of Renaissance technology is typically told as a story of individual engineers such as Leonardo da Vinci, Francesco di Giorgio, and others, as if they worked as independent free agents; and the history of Renaissance politics and culture is typically related to the shifting fortunes of the varied city-states and noble courts and individuals, like Machiavelli. We will see, however, that technology, politics, and culture were actually never far apart. For a time, thoroughly mixing these domains, Machiavelli and Leonardo even collaborated on a grand engineering plan to improve the commercial and military prospects of Florence by moving its river.

This chapter locates Renaissance technologists squarely within the system of court patronage. We will see that the papal court in Rome sponsored or employed such landmark technological figures as Alberti, Leonardo, and Biringuccio. Leonardo's career as an engineer is inseparable from his work for the Medici family, the Sforza court, and the Borgia clan. The pattern of court-sponsored technologies extended right across Europe (and for that matter beyond Europe[2]). Royal courts in France, Spain, and England supported innovations in shipbuilding and silk weaving. Even the well-known history of moveable-type printing needs to be reexamined in the light of pervasive court sponsorship of technical books and surprisingly wide court demand for religious publications. Characteristically, Leonardo and his fellow Renaissance-era technologists had surprising little to do with improving industry or making money in the way we typically think of technology today. Instead, Renaissance-era courts commissioned them for numerous technical projects of city-building, courtly entertainment, and dynastic display, and for the means of war. As one Leonardo scholar observes, "It was within the context of the court that the engineer carried out his many duties, first and foremost of a military nature."[3]

The Career of a Court Engineer

The practical world of artisans that nurtured Leonardo as a young man was far removed from the educated world of courtly scholars such as Machiavelli. In the absence of biographical writings, we must guess at much of Leonardo's early life. Documents recording legal disputes with his stepbrothers make it clear that he was the illegitimate son of a low-level lawyer, and that he was raised by his paternal grandparents. The clarity of his handwriting points to some formal education, while a desire to hide his inventions from prying eyes is suggested by his famous backwards or "mirror" writing. At the age of fourteen he was apprenticed in the Florence workshop of sculptor and painter Andrea del Verrocchio. While Italy's thriving universities were the sites for educating scholars, workshops such as Verrocchio's served as the principal sites for educating artisans and craftsmen. In his decade with Verrocchio (1466–76) Leonardo learned the basics of architecture, sculpture, bronze casting, painting, and possibly some mathematics. During these years, Verrocchio had several major commissions, including an ostentatious family tomb for the Medici and a large bronze statue commemorating a Venetian military hero.

As a member of Verrocchio's workshop, Leonardo had a small but visually prominent role in building the Florence cathedral. Officially the Cathedral of Santa Maria del Fiore, it was described by one contemporary as an "enormous construction towering above the skies, vast enough to cover the entire Tuscan population with its shadow . . . a feat of engineering that people did not believe feasible . . . equally unknown and unimaginable among the ancients" (fig. 1.1). In 1420, more than a century after the cornerstone was laid, Filippo Brunelleschi (1377–1446) gained the city's approval for completing its impressive dome and began work. Brunelleschi's design followed the official plans for an eight-ribbed dome, more than 100 feet high, which began 170 feet above the cathedral's floor. The dome measured 143 feet in diameter—then and now the largest masonry dome in the world. Brunelleschi directed that the dome's lower level be built of solid stone and its upper levels consist of a two-layered shell of stone and brick; at its apex stood a 70-foot-high lantern tower to admit light and air. To construct this novel structure, without wooden bracing from underneath, Brunelleschi devised numerous special cranes and hoisting apparatuses that could lift into place the heavy stones and bricks. Brunelleschi, often with Cosimo de' Medici's support, designed numerous other buildings in Florence, including the Ospedale degli Innocenti, a charitable orphanage commissioned by the silk merchants guild which is notable as an early and influential example of a distinctive Florentine style of architecture. Although the cathedral's dome and lantern had been completed (in 1461) before Leonardo came to Florence, there remained the difficult task of placing an 8-foot-high copper sphere at its very top. Verrocchio's workshop built and finished the copper sphere and, using one of Brunelleschi's cranes that was still in service, placed it to stand a dizzying 350 feet above the city's streets.[4]

The cathedral project occupied Verrocchio's workshop from 1468 to 1472 and made a distinct impression on the young Leonardo. Forty years later he wrote, "Keep in mind how the ball of Santa Maria del Fiore was soldered together."[5] At the time Leonardo made numerous drawings of the varied hoisting machines that Brunelleschi had designed to lift the heavy stones, of the revolving crane used to position them, and of several screwjacks and turnbuckles. In addition to his work as an architect and sculptor, Brunelleschi was a pioneer in geometrical perspective (discussed below), especially useful in capturing the three dimensionality of machines in a two-dimensional drawing. From Leonardo's notebooks it is clear that he mastered this crucial representational technique. To depict Brunelleschi's heaviest hoisting machine, an ox-powered, three-speed winch with a system of reversing ratchets that allowed workers to raise a stone and then carefully lower it into place without physically reversing

FIG. 1.1. Dome of the Florence Cathedral.

Renaissance-era building projects frequently mobilized advanced technology to create impressive cultural displays. To complete the Florence
cathedral Brunelleschi erected the largest masonry dome in the world,
measuring 143 feet in diameter. The octagonal, ribbed dome was formed
in two layers, using bricks laid in a spiral herringbone pattern. During his
apprenticeship, Leonardo da Vinci helped construct and then set in place
the large copper sphere at the top. William J. Anderson, *The Architecture of the Renaissance in Italy* (London: Batsford, 1909), plate 4. Courtesy of Galvin Library, Illinois Institute of Technology.

the animals, Leonardo sketched a general view of the hoist along with a
series of views of its most important details. These multiple-view drawings, done in vivid geometrical perspective, are a signature feature of his
notebooks.

Leonardo's career as an independent technician began as an offshoot
of an assignment from Lorenzo de' Medici. Lorenzo had directed Leo-

nardo to carry a certain gift to Ludovico Sforza, the new and self-installed ruler of the duchy of Milan. Leonardo transformed this charge into a full-time job as the Sforza court's engineer. It turned out to be a fortunate move for him, and he spent nearly two decades there (1482–99). Milan was twice the size of Florence, and the Sforza family presided over an active and powerful court. In a famous letter to his new patron, Leonardo proclaimed his engineering talents and highlighted their uses in war. "I will make bombards, mortars, and light ordnance of fine and useful forms, out of the common type," he stated. He also had promising designs for "covered chariots, safe and unattackable," "covered ways and ladders and other instruments," a way to "take the water out of trenches," "extremely light, strong bridges," and "many machines most efficient for attacking and defending vessels." After enumerating no fewer than nine classes of weapons ("I can contrive various and endless means of offense and defense"), Leonardo then mentioned civilian engineering. "In time of peace I believe I can give perfect satisfaction and to the equal of any other in architecture and the composition of buildings public and private; and in guiding water from one place to another."[6] Yet even the guiding of water, as we will see, was hardly an innocent task.

Ludovico would be formally recognized as duke of Milan some years later (he had seized power by pushing aside his ten-year-old nephew), and then Leonardo would become known as *ingeniarius ducalis* (duke's engineer). But when Leonardo arrived, in the 1480s, Ludovico's claim to the duchy of Milan was shaky. In an effort to create legitimacy for his own regime, Ludovico planned a huge bronze statue, more than 20 feet in height, to commemorate his deceased father, the former and legitimate ruler of Milan. In his letter to the duke, Leonardo shrewdly offered to "undertake the work of the bronze horse, which shall be an immortal glory and eternal honour to the auspicious memory of the Prince your father and of the illustrious house of Sforza."[7] Leonardo, it seems, knew exactly what was expected of him. From this time forward, Leonardo fashioned a career as a court engineer with a strong military slant. His notebooks from Milan are filled with drawings of crossbows, cannons, attack chariots, mobile bridges, firearms, and horses.

Leonardo's drawings create such a vivid image in the mind that it is not easy to tell when he was illustrating an original invention, copying from a treatise, recording something he saw firsthand, or, as was often the case, simply exercising his fertile imagination. Among Leonardo's "technological dreams"—imaginative projects that were beyond the realm of technical possibility—are a huge human-powered wheel apparently meant to drive the winding and firing of four oversize crossbows, and an immense wheeled crossbow at least 40 feet in breadth. A seemingly impossible horse-drawn attack chariot with rotating knives might actually have seen the battlefield, for there is a note by Leonardo that the rotating knives "often did no less injury to friends than to enemies." Pirate raids on the seaport of Genoa prompted Ludovico Sforza to direct Leonardo to design means "for attacking and defending vessels." Leonardo envisioned a submarine, a sealed, watertight vessel temporarily kept afloat by air in leather bags. But he put the lid on this invention to prevent (as he put it) "the evil nature of men" from using such a vessel to mount unseen attacks on enemy ships.[8]

Another "technological dream" with obvious military potential was Leonardo's study of human-powered flight. In the mid 1490s, Leonardo attempted to apply his philosophical stance that nature was mechanically uniform and that humans could imitate the natural "equipment" of flying animals. He devised several devices for transforming a human's arm or leg motions into the birdlike flapping of a mechanical wing. Several of these appear absurdly clumsy. In one, the pilot—in midflight—continually rewound springs that mechanically flapped the contraption's wings. In another, an especially stout design, there were heavy shock absorbers to cushion crash landings. At least one of these wonders took off from the roof of the Sforza palace. Leonardo's instructions include the note, "You will try this machine over a lake, and wear a long wineskin around your waist, so that if you should fall you will not drown."[9]

Leonardo also worked on the expensive, high-technology military hardware of the Renaissance: gunpowder weapons. Indeed, because the Sforzas already employed a court architect who oversaw civilian building projects, Leonardo devoted himself more soundly to military projects. In his notebooks, one finds full engagement with gunpowder weapons. There are characteristic exploded-view drawings for wheel-lock assemblies (to ignite the gunpowder charge), a water-driven machine for rolling the bars to be welded into gun barrels, and a magnificent drawing of workers guiding a huge cannon barrel through the midst of a bustling foundry (fig. 1.2). Still, as Bert Hall, the leading expert on Renaissance gunpowder technology, has emphasized, the "military revolution" we often associate with gunpowder weapons was a slowly developing affair, given the prohibitively high cost of gunpowder, the laughable state of firearm accuracy, and the surprising deadliness of crossbows, pikes, and

FIG. 1.2. Leonardo and the Military Revolution.

Cannons were the expensive high-tech weapons of the court era. In the late thirteenth century the Chinese invented gunpowder weapons capable of shooting projectiles, and gunpowder cannons were in use by Europeans by the early fourteenth century. In this drawing Leonardo captured the bustle of a busy cannon foundry, using geometrical perspective. The prohibitively high cost of high-nitrate gunpowder, however, constrained the wide use of gunpowder weapons for many decades to come. The Royal Collection, © 2003 Her Majesty Queen Elizabeth II, RL 12647.

battle axes.[10] (A classic turning point, considered in chapter 2, is Maurice of Nassau's integration of firearms into battlefield tactics in the early 1600s.) Leonardo also devised numerous means for building and defending fortifications. He illustrated several devices meant to knock down the ladders that an attacking force might place to scale a fortified wall. While certainly not such exciting subjects as muskets or cannon, the varied means for attacking or defending a fortification were at the core of Renaissance-era warfare.

Beyond the military projects that occupied Leonardo in Milan it was entirely characteristic of the court era that labor-saving or industrial technologies were little on his mind. Fully in tune with his courtly patrons, Leonardo focused much of his technological creativity on dynastic displays and courtly entertainments. For the oversize statue commemorating Sforza's ducal father and legitimizing his own rule in Milan Leonardo built a full-scale clay model of the horse-mounted figure. The story is told that Leonardo would leave "the stupendous Horse of clay" at midday, rush over to where he was painting the *Last Supper* fresco, "pick up a brush and give one or two brushstrokes to one of the figures, and then go elsewhere." The immense mass of bronze set aside for the monument was in time diverted to make cannons.[11]

Sforza also charged Leonardo with creating court culture directly. Leonardo took a prominent role in devising the lavish celebration of the marriage in 1489 between a Sforza son and princess Isabella of Aragon. Marriages between noble families were serious affairs, of course. The marriage celebration offered a prominent occasion for proclaiming and cementing alliances between the two families, and hosts often commissioned allegorical performances that "were at the same time ephemeral court entertainments and serious occasions for political propaganda." Leonardo's contribution to the marriage celebration mingled the heavens and the earth. Traditionally, scholars have understood *Il Paradiso,* created and staged for the Sforza fest, as a human-centered dance spectacle that included "seven boys representing the seven planets" circling a throne and that featured the unveiling of paradise. At the finale, the figure of Apollo descended from on high to present princess Isabella with the spectacle's text. It is often suggested that Leonardo chafed at having to design theatrical costumes, yet scholars have recently found evidence indicating that Leonardo also built moving stage platforms and settings—and perhaps even an articulated mechanical robot for these festivities. An eyewitness account describes the set of planets as a mechanical model built by Leonardo. Leonardo's notebooks have designs for at least two other moving-set theatricals, one featuring the mechanical elevation of an allegorical "star" while another offered a glimpse of hell complete with devils, furies, and colored fires. The mechanisms would have been complex spring- or cable-powered devices, constructed mostly of wood. Leonardo's much celebrated "automobile" was most likely also for courtly entertainments. These theatrical and courtly "automata" represented a culturally appropriate expression of Leonardo's interest in self-acting mechanisms.[12]

His fascination with self-acting mechanisms is also evident in Leonardo's many sketches of textile machines found in the surroundings of Milan. These automata have led some breathless admirers to call him the "prophet of automation," yet it seems more likely that he was simply recording the interesting technical devices that he saw in his travels. (This is certainly the case in his drawings of Milan's canal system.) Quite possibly he was sketching possible solutions to problems he had spotted or heard about. One of these drawings that we can certainly identify as Leonardo's own creation (ca. 1493–95) was a rotary machine to make sequins, used for ornamenting fine gowns or theatrical costumes. Again, Leonardo expressed his mechanical talents in culturally resonant ways, just as inventors in later eras would focus on self-acting mechanisms for industry and not for courts.

Leonardo's comfortable life with the Sforza court ended in 1499 when France invaded the region and ousted Sforza. "The duke lost his state, his property, and his freedom, and none of his works was completed for him," recounted Leonardo. In the chaotic eight years that followed, Leonardo acted essentially as a mercenary military engineer. He traveled widely, with the application of military technologies never far from his mind, and worked all sides of the conflicts that were rending Italy at the time. (Even while working for the duke of Milan and in the process of designing a warm-water bath for the duchess, Leonardo recorded in his notebooks "a way of flooding the castle" complete with drawings.) In Venice, which was reeling under attack by the Ottoman Turks, he suggested improvements to that city's all-important water defenses; while several years later, armed with the special knowledge gained in Venice, he would advise the ruler of Milan on how to flood Venice. In Florence once again during 1500–1501 he failed to gain any technical commissions (the city had its own engineers) and so he devoted himself to painting and the study of mathematics.[13]

In the summer of 1502 the infamous Cesare Borgia tapped Leonardo to be military engineer for his campaign in central Italy. It is impossible to make a simple reckoning of the shifting alliances that shaped Leonardo's career during these tumultuous years. The Sforza family had originally gained control of Milan with the support of the Florentine Medici. The French invasion of Milan, which cost Leonardo his job at the

Sforza court, was aided by Alexander VI (1492–1503), the so-called Borgia pope, none other than Cesare Borgia's father, who in effect traded Milan to the French in exchange for their backing of his rule as pope and their support for his son's military venture. We can be certain that while accompanying Cesare Borgia's wide-ranging military campaign Leonardo witnessed the sack of Urbino, where the Montefeltro family had collected a famous library and an illustrious court (of which more below). In addition to its political consequences, the Borgia campaign furnished the occasion when Leonardo first met Machiavelli and when Machiavelli found his "model" prince in Cesare Borgia, the Duke of Valentino.

Returning to war-torn Florence in 1503 Leonardo gained from that city two commissions characteristic of the era. With the Medici family temporarily exiled, the struggling republican government used the fine arts and public works to help secure its political legitimacy. The city kept a squad of historians busy chronicling its glories. "Florence is of such a nature that a more distinguished or splendid city cannot be found on the entire earth," gushed one such account. In the same celebratory vein, the republic commissioned Leonardo to paint a fresco commemorating the Florentines' victory over Milan in the battle of Anghiari in 1440. The second commission, a giant hydraulic-engineering scheme hatched with Machiavelli's behind-the-scenes assistance, took aim at the enemy town of Pisa. The two men planned to move the Arno River—at once devastating the downstream town of Pisa and improving Florence's access to the sea. Two thousand workers began digging a diversion canal, but when the results were unsatisfactory the scheme was halted. During this unsettled period, Leonardo also served as a consultant on military fortifications for Florence's ally, the lord of Piombino, and on painting and architecture for French-held Milan. Leonardo's continual travels between Florence and Milan brought work on the *Battle of Anghiari* fresco to a crawl. Perhaps even Leonardo experienced some discord when simultaneously working for Milan while memorializing Milan's defeat.[14]

Leonardo's famous anatomical studies began during a lull in the military action. After 1507 Leonardo's notebooks record many detailed sketches of the muscles, bones, and tendons of the human body; his access to the city hospital in Florence made his empirical investigations of high quality. Leonardo also conceived of a series of elaborate analogies between geometry and mechanics, anatomy and geology, and the human body and the cosmos. He also outlined a theoretical treatise on water, but keeping in mind the Leonardo-Machiavelli scheme to destroy Pisa by diverting its river, it is difficult to see this treatise as a "pure" theoretical study.

In French-held Milan once again from 1508 to 1513, Leonardo gained the patronage of Louis XII, king of France, who occasionally directed Leonardo toward a specific technical project but mostly provided him general patronage, giving him ample time for theoretical investigations. Yet Leonardo's second Milan period ended when the Sforza family *recaptured* the city and drove out his newly acquired French patron. Leonardo escaped to Florence, where the Medici family had returned to power, then went on to Rome in the service of Giuliano de' Medici, the brother of Pope Leo X (1513–21). In the next two years Leonardo worked on several specific commissions from Giuliano, who flourished as his brother dispensed patronage his way. While in Rome Leonardo mapped the malaria-ridden marshes around Terracina, on the coast south of Rome, with the aim of draining them, a public health project given to Giuliano by his pope-brother. A Leonardo drawing of a rope-making machine is even stamped with the Medici's diamond-ring symbol. But when Giuliano died in 1516, Leonardo once again found himself without work or patron. At least Leonardo saw the situation plainly, writing, "The Medici made me and ruined me."[15]

Leonardo spent the last three years of his life at the royal court of France. The new French king, François I, gave him the grand title of "the King's foremost painter, engineer and architect." Officially Leonardo was to design a new royal palace and the surrounding canals and waterworks, but these plans were never realized. Instead, Leonardo became in true fashion a distinguished figure at court. The king, wrote one courtier, "was extremely taken with his great virtues, [and] took so much pleasure in hearing him speak, that he was separated from him only for a few days out of the year. [The king] believed that there had never been another man born in the world who knew as much as Leonardo, not so much about sculpture, painting, and architecture, as that he was a very great philosopher."[16]

In addition to his duties as a great philosopher and eminent expert, Leonardo created more courtly entertainments. His friend Francesco Melzi described one automaton that Leonardo built to honor the French king. A lion with a bristling mane, it was led by a hermit. On its entrance, we are told, women in the audience drew back in terror; but when the king touched the lion three times with a magic wand handed to him by the hermit, the lion-automaton broke open and spilled at the king's feet a mound of fleur-de-lys. Such events were famously packed with symbols to test the audience's savvy. Everyone understood the flower as a classic symbol of the French royal house; the lion was a prominent heraldic symbol of a rival court.[17] To the end, Leonardo understood how to adapt his considerable talents to court desires. He died, in France, on 2 May 1519.

THE SPECIAL CHARACTER of technological creativity in the Renaissance, as we have seen, resulted from one central fact: the city-states and courts that employed Leonardo and his fellow engineers were scarcely interested in the technologies of industry or commerce. Their dreams and desires focused the era's technologists on warfare, city building, courtly entertainments, and dynastic displays. A glance at Leonardo's technical notebooks confirms that he was an outstanding engineer, architect, and artist. But a closer examination of them, and the notebooks of his contemporaries, reveals that he was not the solitary genius imagined by some authors. For instance, the Renaissance engineers who sketched Brunelleschi's hoisting machinery for the Florence cathedral include, besides Leonardo, Francesco di Giorgio, Buonaccorso Ghiberti, and Giuliano da Sangallo. And while we have a contemporary's quip that Leonardo "was as lonely as a hangman,"[18] we also know that he met personally with, borrowed ideas from, and likely gave inspiration to a number of fellow court engineers. The intellectual resources and social dynamics of this technological community drew on and helped create Renaissance court culture.

Foremost among these intellectual resources was the distinctive three-dimensionality and depth of Renaissance art and engineering. This owed much to Leon Battista Alberti, renowned as the "father of perspective." Alberti (1404–72) was born into a prominent Florentine merchant-banking family, studied canon and civil law at the University of Bologna, and worked for three decades in the administrative offices of the pope's court. As Anthony Grafton has recently made clear, Alberti was a courtier of courtiers. Even more than Leonardo's, his career was bound up with the court system. In his latter years, to secure a place at the Ferrara court, he wrote what amounted to a how-to manual for succeeding in the court system—with examples conveniently drawn from his own life! He was experienced at the game. His patrons, besides the papal curia, included the princely courts at Ferrara, Rimini, Mantua, and Urbino, as well as the great Florentine builder Giovanni Rucellai. In his astonishing career, he was variously a humanist writer and rhetorician, architectural designer and consultant, literary critic, sculptor, mapmaker, and a leading theorist of painting. "He is the sort of man who easily and quickly becomes better than anyone else at whatever pursuit he undertakes," thought one contemporary.[19] If he had done nothing else, Alberti's learned treatises on architecture and on practical mathematics might prompt latter-day readers to label him an engineer.

Linear perspective, which he turned into his most far-reaching theoretical achievement, was one thing that Alberti did *not* invent. Leading Florentine artists such as Masaccio were already practicing something like linear perspective a decade or more before Alberti's famous treatise *On Painting* (1436). During the same time Brunelleschi, at work on the Florence cathedral, staged dramatic public events that popularized the new technique. Positioned on the steps of the cathedral, Brunelleschi painted small precise "show boxes" of the prominent Baptistry across the square, and dazzled passers-by when they had difficulty telling the difference between his painting of the scene and the real thing. He did the same for the Piazza della Signoria, site of the city's government. Alberti, too, later painted show boxes in which (as he wrote) the viewer could see "huge mountains, vast provinces, the immense round of the sea surrounding them, and regions so distant from the eye that sight was dimmed ... they were such that expert and layman alike would insist they saw, not painted things, but real ones in nature."[20]

With his treatise on painting, Alberti turned the practice of perspective into a structured theory. In what is now a commonplace of drawing, he directed artists to treat the two-dimensional picture plane on which they worked, whether it was a wall, panel, or canvas, as if it were a window in which a three-dimensional scene appeared. The classic exercise illustrated and popularized by Albrecht Dürer (1471–1528) is to view an object through a nearby pane of glass, or a mirror, that is ruled into squares, and then to transfer what you see, square by square, onto a piece of paper or canvas that is similarly ruled. Dürer's most famous "object," illustrating his 1525 treatise on geometry and perspective and reproduced widely ever since, was a naked woman on her back, suggesting that perspective was not merely about accurately representing the world but about giving the (male) artist power over it. Whatever the object, vertical lines will remain vertical, while receding horizontal lines will converge toward a vanishing point at the drawing's horizon. Parallel horizontal lines, such as on a tiled floor, must be placed at certain decreasing intervals from front to back. Finally, for maximum effect, the viewer's eye must be precisely positioned. In fact, the "show boxes" directed the observers' point

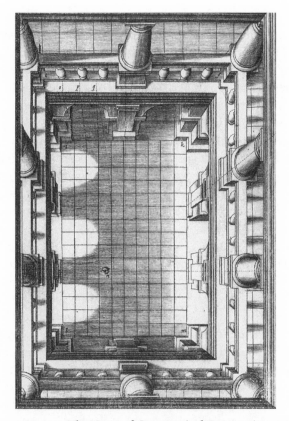

FIG. 1.3. The Gaze of Geometrical Perspective.

Painters, architects, and engineers used geometrical perspective to depict their ideas with precision and persuasion, gaining commissions from the courts and in the process creating culture. Geometrical perspective not merely represented the world but also changed people's ways of thinking. The penetrating eyeball at the symbolic center of this geometrical-perspective drawing hints at the power of the artist's gaze. Joseph Moxon, *Practical Perspective* (London, 1670), plate 36. Courtesy of Illinois Institute of Technology Special Collections.

of view by locating the peephole at the picture's vanishing point, from which they looked at the image reflected in a mirror on the box's far side. Alberti showed how the exact use of perspective can trick the mind's eye into seeing a three-dimensional original.

The geometrical ideas and devices described in Alberti's writings were widely adopted by artists, mathematicians, cartographers, and engineers. And the entertaining show boxes did not harm his flourishing court career. Leonardo, in addition to studying many of Alberti's technical innovations and expanding on his method of empirical investigation, directly incorporated passages of Alberti's writings on painting into his own treatise on the subject.[21] Leonardo even copied many of Alberti's distinctive phrases. It is Alberti's ideas we are reading when Leonardo writes that the perspective picture should look as though it were drawn on a glass through which the objects are seen; or that the *velo,* or square-marked net, locates the correct positions of objects in relation to one another; or that one should judge the accuracy of a picture by its reflection in a mirror. Leonardo even echoed Alberti's dictum that the only "modern" painters were Giotto and Masaccio.

Leonardo's membership in a community of engineers has only recently been appreciated properly. In the absence of an "engineer's guild" listing its members, and because of the pressing needs for military secrecy, we will probably never know all the engineering colleagues with whom Leonardo schemed and dreamed. We do know that chief among Leonardo's contemporaries was Francesco di Giorgio (1439–1501) and that the two men met and in time shared a great deal of technical material. Born in Sienna, Francesco studied painting, sculpture, and architecture. He began his career as a sculptor, then worked as an engineer on his native city's water supply (1469). Francesco became the era's leading military engineer. His career, like Leonardo's and Alberti's, is inseparable from the court system.

Beginning in 1477, Francesco was among the prominent cultural figures that the Montefeltro family welcomed to their fortress-court at Urbino. Court artists there, such as Piero della Francesca, painted the noble family (which numbered one of the Sforza sisters among its members) and authored treatises on painting that were dedicated to the duke and placed in his library. That famous library held some 1,100 richly bound books reportedly costing the whopping sum of 30,000 ducats, about half a year's worth of the payment the Duke of Urbino, a *condottiere* (soldier of fortune) like Sforza, received from his military services. Francesco became the chief designer for the duke's ambitious palace expansion (its scale might be estimated by its stables for 300 horses), and he built a series of military fortresses throughout the state of Urbino. His first treatise on technology—with drawings of such technical devices as artillery, pumps, water-powered saws, methods for attacking fortified

walls, catapults, and siege equipment, as well as tips on defending fortifications in an era of gunpowder—was dedicated to the duke (and with
its ascribed date of ca. 1475 probably secured his employment by the duke
two years later).

Francesco stayed at the Montefeltro court some years after the duke's
death, until 1487, when he once again took up work as chief city engineer
in Sienna. Also during his post-Urbino years, he traveled frequently to
Naples and there advised the Duke of Calabria on military matters. Not
by accident, the second edition of the abovementioned treatise on technology was dedicated to this duke while its third edition was illustrated
by the duke's court painter. When he was called in to advise on the construction of the cathedrals of Pavia and Milan in 1490, Francesco met
Leonardo in person. Francesco had long left the Montefeltro court by the
time Leonardo assisted Cesare Borgia with its capture in 1502.

Close study of the two men's notebooks has revealed that Francesco
was one source of designs for machines and devices that had previously
been attributed to Leonardo alone. For instance, the devices that
Francesco described in his treatise on military architecture are in many
cases more sophisticated than similar devices in Leonardo's notebooks.
We know that Leonardo read and studied Francesco's treatise, since there
is a manuscript copy of it clearly annotated in Leonardo's distinctive
handwriting. It seems Leonardo simply borrowed from a well-regarded
author.

In a curious way, the presence of Leonardo's voluminous notebooks
has helped obscure the breadth and depth of the Renaissance technical
community, because researchers overzealously attributed all the designs
in them to him. At his death in 1519, Leonardo left his notebooks and papers to his "faithful pupil" Francesco Melzi, who kept them at his villa outside Milan for fifty years. Eventually Melzi's son peddled the papers far
and wide. Scholars have found caches of Leonardo's papers close to his
areas of activities in Milan, Turin, and Paris and as far afield as Windsor
Castle, in England. In Madrid a major Leonardo codex was found as recently as 1966. Scholars believe that about one-third (6,000 pages) of
Leonardo's original corpus has been recovered; these papers constitute
the most detailed documentation we have on Renaissance technology.
Unfortunately, only in the last twenty years have scholars abandoned the
naïve view that each and every of Leonardo's notebook entries represents
the master's original thinking. In his varied career Leonardo traveled
widely and often sketched things that caught his attention. His notebooks

record at least four distinct types of technical projects: his specific commissions from courtly patrons; his own technological "dreams," or devices that were then impossible to build; his empirical and theoretical studies; and devices he had seen while traveling or had heard about from fellow engineers; as well as "quotations" from earlier authors, including Vitruvius.

Scholars have intensely debated what should count as Leonardo's own authentic inventions. These probably include a flying machine that was a completely impractical mechanical imitation of a bat, some textile machines that remain tantalizing but obscure, and a machine for polishing mirrors. Characteristic of the court culture of Renaissance technology, the textile machine that is most certainly Leonardo's own was the one for making sequins to decorate fancy garments. Leonardo's "inventions" that should be attributed to others include an assault chariot, endless-screw pumps, several lifting apparatuses, and other pump designs that clearly were known in ancient Rome. Leonardo also copied drawings for a perpetual motion machine.

Perhaps the most distinctive aspect of Leonardo's career was his systematic experimentation, evident in his notebooks especially after 1500. He wrote:

> In dealing with a scientific problem, I first arrange several experiments, since my purpose is to determine the problem in accordance with experience, and then to show why the bodies are compelled so to act. That is the method which must be followed in all researches upon the phenomenon of Nature. . . . We must consult experience in the variety of cases and circumstances until we can draw from them a general rule that [is] contained in them. And for what purposes are these general rules good? They lead us to further investigations of Nature and to creations of art. They prevent us from deceiving ourselves or others, by promising results to ourselves which are not to be obtained.[22]

Some objects of Leonardo's systematic investigations were gears, statics, and fluid flow. He made a survey of different types of gears, designing conical and helical gears of wood. He also investigated the causes of crumbling in walls and found that the direction of cracks in walls indicated the source of strain. In studying the resistance of beams under pressure, he arrived at the general principle that for square horizontal beams supported at each end, their resistance varied with the square of their side and inversely with their length—a fair approximation. Fluids were more challenging. To study the flow of water he built small-scale models involving colored water. He also sought general guidelines for placing dams in rivers. Indeed, it may have been with the failed Arno River scheme in mind that Leonardo reminded himself about "promising results to ourselves which are not to be obtained."

The influence of Renaissance engineers on Europe was substantial. The noted medieval historian Lynn White wrote, "Italian engineers scattered over Europe, from Madrid to Moscow and back to Britain, monopolizing the best jobs, erecting wonderful new machines, building palaces and fortifications, and helping to bankrupt every government which hired them. To tax-paying natives they were a plague of locusts, but rulers in the sixteenth century considered them indispensable. Their impact upon the general culture of Europe was as great as that of the contemporary Italian humanists, artists, and musicians."[23]

Gutenberg's Universe

The invention of moveable type for printing led to an information explosion that profoundly altered scholarship, religious practices, and the character of technology. There is much about the printing revolution that conjures up images of enterprising capitalist printers breaking free of tradition-bound institutions. All the same, courts across Europe created large-scale markets for printed works and shaped the patronage networks for writings about technology. The first several generations of printers as well as the best-known early technical authors were, to a surprising extent, dependent on and participants in late-Renaissance court culture.

Printing was a composite invention that combined the elements of moveable metal-type characters, suitable paper, oil-based ink, and a wooden press. Inks and presses were commonly available in Europe, while paper came to Europe by way of China. Paper was being made in China by the third century A.D. from the fibers of silk, bamboo, flax, rice, and wheat straw. Papermaking spread to the Middle East by Chinese papermakers taken prisoner by Arabs in 751. From Samarkand, the great Arabic center of astronomy and learning, the "paper route" passed through Baghdad (793), Cairo (900), Fez (1000), Palermo (1109), Játiva (1150), Fabriano (1276), and Nuremberg (1390). Another route of transmission passed along the coast of Northern Africa, through Spain (beginning of thirteenth century), and into France.

Moveable type was also "first" developed in the Far East, centuries before Gutenberg. Printing from carved stone or wood plates was well advanced in China at the turn of the first millennium; a set of 130 volumes of Chinese classics was printed in 953, and a key Buddhist text was printed using 130,000 plates in 982. The first truly moveable type is credited to Pi Sheng (1041–48), who engraved individual characters in clay, fired them, and then assembled them on a frame for printing. This method was reinvented and improved around 1314 by Wang Cheng, a government magistrate and prolific compiler of agricultural treatises. He commissioned artisans to carve a set of 60,000 specially designed characters and used them to publish up to 100 copies of a monthly gazette. During the Ming Dynasty (1368–1644) moveable wooden characters were used to publish the official court gazette. In Korea characters were first cast from metal (lead and copper) in 1403 and used to print numerous works during that century. The Korean characters spread to China (end of the fifteenth century) and to Japan (1596), and yet metal type did not entirely displace woodblock printing. Traditional woodblock printing persisted, since metal type could not print on two sides of the thin mulberry-tree paper. The further difficulty of moveable type printing in Chinese is evident in Wang Cheng's extraordinarily large set of carved characters.

While it was a vector for the "paper route," the Arab world halted the spread of Asian printing to the West. Islam permitted handwriting the words of Allah on paper but for many years forbad its mechanical printing. The first Arabic-language book printed in Cairo, Egypt, did not appear until 1825.[24]

"The admirable art of typography was invented by the ingenious Johann Gutenberg in 1450 at Mainz," or so stated the son of Gutenberg's partner in 1505. Remarkably enough, the only direct evidence to assess this claim are two legal documents from 1439 and 1455. We know surprisingly little about Gutenberg himself. For almost ten years of his life (1429–34 and 1444–48) not even his city of residence is known with certainty. And while, for instance, tax records tell us that Gutenberg in 1439 had a wine cellar storing 2,000 liters, no one can be sure if he was married to the legendary Ennelin—who may have been his wife, his lover, or a Beguine nun. It is clear that Gutenberg was born into a wealthy and established family of Mainz, sometime around 1400—by printers' fable it was St. John the Baptist's Day, 24 June 1400. His father and several other family members were well connected with the church mint at Mainz, and from them, it seems, Johann gained his knowledge of metal casting, punch-cutting, and goldsmithing.

In 1434 Gutenberg moved to Strasbourg, near the Rhine River in northeast France, and by 1436 he was engaged in experiments on printing. During his residence in Strasbourg he also practiced and taught gold-smithing, gem cutting, and mirror making. In 1438 he agreed to convey "the adventure and art" of printing to two of his mirror-making associates; in effect they paid Gutenberg 250 guilders to create a five-year partnership. However, one of the two died, and his brothers sued Gutenberg in 1439 to be taken into the partnership. The court records indicate little more than that Gutenberg owned a press, had bought lead and other metals, and from them had cast what were called "forms" (a word used to describe the molding or casting of iron). A goldsmith testified that Gutenberg had paid him 100 guilders in 1436 for "that which pertains to the use of a press." Secrecy was an overriding concern of Gutenberg's. When his partner was dying, Gutenberg had all the existing "forms" melted down in his presence. He also directed a helper to take apart an object with "four pieces" held together by two screws, quite likely an adjustable mold for casting type. More than this we will probably never know. To finance these experiments, Gutenberg borrowed large sums of money.[25]

Gutenberg left Strasbourg in 1444, but it is unclear what he did and where he went until his return to his native Mainz four years later. Money surely mattered. Shortly after his reappearance in Mainz, he secured a 150-guilder loan guaranteed by a relative. In 1450 he borrowed 800 guilders from a merchant named Johann Fust, and soon enough an additional 800 guilders with the proviso that Fust was to become a profit-sharing partner in "the works of the books" (namely the printing of the Latin Bible). These "works" comprised perhaps six presses and twenty printers and assistants. In 1455, generating the second court dossier, Fust sued Gutenberg for the repayment of the two loans including interest, a total of over 2,000 guilders. It is often assumed that this court action wiped out Gutenberg's finances and deprived him of his livelihood, but in fact the court required Gutenberg only to repay the first 800 guilders.

Gutenberg retained his half-share in the Bible sales and revenue from other printing jobs, and a recent researcher concludes that he not only repaid Fust but also regained possession of the printing equipment held as collateral. Fust set up a new and highly successful partnership with Peter Schöffer, who had also worked with Gutenberg; between Fust's death in 1466 and his own in 1503, Schöffer published at least 228 books and broadsides. For his part, Gutenberg continued in the printing trade and employed at least two workmen, who became successful printers in their own right after his retirement in 1465. In that year, as the Archbishop of Mainz formally declared, "by special dispensation have we admitted and received

him as our servant and courtier." The honor gave Gutenberg a financially secure retirement and, until his death in 1468, the privileges of a nobleman.[26]

Gutenberg's principal inventions were the adjustable mold for casting
type and a suitable metal alloy for the type. Understandably, the two law-
suits reveal few technical details of the type-molding apparatus, but a
Latin grammar and dictionary printed in Mainz in 1460 (by whom is not
clear) points out that its successful printing resulted from the "marvellous consistency in size and proportion between patterns and moulds."
We can see why such consistency would be important in type that would
be assembled in lines and locked into a frame when placed on the printing press. The letters "M" and "W" are roughly the same width, but "i"
and "W" are not. Making different size letters of exactly the same depth
and height-on-paper required an adjustable type mold. Gutenberg also
struggled to find a type metal that would be easy to melt and mold, yet
hard enough to be durable. Again, while the exact composition of Gutenberg's own alloy is not known, the first printed book that describes type-
casting, Biringuccio's *Pirotechnia* (1540), of which more later, states, "the
letters for printing books are made of a composition of three parts fine
tin, an eighth part of lead, and another eighth part of fused marcasite of
antimony." Analysis of samples from the printshop of Christopher Plantin from 1580 are 82 percent lead, 9 percent tin, and 6 percent antimony,
with a trace of copper.[27]

Printing traveled quickly. By 1471, printing establishments had sprung
up in a dozen European cities as far away as Venice, Rome, and Seville. By
1480, there were two dozen printing cities in northern Italy alone, while
printers had moved east to Cracow and Budapest and north to London
and Oxford. By 1500, there were printers as far north as Stockholm and
as far west as Lisbon.

The printing press made a little-known German theology professor
named Martin Luther into a best-selling author and helped usher in the
Protestant Reformation. The Reformation is officially dated from 31 October 1517, the day Luther tacked his Ninety-Five Theses onto the church
door at Wittenberg, Germany. His Theses were written in Latin, not the
language of the common people, and church doors had been "the cus-
tomary place for medieval publicity." Yet, printers sensed a huge market
for his work and quickly made bootleg copies in Latin, German, and other
vernacular languages to fill it. It was said that Luther's theses were known
across Germany in two weeks and across Europe in a month. Luther wrote
to Pope Leo X, another Medici pope and brother of Leonardo's onetime

patron: "It is a mystery to me how my theses . . . were spread to so many places. They were meant exclusively for our academic circle here." Within three years (1517–20) enterprising printers had sold over 300,000 copies of Luther's writings. "Lutheranism was from the first the child of the printed book, and through this vehicle Luther was able to make exact, standardized and ineradicable impressions on the minds of Europe," writes historian Elizabeth Eisenstein. Eventually, Luther himself hailed printing as "God's highest and extremest act of grace, whereby the business of the Gospel is driven forward."[28]

The Catholic Church responded to the specific theological arguments Luther had raised but could not stop the spread of the printed word. And in this instance, the medium formed the message. The Protestant movement's emphasis on individuals' reading the Bible themselves required a massive printing effort. Whatever their personal beliefs, printers thus had material reasons to support Protestantism. Even the Catholic Church unwittingly helped the cause. Beginning in 1559, the Church issued its notorious *Index of Prohibited Books*, a terror to free-thinking authors and publishers in Catholic countries. But for Protestant-leaning authors it amounted to free publicity, and for printers in Protestant countries it amounted to a conveniently compiled list of potentially best-selling titles.[29] Machiavelli and Galileo were among the well-known figures featured in the *Index*.

It was in part because of the fumbling attempts at repressing printing in the Catholic countries of southern Europe that printing flourished in northern Europe. During the forty years from 1500 to 1540 no fewer than 133 printers produced more than 4,000 books in the Netherlands, whose publishing trade was then centered in the city of Antwerp. Only Paris surpassed Antwerp in population and commercial activity, and it was in Antwerp that Christopher Plantin (c. 1520–89) transformed the craft of printing into an industry. Born in France and apprenticed as a bookbinder in Paris, Plantin moved to Antwerp and by 1550 shows up as "boeckprinter" in that city's guild list. He established his own independent press in 1555. Ten years later, when other leading printers had between two and four presses, Plantin had seven; at the height of his career in 1576 he had a total of twenty-two presses and a substantial workforce of typesetters, printers, and proofreaders.

Although it is tempting to see printers as proto-capitalists—owing to their strong market orientation and substantial capital needs—their livelihood owed much to the patronage and politics of the court system.

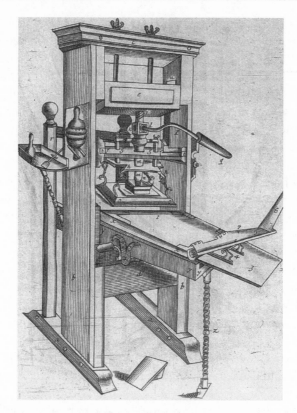

FIG. 1.4. Early Dutch-Style Printing Press.

Christopher Plantin's secret weapon? The pioneering authority on early
printing, Joseph Moxon, identified this as a "new fashioned" printing
press invented in Amsterdam and widely used in the Low Countries. It
is similar to the seventeenth-century printing presses on exhibit today
at the Plantin-Moretus Museum (Antwerp). Joseph Moxon, *Mechanick
Exercises* (London, 1683), plate 4. Courtesy of Illinois Institute of Technology Special Collections.

Plantin's first book, printed in 1555, was an advice manual for the up-
bringing of young ladies of the nobility, printed in Italian and French.
Plantin benefited from the goodwill and patronage of Cardinal Granvelle
and of Gabriel de Zayas, secretary to King Philip II of Spain. With Philip
II's patronage, including a healthy direct subsidy, Plantin in the 1560s conceived and printed a massive five-language Bible in eight volumes. During 1571–76, "his period of greatest prosperity," Plantin kept his many

presses busy printing Bibles and liturgical books for the king of Spain. In these years his firm sent to Spain no fewer than 18,370 breviaries, 16,755 missals, 9,120 books of hours, and 3,200 hymnals. Botanical works, maps, dictionaries, and literary and scholarly works of many kinds rounded out his catalogue. "Plantin's firm in Antwerp thus gleaned a lion's share of rewards from the expansion of an overseas book trade in Spain's golden age," writes Eisenstein. "Plantin himself became not only Royal Typographer but chief censor of all his competitors' output while gaining a monopoly of all the liturgical books needed by priests throughout the far-flung Habsburg realms."[30]

When the political winds shifted, Plantin quickly changed his tack. In 1578, when the Spanish were driven from Antwerp temporarily, Plantin became official printer to the new authorities, the States General, and soon printed many anti-Spanish tracts. But when in 1585 the Spanish threatened to take Antwerp again, and this time for good, Plantin opened a second office in Leiden and for two years served as printer to the newly founded University of Leiden. As Eisenstein observes, "Plantin's vast publishing empire, which was the largest in Europe at the time, owed much to his capacity to hedge all bets by winning rich and powerful friends in different regions who belonged to diverse confessions."[31]

Plantin's massive output suggests the huge scale of book production at the time. In the first fifty years of printing (1450s-1500) eight million books were produced in Europe. The economics of the new technology were impressive. For instance, in 1483 a certain Florence printer charged about three times more than a scribe would have for reproducing Plato's *Dialogues*, but the printer made more than a thousand copies while the scribe would have made just one. Moreover, handwritten scribal copies contained idiosyncratic alterations from the original, while the printer's copies were identical to each other, an additional dimension of quality. This economy of scale sharply reduced the cost of books, which meant that one scholar could have at hand multiple copies from several scholarly traditions, inviting comparison and evaluation. Eisenstein writes, "Not only was confidence in old theories weakened, but an enriched reading matter also encouraged the development of new intellectual combinations and permutations."[32] In this way, the availability of vastly more and radically cheaper information led to fundamental changes in scholarship and learning. Printing lent a permanent and cumulative character to the fifteenth-century Renaissance. The same was true for many other technologies.

Technology and Tradition

We tend to think of technology as cumulative and irreversible, permanent and for all time, but it has not always been so. Technologies have ebbed and flowed with the cultures that they were intrinsically a part of. The Middle Ages became known as the "dark ages" because Western Europeans had "lost" contact with the classical civilizations of Greece and Rome (whose knowledge of science, medicine, and technology resided in the libraries of the Islamic world). Not until the incremental technical changes of the late Middle Ages were combined with the innovation of three-dimensional perspective, courtesy of Alberti, Leonardo, and others, and with the possibility of exact reproduction through printing did technology become a cumulative and permanent cultural element. This cultural shift, which we understand to be characteristic of the "modern" world, deserves to be better understood.

Transfer of technology before the Renaissance could be hit-or-miss. Machines invented in one time, or place, might well need to be rediscovered or even reinvented. Indeed, something very much like this occurred after the great technological advances of Song China (960–1279). The list of technologies invented in China during these years is formidable indeed, including not only paper and printing, but also gunpowder weapons, magnetic compasses, all manner of canals, locks, and hydraulic engineering, as well as the industrial-scale production and consumption of iron. China's ironmasters in the northern Heibei and Henan regions were using coke to smelt iron ore fully five centuries before the English industrial revolution. Chinese silk-reeling mills, farm implements (made of iron), and bridges are among the numerous other advanced technical fields where, as historian Arnold Pacey writes, "techniques in use in eleventh-century China . . . had no parallel in Europe until around 1700."[33]

Yet these pioneering Chinese technologies were not reliably recorded with the rigorous geometrical perspective that allowed Renaissance engineers to set down their ideas about the crucial workings of machines. Chinese drawings of silk-reeling mills, for example, are often so distorted that to Western eyes, accustomed to geometrical perspective, it is very difficult to tell how they should be built. The definitive *Science and Civilisation in China* series offers numerous instances of inaccurate and/or incomplete illustrations of textile machines. In the volume on textile spinning, it is clear that hand scrolls (themselves works of fine art) invariably have the "best" illustrations of textile technology while the post-Song-era encyclopedias and technical books are plagued by incomplete and misleading illustrations. In fact, the progressive corruption of images of silk-reeling machinery in successive encyclopedia editions of 1313, 1530, and 1774 rather clinches the point. In the wake of political disruptions after 1279 Song China's technical brilliance was lost not only to the Chinese themselves but also to the West, whose residents formed an inaccurate and incomplete view of China's accomplishments.[34]

Eugene Ferguson, a leading engineer-historian, has brilliantly shown how quickly technical drawings might be corrupted, even in the West. He compares a series of drawings made by Francesco di Giorgio around 1470 with copies made in the 1540s by artist-scribes in Sienna who had been specially trained to copy mechanical drawings.[35] The results are startling. Francesco's original perspective drawings of hoisting cranes and automobiles are workable, if not always brilliantly designed. But the scribes' copies are something else; many mechanical details are missing or distorted. For instance, a hoisting crane's block-and-tackle system, useful for gaining mechanical advantage in lifting, was reduced in the copy to a single pulley, which provides no mechanical advantage. Similarly, Francesco's original versions of a bar spreader and bar puller resembled modern turnbuckles in that the *right*-handed threads on one side of a square nut and the *left*-handed threads on the other served to push apart or pull together the *ends*. In the copyist's version, however, left-handed threads are shown throughout, creating a device that would simply move the *nut* toward one end or the other. While Francesco's originals rigorously observe the conventions of three-dimensional perspective (so that parallel lines running right-to-left are really parallel while lines running back-to-front converge to a vanishing point at the rear of the drawing), the copyists' versions often have no vanishing point.

For some of these devices, a knowledgeable technical worker might have been able to "see through" the warped perspective of the copied drawing and "fix" the errors. But now imagine that the copyist's version was again copied by hand and that additional distortions were introduced in the process. In time, recreating Francesco's original devices would become hopelessly difficult. His designs, like those of the Chinese technologists, would effectively be "lost." The point is that before the combination of printing and geometrical perspective, inventions made in one generation might not be available to successive generations or for that matter beyond the close circle of colleagues sharing notebooks or craft practices. In these circumstances, a disruption to the social order, like the fall of a ruler and destruction of his library, would entail a disruption in the technological tradition. Technological change could not be permanent and cumulative.

In these terms a permanent and cumulative tradition in technology, enabled by the invention of printing and perspective, appeared first in central Europe's mining industry. Vannoccio Biringuccio's *Pirotechnia* (1540), Georgius Agricola's *De re metallica* (1556), and Lazarus Ercker's *Treatise on Ores and Assaying* (1580) are brilliantly illustrated printed books that even today convey the details of distant technologies. These three volumes have several striking similarities, including once again surprisingly close connections to late-Renaissance court culture, even though these authors came from diverse social backgrounds. Biringuccio was a supervisor of silver and iron mines with extensive practical experience, Agricola was a university-trained humanist, and Ercker was an upwardly mobile mining supervisor who married into a prominent family. As Pamela Long has shown in her studies of early-modern technical writing, each author was associated with the mining industry of central Europe at a time when technology-won efficiencies were urgently needed. Costs were rising because an earlier mining boom (of 1460 to 1530) had already tapped the obvious veins of silver, gold, and copper, while prices were falling due to the influx of Spanish gold and silver stripped from the New World. Thus, wealthy investors and the holders of royal mining rights eagerly consumed information about mining technology that promised substantial economic returns. As Long writes, "the great majority of 16th-century mining books were written by Germans in the regions of the empire where the capitalist transformations of mining were most pronounced—the Harz Mountains near Goslar, the Erzgebirge Mountains in Saxony and Bohemia, and the Tyrolian Alps to the south."[36] Biringuccio, an Italian, had visited mines in this region and wrote to urge his countrymen to make comparable investments in technological changes in their own copper mines.

Each of these three authors flourished by gaining the courtly favor of "prince-practitioners," as Long calls them, who had a special interest in mining and metal technology. Biringuccio (1480–1539) worked for Italian princes, including the Farnese of Parma and Alphonso I d'Este of Ferrara, for the ruling Petrucci family of his native Sienna, and for the Florentine and Venetian republics; and at the time of his death was director of the papal foundry and munitions plant in Rome. Agricola (1494–1555) was born in Saxony during a great expansion of that region's silver mines, graduated from Leipzig University, studied medicine in Italy, and served as town physician in the Erzgebirge region. He turned his close observations of mining and careful investments to his personal enrichment (by 1542 he was among the twelve wealthiest inhabitants of Chemnitz, Saxony) and to courtly recognition by the Saxon prince Maurice, who gave him a house and land in 1543 and three years later made him burgomaster and later councilor in the court of Saxony. His famous *De re metallica* is dedicated to the Saxon princes Augustus and Maurice.

The same Saxon princes figured in the career of Lazarus Ercker (ca. 1530–94). After attending the University of Wittenberg and marrying into a prominent family, he was named assayer at Dresden by the elector Augustus. Within a year, he dedicated a practical metallurgical handbook to Augustus, who named him general assay master for Freiberg, Annaberg, and Schneeberg. Then, Ercker aligned himself with a new prince-practitioner, Prince Henry of Braunschweig, who named him assay warden of the Goslar mint. After dedicating his second book, on minting, to Henry's son, Julius, duke of Braunschweig-Wolfenbüttel, Ercker was named master of the Goslar mint. Moving to Bohemia in the mid-1560s, Ercker was made control assayer in Kutná Hora (Kuttenberg) through the influence of his second wife's brother. His wife Susanne served alongside Ercker as "manager-mistress" of the mint for many years. His *Treatise on Ores and Assaying*, first published in 1574, was dedicated to the Emperor Maximilian II, from whom he received a courtly position. Ercker was named chief inspector of mines by Maximilian's successor, Rudolf II, who also gave him a knighthood in 1586.[37]

Each of these three authors praised the values of complete disclosure, precise description, and openness often associated with the "scientific revolution." These books detailed the processes of mining, smelting, refining, founding, and assaying. Biringuccio and Agricola used extensive illustrations to convey the best technical practices of their time (see fig. 1.5). Biringuccio's *Pirotechnia* features 85 wood-block engravings while Agricola's *De re metallica* has more than 250. Agricola hired illustrators to make detailed drawings of "veins, tools, vessels, sluices, machines, and furnaces . . . lest descriptions which are conveyed by words should either not be understood by the men of our times or should cause difficulty to posterity."[38]

Even the less practical illustrated volumes known as "theaters of machines" published by Jacques Besson (1578) and Agostino Ramelli (1588) reflected the support of aristocratic and royal patrons. These books were

FIG. 1.5. Agricola's Lifting Crane.

Agricola hired illustrators to make detailed drawings, like this vivid illustration of a two-directional lifting crane powered by 36-foot-high waterwheel, "lest descriptions . . . conveyed by words . . . not be understood." Georg Agricola, *De re metallica* (1556; reprint, Berlin: VDI Verlag, 1928), 170. Courtesy of Illinois Institute of Technology Special Collections.

something like published versions of Leonardo's technological dreams, in that they recorded often highly imaginative designs that no one had built. Besson was engineer at the court of France's King Charles IX. Ramelli, an Italian military engineer, was also in the service of the king of France when his book—*Diverse and Ingenious Machines*—was published. While Besson's book had 60 engraved plates, Ramelli's book featured 195 full-page copperplate engravings of grain mills, sawmills, cranes, water-raising machines, military bridges, and catapults. Ramelli devised more than 100 different types of water-raising machines. Eugene Ferguson observes that Ramelli "was answering questions that had never been asked and solving problems that nobody . . . [had] posed. There is no suggestion that economic forces induced these inventions. The machines were clearly ends, not means." Yet many of Ramelli's machines, including a sliding vane water pump that is original to him, were eventually realized, as the problems of friction and materials were overcome. Ferguson concludes, "The seeds of the explosive expansion of technology in the West lie in books such as these."[39]

The scientific revolution was also surprisingly dependent on printing technology and courtly patronage networks. Tycho Brahe, a young Danish nobleman who taught himself astronomy through printed books in the 1560s, had access not only to accurate printed versions of all of Ptolemy's work (including a new translation of the *Almagest* from the Greek), but also to tables that had been recomputed on the basis of Copernicus's work, to printed sine tables, trigonometry texts, and star catalogues. Tycho could directly compare star tables computed from Copernicus and Ptolemy. On the island of Hveen, a gift from the Danish king, Tycho's observatory, Uraniborg, had no telescope; but it did have a well-stocked library, fifty assistants, and a busy printing press.

Galileo was not only a first-rate scientist but also a prolific popular author and imaginative scientist-courtier. His career began with a landmark study of mechanics that owed much to his observations of large hoisting machines at the Arsenal of Venice, and he secured his fame with another technological breakthrough—the telescope. Among the astronomical discoveries he reported in *Starry Messenger* (1610), something of a best-seller across Europe, were the four moons of Jupiter. Galileo named them the "Medicean stars" in a frank bid for the favor of Cosimo de Medici, the namesake and latter-day heir of the great Florentine powerbroker. In 1616 the Catholic Church placed Copernicus's work on its *Index* of prohibited works, while Galileo's own *Dialogue on Two World Systems* (1633), which left little doubt about his sympathies for the Copernican system, also made the *Index*. Because of his open support of the sun-centered cosmology, Galileo was placed under house arrest, where he worked on the uncontroversial topic of classical mechanics. Yet even this treatise had to be smuggled out of Italy by a Dutch printer, Louis Elsevier, and printed in Leiden as *Discourses on Two New Sciences* (1638). "I have not been able," wrote Galileo in 1639, "to obtain a single copy of my new dialogue. . . . Yet I know that they circulated through all the northern countries. The copies lost must be those which, as soon as they arrived in Prague were immediately bought by the Jesuit fathers so that not even the Emperor was able to get one."[40]

THE DESIRES AND DREAMS of Renaissance courts and city-states defined the character of the era's technology and much of the character of its culture. Leonardo, Francesco, Alberti, and other engineers of the Renaissance era worked on war making, city building, courtly entertainments, and dynastic displays because that is what courtly patrons valued and that is what they paid for. We can easily broaden our view to include Italy's famous luxury glass and fancy silk industries and the impressive state-directed Arsenal of Venice without significantly changing this basic picture of court dominance. In a major study of the silk industry, Luca Molà writes, "While in the fourteenth and in the first half of the fifteenth century merchant oligarchies ruling over Italian cities dedicated considerable energy to the development of silk manufacturing, later on champions of the new industry were to include the major princely families of the peninsula (such as the Visconti, the Sforza, the Gonzaga, the Este, the Medici, the Savoia, the della Rovere), more than one pope, and the monarchs of Spain, France, and England."[41] Courts in China, Turkey, India, the Persian empire, and Japan also during these years, in varied ways, channeled technologies toward court-relevant ends.[42]

The patrons of Renaissance technologies, especially when compared with those of eras discussed in later chapters, were not much concerned with labor-saving industrial technologies or with profit-spinning commercial ones. Similarly, the early generation of moveable type printing was deeply dependent on European courts for large-scale printing jobs, while authors of the books that made technology into a cumulative and progressive tradition depended on courtly patronage networks. The pervasiveness of the court system in the Renaissance should not really surprise us, since it was the dominant cultural and political actor at the time, fully analogous to the commercial and industrial institutions as well as the nation-states, corporations, and government agencies that followed with different imperatives and visions for technologies.

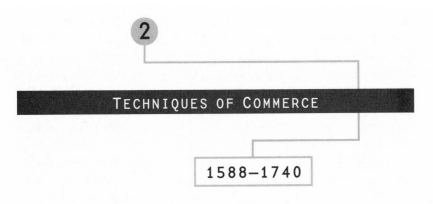

2

TECHNIQUES OF COMMERCE

1588–1740

THE NOBLE COURTS, city-states, and prince-practitioners who employed Renaissance technologists to build cities, wage war, entertain courts, and display dynasties were not using technologies principally to create wealth or improve industries. Rather, they were using their wealth—from land rents, banking, and mercenary activities—to pay for the creation and deployment of technologies.[1] This practice shaped the character of that era's technology and—through the resulting cathedrals and sculptures, urban palaces and rural villas, court automata and printed books—the character of Renaissance society and culture as well. We have seen how this occurred in the growth and expansion of the mining industry in central Europe around 1600 and in the profusion of court-sponsored technology books on mining.

The imperatives of creating wealth reshaped the content and purpose of technology during the great expansion of commerce across Europe. In Venice, Florence, and other Italian city-states commercial activities began expanding even before the Renaissance, of course, but the commercial era was fully realized a bit later in Antwerp, Amsterdam, and London. Each of these three cities was the node for far-flung trading networks, constructed when commercial traders following up on the pioneering "voyages of discovery" created maritime trading routes to Asia, Africa, and the New World. Even though no single year marks a shift from one era to another, the influence of Renaissance-era courts was on the wane by around 1600 while the influence of commerce was distinctly rising.[2] It is important to emphasize the historical distinctiveness of the commercial era and to avoid reducing it to an "early" but somehow failed version of industrial capitalism. As we shall see, the era of commerce was thoroughly capitalistic but not industrial in character. The imperatives of commerce included carrying goods cheaply, processing them profitably, and funding the means for shipping and trading. Technologies such as innovative ship designs, import-processing techniques, and a host of financial innovations reflected these commercial impulses, just as attack chariots, court automata, and princely palaces expressed the court vision of Renaissance patrons of technologies.

The age of commerce, anticipated in Spain and Portugal as well as in China and India, found its fullest expression during the seventeenth-century Golden Age of the Dutch Republic. The Dutch Republic at this time was reminiscent of Renaissance Italy in that all manner of cultural activities flourished to "the fear of some, the envy of others and the wonder of all their neighbors." Indeed, the many parallels between the older centers in the south of Europe and the rising commercial centers in the north led Fernand Braudel to posit long-term "secular trends" in which each city-state had, as it were, its historical moment at the center. Yet such a view, however appealing as an overarching world-historical theme, suggests a troubling inevitability about historical change and undervalues how the Dutch developed new technologies to capture the leading economic role in Europe and construct a trading empire of unprecedented global scope.[3]

Venice had dominated trade in the Mediterranean, while Spain captured the riches of the New World; but the Dutch Republic became the center of a worldwide trading and processing network that linked slaves from Africa, copper from Scandinavia, sugar from Brazil, and tea and spices from Asia, with maritime, processing, and trading technologies at home. It was an improbable success for a tiny state that lacked manpower, raw materials, and energy sources. "It seems a wonder to the world," puzzled one English writer of the time, "that such a small country, not fully so big as two of our best shires [counties], having little natural wealth, victuals, timber or other necessary ammunitions, either for war or peace, should notwithstanding possess them all in such extraordinary plenty that besides their own wants (which are very great) they can and do likewise serve and sell to other Princes, ships, ordnance, cordage, corn, powder, shot and what not, which by their industrious trading they gather from all the quarters of the world."[4] This chapter examines how citizens of the Dutch Republic shaped technologies in the pursuit of commerce and how commercial technologies shaped their culture.

Technology and Trade

The Rhine River is the central geographical feature that propelled the so-called Low Countries to economic preeminence even before the seventeenth century and continues to shape commercial prospects there today. The Rhine begins with melting ice in the mountains of central Switzerland and runs 800 miles through the heart of western Europe. The river passes in turn through the cities of Basel, Strasbourg, Cologne, Düsseldorf, and Rotterdam before exiting to the North Sea. Its several tributaries connect to a score of commercial centers, including Antwerp, Frankfurt, Stuttgart, and Zurich. Even today, the Rhine's vessel-tonnage nearly exceeds the tonnage carried on the Mississippi River and St. Lawrence Seaway combined. Handling this flow of cargo occupies two of the largest ports in the world, Rotterdam and Antwerp, which organize the multifarious businesses of offloading, processing, and reexporting the commerce of all Europe. Yet, these basic port and processing functions have changed surprisingly little since the Dutch Golden Age, allowing, of course, for the gargantuan scale of today's port complexes. A high level of trading activity occurred early on; by 1550 imports for the Low Countries were already four times those for England and France in per capita terms.

The Dutch Republic as a political and cultural entity took form in the wake of Luther's Bible-printing Reformation. Beginning in the 1570s, Catholic Spain and the Protestant core of the northern Low Countries, the United Provinces, fought for control over the region, then part of the Spanish empire. Spain's dominance over the southern provinces of Flanders and Brabant, the campaign for which included its recapture of Antwerp in 1585, prompted many merchants and craftsmen, including the theologically flexible printer Christopher Plantin (see chapter 1) to set up shop elsewhere during these unsettled years. The tide turned, however, when the English Navy defeated Spain's "Invincible" Armada (1588) and thereby removed a certain threat to the Dutch maritime economy. During the next decade, the Dutch blockaded Spanish-occupied Antwerp, cutting off that rival port from North Sea shipping and striking a blow to its leading role in commerce. The Dutch then forcibly reclaimed from Spain the inland provinces surrounding the Zuider Zee.

During these battles Maurice of Nassau, the Dutch military commander, turned handheld firearms into winning weapons. He transformed the matchlock muskets of the time into formidable battlefield weapons through systematic drilling of his soldiers. Maurice divided his troops into small units and separated the loading, aiming, and firing of a musket into forty-two individual steps, each one activated on the battle-field by a particular shouted command. For instance, steps eleven to six-teen directed each soldier: "hold up your musket and present; give fire; take down your musket and carry it with your rest [pole that supported mus-ket during firing]; uncock your match; and put it again betwixt your fin-gers; blow your [firing] pan." Maurice drilled his soldiers to form revolv-ing ranks in battle: in the protected rear rank, reloading their muskets; in the middle, preparing to fire; in the front, firing a coordinated volley at the enemy; and then falling back in orderly fashion to reload. Drilled to make fewer mistakes, the disciplined Dutch soldiers, when directed, let loose deadly barrages of musket balls. The Spaniards fell in droves.[5]

As a result of Maurice's military victories, the newly formed Dutch Republic in 1609 secured an advantageous truce with Spain that granted it political independence. The new political confederation comprised the maritime provinces (Zeeland, Holland, and Friesland, which had feder-ated thirty years earlier) and the inland provinces to the north. They were assertively Protestant. The southern provinces (more or less today's Bel-gium) remained under the control of Catholic Spain. "This had disastrous effects on the once wealthy economies of Brabant, Flanders, and the Wal-loon provinces," writes economic historian Joel Mokyr. "Spanish merce-naries devastated the land, a huge tax burden crippled the economy, and the relentless persecution of Protestants prompted thousands of highly skilled artisans to flee abroad, where they strengthened their homeland's competitors."[6]

Those fleeing the Spanish-controlled lands included craftsmen and sailors as well as merchants and financiers. Walloon exiles introduced new mills for the "fulling," or finishing, of woolen cloth at Leiden and Rotter-dam, in 1585 and 1591, respectively. Many merchants fleeing the Spanish-dominated lands initially went to northwestern Germany, then settled in Amsterdam. In Amsterdam, Walloon exiles were among the leading share-holders in the Dutch East India Company (discussed below). Walloons and other groups who had suffered religious persecution welcomed the climate of religious tolerance in the Dutch Republic, exceptional in Eu-rope at the time, which extended across the spectrum of Protestant sects and included even Catholics and Jews. Jews became leading figures in Am-sterdam's tobacco and diamond trades.

Louis de Geer (1587–1652) numbered among the emigrant merchants and technologists whose influence was felt from Portugal to Prague, especially in the fields of fortification, urban drainage, mining, and harbor engineering. De Geer, originally from a Liège family, began his career in the Dutch city of Dordrecht trading a mix of goods, then moved to Sweden. De Geer's extensive involvement with Sweden exemplifies the substantial Dutch technical influence in northern Europe as well as a subtle but pronounced shift from court to commerce. Sweden at the time was seeking to develop its rich mineral holdings in iron and especially in copper, much in demand for coinage and cannons. The king of Sweden chartered the Royal Company (in 1619) to win back control of the copper trade, which had slipped to Amsterdam. Since Sweden's state finances were dependent on copper, the trade was a matter of some urgency. This court-sanctioned venture did not succeed, however, and in the next decade control of the Swedish copper industry passed firmly into Dutch hands. Dutch investors before and after the Royal Company's failure provided investment capital to Sweden that required repayment in copper, which tilted the field against the crown and toward commerce.

Louis de Geer played a leading role not only in routing Swedish copper exports to Amsterdam but also in transferring valuable mining and smelting technologies to Sweden. In Stockholm, de Geer created the largest industrial complex in Sweden, making iron and brass, ships and ropes, cannon and cannon balls. The *garmakeriet* method for refining copper was an import from the southern Netherlands, as was the *vallonsmidet* (literally, "Walloon smithy") process for ironmaking. Another Dutchman, with the backing of the Swedish king, set up a famous iron works at Eskilstuna that grew into another of Sweden's major industrial districts. At the peak of its copper industry, around 1650, Sweden accounted for fully half of Europe's total copper production. The only other nation with substantial copper production at the time was Japan, and Dutch merchants controlled its copper trade, too.[7]

The emergence of specialized ship designs in the Netherlands was another early signal that the Dutch understood how to bring technology and trade together in the pursuit of commerce. Most seafaring nations either possessed the raw materials needed for wooden shipbuilding or had ready access to colonies that did, while the Dutch had neither. Among the necessary raw materials for wooden shipbuilders were, obviously, timber for masts and planking; resin, pitch, and turpentine (naval stores) for waterproofing; and rope for rigging. In a sense Dutch shipbuilding began with wheels of cheese, which along with other export goods, including salt, wine, and herring, were shipped off to Norway and the Baltic countries in exchange for timber and naval stores. Dutch economic dominance relied on savvy trading, such as that practiced by the merchants of Dordrecht and Zeeland, the leading exporters of German and French wines, respectively. The real key, though, was in spotting where a low-value import such as salt from Spain, Portugal, or France could be processed into a high-value export. In this respect the salt-refining industry in Zeeland was an important precursor to the distinctive and wide-ranging "traffics" system, discussed below.

An innovative, purpose-built ship secured to the Dutch an early and thorough dominance of the North Sea herring fishery. The full-rigged herring buss was an unusually large fishing vessel—with a crew of fifteen or more—and was virtually a self-contained factory, designed to stay out in open seas for up to eight weeks and through the roughest weather while carrying the salt and barrels and manpower needed to gut, salt, and pack the freshly caught herring right on board (fig. 2.1). By the 1560s the northern provinces already had around 500 herring busses; the largest brought home 140 tons of packed fish. Yet the real distinction of the Dutch herring fishery was not so much its volume of production but rather the consistently high *quality* of the packed herring and their correspondingly high trading value—characteristics that one finds again and again in the Dutch commercial era and that sharply distinguish it from the industrial era that followed.

Use of the factorylike herring busses was just one of the distinctive Dutch maritime activities. The province of Holland at that time had 1,800 seagoing ships. (Amsterdam's 500 ships easily exceeded Venice's fleet of seagoing ships, estimated at 300 at its peak around 1450.) The labor force required just for this shipping fleet was perhaps 30,000 men, not counting the thousands of men needed to build the ships, rigging, and sails and to move the off-loaded goods within the country. This shift of labor out of the agricultural sector, a defining feature of "modern" economies, relied on the large imports of Baltic grain that allowed Dutch rural workers to specialize in exportable goods, like cheese and butter, further fanning the export trade. For a small population—the Dutch Republic had two million at its height in the seventeenth century—such an export-oriented agriculture was indispensable.[8]

The Dutch effected their most brilliant departure from traditional ship designs in the field of commercial trading vessels. "English shipping in the late sixteenth century was, generally speaking, multi-purpose and, being often used for voyages to the Mediterranean, tended to be strongly

FIG. 2.1. Dutch Herring Buss.

Something like a fishing vessel married to a factory, the herring buss carried all the salt, wooden barrels, and labor needed to transform freshly caught North Sea herring into a marketable commodity. Courtesy of The Kendall Whaling Museum, Sharon, Mass.

constructed, well manned, and well armed," writes Jonathan Israel. "Dutch shipbuilders, by contrast, concentrated on low-cost hull forms that maximized cargo space, discarded armaments, and used only the simplest rigging. Dutch shipping was designed for minimal crews and maximum economy."[9] Venice's ships were, even more than Britain's, tailored to the commerce of the pirate-infested Mediterranean (and given the recurrent European wars and the long-running contest with the Ottoman Turks, one country's detestable "pirates" might be another country's heroic privateers). Spanish ship designs were oriented to the essential imperial lifeline of the New World silver trade, where heavily armed sailing ships packed with bullion plied the Atlantic. The same Spanish ship might, between silver runs, be ordered to shuttle goods or ferry soldiers to a European port. The higher costs of moving goods with armed ships, where space for soldiers and weaponry crowded out cargo capacity, hampered the commercial prospects of both Venice and Spain.

The famous Dutch *fluytschip* was in several ways the precise embodiment of commerce. The fluyt, a distinctive cargo ship design which dominated northern European shipping for a century or more, emerged from the city of Hoorn in the mid-1590s (fig. 2.2). This was no graceful clipper ship with sleek lines and tall masts. The fluyt was instead stubby and squat: its sail area was small and its full-rigged masts short compared to its capacious cargo hold. The vessel was ploddingly slow, yet with its excellent handling qualities and the extensive use of pulleys and blocks, a small crew of around a dozen could easily control the yards and sails. Another economy was effected by the use of cheap light pine in the vessel, except for the heavy hull, made of oak to resist salt water corrosion.

The basic fluyt design sacrificed the standard reinforced construction, which every armed ship needed to support the weight and recoil of cannons, in exchange for increased cargo space and simplicity of handling. The resulting ship—an artifact shaped by commerce—was exquisitely well adapted to the peaceful trade of northern Europe, if not for the more dangerous waters of the Mediterranean or the Atlantic. Since transit taxes in the Baltic were levied according to a ship's deck area (rather than its cargo capacity) Dutch fluyt builders logically enough reduced the topdeck's area while retaining a large cargo hold underneath. In fact, surviving models of Dutch fluyts from this era are so narrow-decked, wide-bodied, and round-shaped at the bow and stern that they appear to be whimsically "inflated" versions of real ships; nevertheless, these models agree closely with contemporaneous documentation of the ships' actual shapes and sizes.

Following on these innovations, Dutch shipbuilders practiced a high degree of design specialization during the seventeenth century. In contrast with most other European countries, where military ships and commercial ships were used interchangeably, Dutch shipbuilders segregated military and commercial shipbuilding while segmenting the basic fluyt design. "The output of private shipbuilding yards was so much greater than that of naval yards that the republican Admiralties found it almost impossible to meddle in the affairs of commercial shipbuilders," writes Richard Unger. In the Netherlands, "the design of warships was rather influenced by successful experiments with merchant vessels, the inverse of what happened in the rest of Europe."[10]

The Dutch East India Company (discussed below) had its own shipbuilding yard, where it evolved a distinctive broad-hulled design for the famous East Indiaman. Also known as a *retourschip,* this hybrid vessel was

FIG. 2.2. Dutch Cargo Fluyt.

Line drawing of a fluyt; the original vessel was 100 feet long. Courtesy
of The Kendall Whaling Museum, Sharon, Mass.

a larger and more heavily armed version of the pinnance, itself an armed
version of the fluyt. But while most fluyts were no larger than 500 tons
displacement, the largest *retourschip* displaced 1,200 tons and might be
45 meters long. Meanwhile, Dutch commercial shipbuilders tailored the
basic fluyt design to many special requirements. Whaling fluyts had
double-thick bows to protect against the ice in far northern waters; timber fluyts had special hatches to permit the loading of entire trees; and
the so-called *Oostvaarder* fluyts were specially designed for cheating the
Danish king out of taxes. For a time, until the Danish saw through the
scheme, the *Oostvaarders* had bizarre hourglass-shaped topdecks that
were extra narrow precisely at the high midpoint where the all-important
tax measurement was taken.

The commerce-inspired designs for herring busses and cargo-carrying
fluyts are impressive evidence of the Dutch responsiveness to commerce.
Yet the real distinction of the Dutch was to take a set of innovations and
make them into broad society-wide developments that shaped Dutch
culture not only at the top class of wealthy merchants and investors but

also down through the merchant and artisan classes. Even rural workers milking cows for cheese exports participated in the international trading economy. The distinctive *trekvaarten* network of horse-towed barges provided scheduled passenger service throughout the western region of the Netherlands, enabling merchants and traders to travel to and from a business meeting within the day. The Dutch impulse to broaden the commercial economy manifested itself in uniquely broad ownership of ships and, as we will soon see, of much else.[11]

At other European shipping centers of the time, the owning of a single ship had been divided among three or four investors. Impressed with how multiple-share ownership helped raise money and spread the risk of losses, the Dutch took the practice much further. For the bulk-carrying fluyts, as well as for fishing, whaling, and timber-carrying ships, ownership of sixteenth, thirty-second, or even sixty-fourth shares became common. At his death in 1610, one Amsterdam shipowner left fractional shares in no fewer than 22 ships. Middle-level merchants connected with the bulk trades were typically the collective owners of the fluyts. Technology investments similarly took collective forms, as when twelve Amsterdam brewers shared the cost for mechanizing malt grinding or in the extensive multishare ownership of linseed oil mills in the Zaan district (see below). It is worth emphasizing that, although history remembers the wealthiest merchants who patronized the famous painters and built the finest town houses, the Dutch economy, in shipping and in many other sectors, got much of its vitality from an unprecedentedly broad base of investment and activity. In short order, the Dutch led all the rest of Europe in shipbuilding. A contemporary estimate had it that in the late 1660s fully three-quarters of the 20,000 ships used in the European maritime trade were Dutch, with England and France falling into distant second and third places.[12]

Creating Global Capitalism

While the British East India Company had its heyday somewhat later, presiding, as we will see, over nineteenth-century British imperialism in India, its Dutch counterpart (1602–1798) was quick off the mark. By the middle of the seventeenth century, the Dutch East India Company's economic reach and political influence spread from its singular port at Nagasaki, Japan, clear across the Far East to Yemen in the Middle East. The company's trading patterns—a web of intra-Asian routes in addition to bilateral Asian-European routes—differed fundamentally from the simple bilateral trading practiced by Europeans since the days of the legendary Silk Route. By comparison, the "global" trading empire of which Spain boasted was really mostly an Atlantic one; from the 1530s onward, its showpiece West Indies trade centered on extracting silver from Mexico and Peru. Spanish trading in the Pacific focused on the Philippines, whose trade was bound for Mexico. Goods from the Philippines reached Spain only after they were transshipped across land in Mexico and then put on a second vessel and sent onward across the Atlantic. In the 1620s the Spanish lost their dominance over the Chinese silk trade and were soon expelled from Japan. At the British East India Company significant intra-Asian trade emerged only in the 1830s. At the height of Dutch activities in Asia during the mid-seventeenth century,

> Bengal sugar and Taiwan sugar were sold in Persia and in Europe, while silk from Persia, Bengal, Tongking, and China was sold in Japan and in Europe. One key set of exchanges could be traced as follows: Pepper, sandalwood, and other Southeast Asian goods were sold in China . . . , and gold, tutenag [a semiprecious alloy of copper, zinc, and nickel], raw silk, and silk fabrics were bought. Much of the raw silk and silk fabrics was sold in Japan, and the silver earned, along with the gold and tutenag from China, was taken to India to buy cotton fabrics, which were the principal goods that could be exchanged for spices in Indonesia.

The Dutch—through their East India Company in the Pacific and West India Company in the Atlantic, coupled with the extensive trading in Europe and Africa—in effect created the first truly global economy.[13]

Underlying this global commercial expansion were extensive Dutch innovations in the basic institutions of commercial capitalism, including commodity exchanges, a public exchange bank, and a stock exchange. If the Dutch did not exactly invent capitalism, they created the first society where the principles of commerce and capitalism pervaded the culture. Capitalism is typically traced back to the merchants of Venice, and it is true that at least six other European cities had commodity exchanges during the sixteenth century. But while these exchanges had typically set only regional prices, Amsterdam's exchanges became the world center of commodity pricing and trade flows. The commodity traders' guild began publishing weekly lists of prices in 1585. Within a few years, the Amsterdam commodity exchanges—for grain, salt, silks, sugar, and more—had surpassed their regional rivals and become a set of global exchanges. By 1613 the Amsterdam Exchange was lodged in a new building, where a well-

FIG. 2.3. Delivery of a Futures Contract.

Amsterdam's financial markets dealt extensively with futures contracts in grain, wool, silks, and even the "buying of herrings before they be catched." Courtesy of The Kendall Whaling Museum, Sharon, Mass.

organized guild ensured honest trades by tightly regulating its 300 licensed brokers. Grain merchants soon built a separate grain exchange. By 1634 the commodities exchange's weekly bulletins set authoritative prices across Europe for 359 commodities; the exchange added nearly 200 more in the next half-century.

The growth of the Amsterdam exchanges can be measured not only in their size, scope, and specialization but also in their financial sophistication. For example, futures contracts emerged, speculating on grain that had not yet been delivered and the "buying of herrings before they be catched." In time, Amsterdam merchants were purchasing such varied goods as Spanish and German wools and Italian silks up to twenty-four months in advance of their arrival. Issuing maritime insurance became yet another financial activity linked to global trade. At least until London in the 1700s (see chapter 3), there was simply no rival to Amsterdam in the breadth, depth, and refinement of its financial markets.[14]

The rising volume and complexity of trading in physical commodities depended on sophisticated means of trading money. The Amsterdam Wisselbank, or exchange bank, founded in 1609, was the first public bank outside Italy and was without peer in northern Europe until similar banks were organized in Hamburg and London, in 1619 and 1694, respectively. Before the advent of exchange banks, a merchant needing to pay for a delivery of foreign goods might visit a money-changer to purchase sufficient foreign currency. Where possible, merchants usually preferred so-called bills of exchange, which were in effect a merchant's paper promise to pay the bearer a certain sum of money on demand. However, bills of exchange issued by merchants circulated at a profit-sapping discount that varied with the trustworthiness of the merchant and the trading distance.

The Amsterdam Wisselbank slashed these cumbersome transaction costs. In essence, it provided a means for merchants to quickly and efficiently pay bills, backed by its huge pool of capital. By 1650 a merchant's deposit into the Wisselbank, because of the certainty of getting payment in settling an account, commanded a premium (not discount) of 5 percent. Secrecy surrounded the Wisselbank's accounts; an estimate from the early eighteenth century put its holdings in coin or specie at an astounding 300 million guilders, although modern scholars believe the bank held no more than one-tenth this amount. This deep pool of exchange capital resulted in a ready measure of confidence: Dutch interest rates were half those in England and an even smaller fraction of those in France and Germany. This meant that a Dutch merchant borrowing money to purchase goods paid substantially lower interest charges and pocketed the difference. As one Englishman put the matter in 1672, low Dutch interest rates "hath robbed us totally of all trade not inseparably annexed to this Kingdom."[15]

It would be a mistake, even in the face of this determined financial innovation, to see the Dutch solely as sober paragons of economic calculation. After all, in the 1630s the Dutch fostered the great tulip mania, certainly the most colorful economic "bubble" the world has ever seen. Whatever the intrinsic value of a flower bulb, crazed Dutch traders bid tulip prices to remarkable heights. Already by 1633 an entire house in the shipbuilding town of Hoorn was traded for three rare tulips. More to the point, tulip trading embodied several of the classic Dutch financial techniques, including futures contracts, commodity pricing, and multiple-share ownership. Traditionally tulip bulbs had changed hands only during the months from June to September when the bulbs were "lifted" from the soil after flowering and were available for inspection. The first step toward a futures market in tulips was when traders, sensibly enough, began accepting a future delivery of a particular tulip bulb; a note on a slip of paper promised delivery when the bulb was lifted the following June.

The circulation of these valuable slips of paper seems quickly to have gotten out of hand. It was difficult to know, in the absence of any regulatory scheme, whether a given slip of paper represented a "real" tulip in the soil or merely the hope of a seller's obtaining one in the future. An increasing number of traders cared not at all to possess a physical bulb but simply to buy these notes low and sell them high. Other tulip trading techniques modeled on the commodity exchanges were the elaboration of weight-based pricing and the distinction between "piece" and "pound" goods. The most valuable bulbs were always sold as a "piece" good, after 1635 invariably priced according to their weight in "aces" (around one-twentieth of a gram), while aggregate lots of the less valuable bulbs were traded as "pound" goods similar to the practice in trading commodified grain. Finally, in a move that made tulips more like ships than flowers, at least one Amsterdam grower sold a half share in three expensive bulbs to a hopeful customer.

These financial practices, fanned by greed and fear, pushed tulip prices to a sharp peak in February 1637. In the first week of that month an exceptionally fine harvest of tulip bulbs, owned outright by the surviving children of an Alkmaar tulip grower named Wouter Winkel, was auctioned with great fanfare. The prize in the collection was a mother bulb, including a valuable offset, of the variety Violetten Admirael van Enkhuisen, for which a private buyer paid 5,200 guilders, while the most valuable bulbs auctioned publicly were two Viceroys that fetched 4,200 and 3,000 guilders. Dozens of other bulbs, both piece and pound goods, commanded record prices. (At the time, 5,000 guilders would purchase a large townhouse in Amsterdam while the lesser sum of 3,000 guilders, as one contemporary pamphleteer enumerated it, would purchase 2 lasts of wheat [around 24 tons], 4 lasts of rye, 4 well-fed oxen, 8 well-fed pigs, 12 well-fed sheep, 2 oxheads of wine, 4 tons of eight-guilder beer, 2 tons of butter, 1,000 pounds of cheese, 1 bed with accessories, 1 stack of clothes, and 1 silver chalice. And a boat to load the goods into.) Altogether, Winkel's prime lot of tulips brought in the stupendous sum of 90,000 guilders.[16]

One might guess that tulip traders at the height of the mania, while calculating and profit-minded, were far from sober. While tulip trading in earlier years had been the province of wealthy connoisseurs, the run-up in tulip prices in the 1630s drew a large number of fortune seekers into the trade. Instead of the tightly regulated commodity exchanges or the Amsterdam Stock Exchange, which opened in 1610 and oversaw a set two-hour trading day, the trading floor for tulips consisted of the back rooms of inns, where trading might extend from ten o'clock in the morning until the early hours of the next day. With wine for the wealthy, and beer and cheap spirits for everyone else, Dutch taverns had long-established traditions of heavy social drinking. "All these gentlemen of the Netherlands," complained one Frenchman, "have so many rules and ceremonies for getting drunk that I am repelled as much by the discipline as by the excess." Traditionally a center of beer brewing, Haarlem emerged as a center of tulip trading also; its taverns served up 40,000 pints of beer each day, one-third of the city's output—not bad for a population of just 30,000 men, women, and children. The combination of tulip trading and taverns is no coincidence. "This trade," offered one contemporary account, "must be done with an intoxicated head, and the bolder one is the better."[17]

In the same month that the Winkel bulbs drew their astronomical prices, the price of tulips collapsed, wiping out many down-market tulip speculators. Very shortly, the buzz about priceless tulip bulbs decayed into a squabble over worthless tulip futures. The crash had little effect on the mainstream commodity exchanges, the stock exchange, or Wisselbank. During the same years as the tulip mania, speculation was rampant in shares of the United East India Company (Verenigde Oostindische Compagnie, or VOC). The VOC represented a typically Dutch solution to the problems of importing spices from the East Indies and contesting the dominance of Portugal and Spain. Beginning in the 1590s, Dutch merchants had flocked into the lucrative East Indian trade. The states of Holland and Zeeland offered them not only free import duties but also free firearms. By 1601 there were fourteen separate Dutch fleets comprising sixty-five ships, all plying East Indian waters. The economic consequences of this free-for-all, however, were ominous. The prices Dutch traders paid for pepper in the East Indies had doubled over the previous six years, the prices they charged to Dutch consumers back home had fallen, and their profits had tumbled.

This ruinous competition came to an end when, at the merchants' request, the Dutch federal government assembled the directors of the numerous trading companies at The Hague. In 1602 the conference created the VOC. It was a state-supported, joint-stock trading monopoly, with a decentralized federal structure. The VOC's four "chambers" raised separate capital and kept separate books and chose separate directors. And even though Amsterdam claimed half the total votes of the VOC's board of directors, because of that city's leading role in financing the existing East Indies trade, the prospect of Amsterdam's having such a large voting bloc went against the pervasive preference for federalism and decentralized power. Amsterdam got eight of seventeen voting seats. Four seats went to Zeeland, two each to the North Quarter (Hoorn and Enkhuizen) and South Holland (Delft and Rotterdam), with the last seat rotating among the three *non*-Amsterdam partners. The VOC's founding capitalization of 6.4 million guilders was comparable to the value of 1,000 large town houses. Over half of this total capitalization (57 percent) came from the Amsterdam chamber. While the richest of Amsterdam's native merchant elite invested sums of 12,000 to 30,000 guilders, Walloons exiled from the southern Netherlands made the very largest individual investments, contributing sums of from 18,000 to 85,000 guilders.[18]

The VOC was much more than a harmless Dutch spice-trading cartel. It was in fact intended to break Spain and Portugal's political and trading dominance in Asia, and the VOC from its founding was granted wide-ranging commercial, military, and political responsibilities. The Dutch federal government gave it warships and an annual subsidy, raised to 200,000 guilders in the mid-1610s. The warships worked wonders. "The Dutch ships were very large and tall, and each carried more than a thousand guns and small arms; they blocked the whole strait," wrote one beleaguered Chinese official. Within three years of its founding, the VOC wrested control of the legendary Spice Islands from the Portuguese and thereby cornered the world trade in cloves, mace, and nutmeg. Further gains in Asia—effecting a network of alliances with local rulers in southern India and the Spice Islands—pushed VOC share prices on the Amsterdam Exchange above 200, or double their face value. However, the independence-making truce between the Dutch Republic and Spain in 1609, because it implied a resurgence of trade and influence by Spain and Portugal, sent VOC shares down sharply; they did not recover for years.[19]

The key to the lucrative spice trade, as the European rivals understood, was in Asia itself. India in the preindustrial era had a thriving cotton industry, and its cloth could be traded for spices. On the southeast coast of India and on the innumerable islands of what is now Indonesia, each of the trading countries sought to establish trading alliances; and when these alliances were betrayed, they tried unarmed trading "factories" (warehouse-like buildings where "factors"—traders—did business). When these were attacked, they built heavily fortified factories garrisoned with soldiers. The Dutch, backed by its fleet of forty-odd warships, erected a massive stone fortress at Pulicat, in the southeast of India, that became the leading European factory in the region and the centerpiece of Dutch control of the Indian cotton trade. By 1617 there were twenty Dutch fortress-factories across Asia. The Portuguese too built fortress-factories in India, at Goa and Ceylon; the Spaniards, also, in the Philippines; while the English went after the Spice Islands. These "factories" sometimes finished cotton cloth imported from the countryside and as such can be considered the first cotton-textile factories in the world.[20]

Beginning in the 1620s, the VOC created a vast intra-Asian trading network. Not gold or silver bullion but goods were the currency of exchange. "Shipping fine spices, Chinese silks and porcelain, and Japanese copper to India, the Company purchased cotton textiles [in India] . . . which it then sold in [Indonesia] for pepper and fine spices. In the same way, spices, Chinese silks, Japanese copper, and also coffee from Mocha helped pay for the VOC's purchases of silk and drugs [opium] in Persia. Pepper and spices also supplemented silver in the Company's purchases of Chinese wares, on Taiwan." When in the 1630s the Japanese government restricted its merchants' conduct of foreign trade (one of several harsh measures to preserve its court-based culture), the Dutch took over Japan's active silk and copper trades. As of 1640, the Dutch trading factory at Nagasaki was the sole remaining European trade link with all of Japan. VOC share prices, scarcely registering the collapse of the tulip mania, climbed all the way to 500.[21]

It was also in the go-ahead decade of the 1620s that a second Dutch overseas company was organized. The strategic aim of the West India Company, created in 1621–24, was to secure Atlantic-based trading routes between West Africa and the West Indies (from Brazil to Cuba). Like the VOC it was a state-sanctioned trade monopoly federated among five regional "chambers." Even more than the VOC, the West India Company drew investors from outside Amsterdam to supply its 7.1 million guilders in starting capital. Such inland cities as Leiden, Utrecht, Dordrecht, and Groningen together invested substantially more than the city of Amsterdam. Amsterdam once again had eight votes, but this time it was out of an expanded total of nineteen directors. While the VOC dealt with spices and cotton, the West India Company traded in slaves and sugar.[22]

The West India Company focused initially on West Africa and the trade of Swedish copperware for Guinea gold. In this way copper became part of a far-flung trading network. Whereas in the early 1620s 2 ounces of African gold traded for 70 or 80 pounds of copperware, the West India Company's competition-damping monopoly allowed it to exchange just 35 pounds of copper for the same amount of gold. (From an African perspective, gold was thus worth half as much in copperware.) In its first thirteen years, the West India Company imported 12 million guilders in Guinea gold, handsomely exceeding its starting capital; in addition, it raided more than 500 Spanish and Portuguese vessels. The company in one stroke netted a whopping 11 million guilders when in 1628 its ships captured an entire Spanish silver fleet off Cuba, consequently paralyzing the silver-dependent Spanish imperial economy. In response to these gains, the company's share price around 1630 topped even the VOC's. For a time it looked as if the company's booty would actually pay for its heavy expenditures for 220 large warships, a vast store of guns and munitions, and a heavy burden of wages and salaries.

The West India Company planted a profitable sugar colony on the northeast coast of Brazil, on land wrenched from Portugal in 1635. Cultivating sugar demanded a large labor force, which the Dutch West African trade was well positioned to supply in the form of slaves. "The Dutch now controlled the international sugar trade for the first time and, as a direct consequence, also the Atlantic slave trade. Plantations and slaves went together," writes Israel. During the decade 1636–45 the West India Company auctioned nearly 25,000 African slaves in its Brazilian territory, while the price of sugar in Amsterdam slid downward 20 percent because of the new supply. But in 1645 the Portuguese swarmed up from southern Brazil, staged a series of raids that forced the Dutch back into their fortifications, and effectively ended Dutch sugar planting on the mainland. When Dutch shipments of Brazilian sugar came to a close in 1646, sugar prices in Amsterdam climbed 40 percent in one year. As a sugar colony, Netherlands Brazil never recovered; and West India Company shares collapsed. Dutch sugar and slave traders shifted their focus to the islands of the Caribbean. From 1676 to 1689, the Dutch shipped 20,000 black slaves from West Africa to the West Indies; most of them were deposited at Curaçao (off the coast of Venezuela), whence these hapless souls were parceled out across the sugar-cultivating region.[23]

"The Great Traffic"

The material plenty of the Dutch Golden Age was created from a heady mix of shipping, financing, trading, warring, and slaving that generated

FIG. 2.4. Secrets of Dutch Shipbuilding Revealed.

Title page from a book published in Amsterdam in 1697. Courtesy of
Kendall Whaling Museum, Sharon, Mass.

a constant stream of raw materials coming into the Dutch Republic. Yet
however impressive or sordid these trading activities seem today, they
were not the mainstay of the Dutch commercial economy. Dutch preeminence came through the targeted processing and selective reexporting of
the traded materials. "A more or less unique sector emerged, often referred to as *trafieken* (traffics) as opposed to *fabrieken* (manufactures),"
writes Joel Mokyr. Among the "traffics" with links to the maritime sector were sugar refining, papermaking, brewing, distilling, soap boiling,
cotton printing, and tobacco processing, as well as the complex of activities related to shipbuilding (fig. 2.4). Other highly specialized activities in
which the Dutch gained global dominance include processing dyes and
glazes, cutting and polishing diamonds, grinding glass lenses, refining
whale oil, bleaching linens, and dyeing and finishing broadcloth. The
making of highly precise nautical compasses, maps, and chronometers re-

inforced Dutch maritime dominance. Dutch control over the Swedish copper trade led at least four Dutch cities to set up copper mills, often with skilled artisans or equipment from Germany. For each of these traffics, mastering special techniques and attaining superior quality were more important than achieving high levels of output. Indeed, high wages, relatively low volumes, and high-quality production typified the traffics, in sharp contrast with early industrial technologies, which emphasized low wages, high volumes, and low-quality production (see chapter 3).[24]

The "great traffic" was surprisingly wide reaching. Amsterdam was the largest city in the Dutch Republic but not by much. Its population just exceeded Haarlem and Leiden together, and these two towns often teamed with others to outvote Amsterdam on matters relating to foreign affairs and trade. The varied stream of imported goods strongly shaped which cities and which industries would thrive. For instance, although Amsterdam lobbied for a peace settlement with Spain from the 1620s on—because the war harmed its European trade—the country as a whole pursued a policy of prolonging the war, following the lead of Haarlem and Leiden, which were in league with several inland provinces. The textile towns of Haarlem and Leiden believed that prolonging war helped them compete with their chief rival, Spanish-occupied Flanders, while the inland towns favored war owing to their heavy investments in the two colonial trading companies. Amsterdam, writes Israel, "was merely the hub of a large clustering of thriving towns, all of which directly participated in the process of Dutch penetration of foreign markets."[25]

Textile towns such as Haarlem and Leiden, no less than the better-known commercial centers, benefited handsomely from the arrival of exiled Spanish Netherlanders. In the years after 1585 Haarlem prospered with the arrival of Flemish linen-bleachers, who found a ready supply of water and soon began bleaching "to the highest standards of whiteness then known in Europe." The population of Leiden more than doubled from 1581 to 1600. Initially Leiden focused on lower grades of cloth (says and fustians) produced from cheap imports of Baltic wool. In the 1630s Leiden dramatically expanded its output of high-value lakens, using high-quality wool from Spain (see fig. 2.5). "It seems likely that this was in part due to the important technical innovations which were soon to make Leiden fine cloth famous throughout the globe and which gave it its characteristic smoothness of texture," writes Israel. Another high-grade Leiden cloth that experienced rapid growth was camlet, made from

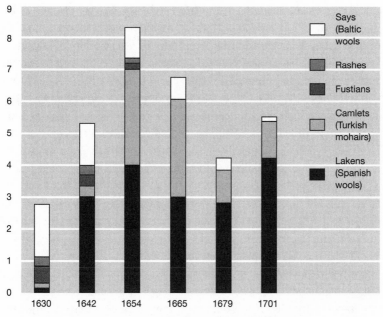

Says (Baltic wools
Rashes
Fustians
Camlets (Turkish mohairs)
Lakens (Spanish wools)

FIG. 2.5. Leiden's High-Grade Textiles.

In the early seventeenth century, Leiden's textile industry made middle-quality fabrics, such as says (from Baltic wools). By midcentury they increasingly focused on the upmarket camlets (from Turkish mohairs) and lakens (from superfine Spanish wools). Data from Jonathan Israel, *Dutch Primacy in World Trade, 1585–1740* (Oxford: Clarendon Press, 1989), 195, 261.

expensive Turkish mohair. By the mid-1650s two-thirds of Leiden's 36,000 textile workers were employed in the making of lakens and camlets; the value of the city's textile output was 9 million guilders. The Dutch at that time took in at least three-quarters of Spain's wool exports and the majority of Turkey's mohair yarn. Significantly, the principal markets for Dutch fine cloth were in just the countries that had produced the raw materials—France, Spain, and Turkey.[26]

In 1670, when Venice tried to reestablish its fine cloth industry, importing Dutch methods was the only way it could achieve the lightness and smoothness of Leiden fine cloth, which had become the standard demanded by the principal consumers. French observers of the time noted that the Dutch methods used a third less wool and less labor to produce the same amount of fine cloth. Even lakens from Brabant, once the chief rival, lacked the characteristic smooth finish of the Dutch product. "The technical factor was crucial to both the continuance of Dutch dominance of the Spanish wool trade and the dominance of fine-cloth production itself," states Israel. Other Dutch textile centers that grew to international prominence included the silk industry of Amsterdam and Haarlem (which gained the skills of exiled French Protestants in the 1680s), the linen-weaving industry in the Twente and Helmond regions, and the sail-canvas industry in Haarlem, Enkhuizen, and the Zaan district.[27]

It must have been especially galling for the Spanish wool manufacturers to see all that wool slip through their grasp. From the late Middle Ages on, a massive "herding economy" active through much of central Spain tended the famous Merino sheep that grew superfine wool. In the late fifteenth century, the Spanish crown granted merchants in the northern town of Burgos control over the export trade in wool. Using ships from the north coast of Spain, they sent large sacks of washed, but otherwise unfinished wool to England, France, and the southern Netherlands. The Burgos merchants adopted many of the classic Italian financial techniques, such as double-entry bookkeeping, insurance underwriting, and advance purchasing. Domestic wool manufacturers in Segovia, Cordoba, Toledo, Cuenca, and other centers organized regional "putting-out" networks to spin and weave the wool, mostly for domestic markets. The Spanish wool trade flourished until around 1610. At the time when the Dutch economy took off, Spain's wool industry was hampered by high labor costs, creeping inflation, and the absence of technical innovation. A French traveler to Segovia, noting the city's past wealth from "the great commerce in wool and the beautiful cloth that was made there," observed in 1660, "the city is almost deserted and poor."[28]

The rise and fall of Dutch sugar-refining shows how dependent the great traffic was on foreign trade. As noted above, Dutch sugar refining expanded in the 1630s when the Dutch briefly controlled northeast Brazil, but the loss of that colony in 1645 forced the Dutch sugar traders to fall back on the Caribbean export trade. By 1662 sugar refining was firmly centered in Amsterdam; the city had fifty or more sugar refineries, around half of Europe's total. In addition, there were another dozen sugar refineries located in other Dutch cities. But sugar was too important to be monopolized by any small country. The Dutch lost their largest market for processed sugar when France imposed restrictive trade policies during the 1660s. By the end of the century, the Dutch had lost leadership in the Caribbean sugar trade to England, with its growing commercial activities and its insatiable domestic hunger for sugar. Even so, the English frequently reexported sugar for processing in Amsterdam. (They did the same with tobacco.)[29]

After 1650 the Dutch traffics expanded in scope. Delft made pottery that simulated the finest Chinese blue porcelain. Gouda specialized in white clay tobacco pipes. And the Zaan, a vast inland estuary stretching from Amsterdam to Haarlem (now filled in) evolved into a major proto-industrial district. Already a shipbuilding center, the site of numerous timber-sawing mills, and the world's largest timber depot, the Zaan in the latter part of the seventeenth century became the Dutch center of high-quality papermaking and one of two centers for refining whale blubber into soap, lighting fuel, and other products. It also featured rope- and sail-making concerns and refineries processing animal fats and oils. This district alone had 128 industrial windmills in 1630 and perhaps 400 by the end of the century.[30]

Surveying the improbable commercial success of the Dutch Republic, one Englishman observed in 1672 that the Dutch had achieved a universal reputation for the "exact making of all their native commodities," and that they gave "great encouragement and immunities to the inventors of new manufacturers." Not only had Dutch traders captured commercial control over many key raw materials, including Spanish wool, Turkish mohair yarns, Swedish copper, and South American dyestuffs; the "traffic" system had also erected a superstructure of processing industries that added value to the flow of raw materials. The Dutch conditions of high wages and labor scarcity put a premium on mechanical innovation, the fruits of which were protected by patents. Another economic role taken on by the Dutch state (at the federal, state, and municipal levels) was the close regulation of industry in the form of setting standards for quality and for the packaging of goods.[31]

Given the commanding heights attained by the Dutch economy during the seventeenth century, many have asked why did it not become the first industrial nation, when that age arrived in the eighteenth century. Three lines of inquiry seem most pertinent: raw materials and energy, international trade, and the traffic industries.[32] Economic historians under the sway of British industrialization's coal paradigm often mention that the Dutch Republic lacked coal. While it is true that the Dutch Republic needed to import many raw materials, it did have numerous sources of industrial energy other than coal, including peat, water power, and wind power. Peat—a biological precursor to coal—was extensively used for

TABLE 2.1. Dutch International Trade and
Traffic Industries, 1661–1702

	1661–1665	1666–1670	1671–1675	1676–1680	1681–1685	1686–1690	1691–1695	1696–1700	1701–1702
Leiden camlets (1,000 pieces)	44	54	54	34		28	24	31	18
Leiden lakens (1,000 pieces)	19	18	19					25	24
Dutch voyages to Baltic	589	774	496	694	1,027	885	573	591	484
Dutch whaling voyages			148	136	212	175	56	135	
Guinea gold imports (1,000 guilders)				551	525	619	503	317	222
Bengali raw silk imports (1,000 pounds)	79	77	121	105			160	185	

Source: Jonathan I. Israel, *Dutch Primacy in World Trade, 1585–1740* (Oxford: Clarendon Press, 1989), 257, 263, 301, 307, 329, 337, 354, 366, 387.

Note: Figures are annual averages for each five-year period.

heating homes and in the manufacturing of bricks, glass, and beer; while imported coal was increasingly used in industrial activities such as sugar refining, soap boiling, distilling, and copper working. Water mills were powered by inland streams and by tides at the shore. The most famous Dutch mills were the windmills. It is estimated that by 1630 there were 222 industrial windmills in the country; they were used to drive saw mills, oil mills, and paper mills. Counting those used for draining land and milling grain, the Dutch had a total of perhaps 3,000 windmills.[33]

International trade overall proved remarkably resilient during turbulent times in the seventeenth century (see table 2.1). In 1672 the English declared war on the Dutch Republic, and the French actually invaded the northern and inland provinces, which brought chaos there as well as to the maritime provinces. For two years the French bottled up Dutch merchant ships in port. Yet as soon as the ships were able to leave port, trade bounced back. Dutch trade with the East Indies actually grew right across the seventeenth century, despite the recurrent wars with France and England. Dutch shipments of raw silk from Bengal remained surprisingly strong, too. But by the 1690s, several other sectors of Dutch shipping reached a breaking point. Whaling voyages decreased, while shipments of West African gold fell. Closer to home, Dutch trade with the Baltic withered under stiff competition from the English and Scandinavian fleets.[34]

These reverses, in tandem with the increasingly protectionist policies of France and England, put great pressure on the "traffics" system. Remarkably, in the years after the 1672 French invasion, there was expansion in several export-oriented industries, including silk, sail cloth, high-quality papermaking, linen weaving, and gin distilling. Silk flourished with the arrival of French Protestant Huguenots in the 1680s. But the traffics system as a whole never recovered from the disruption of the invasion. The textile center at Leiden, for example, experienced several sharp declines in its fine-cloth production between 1667 and 1702. The output of camlets, from fine Turkish mohair, fell steadily across this period (see table 2.1). Growth did not resume in the early eighteenth century, even as the English woolen-textile industry was experiencing great leaps in cheaper-grade production (see chapter 3). At Leiden the production of lakens, from fine grades of Spanish wool, fell steadily, from 25,000 pieces around 1700 to a mere 6,700 pieces in 1750.

The wealth-creating imperatives of traders and merchants, boat-builders and shipowners, sugar refiners and textile makers, and many others—a far more diverse cast than the Renaissance court patrons—altered the character of technology during the era of commerce. While choosing, developing, and using technologies with the aim of creating wealth had been an undercurrent before, this era saw the flourishing of an international (if nonindustrial) capitalism as a central purpose for technology. It is really a *set* of wealth-creating technologies and techniques that distinguishes the Dutch commercial era: no other age and place combined bulbous cargo-carrying fluyts and factorylike herring busses, large port complexes coupled to buzzing inland cities, the array of added-value "traffic" industries, and the elaboration of world-spanning financial institutions, including exchanges for the trading of stocks and commodities, multishare ownership of ships, and futures markets for herrings, woolens, and for a time tulips.

These technologies not only set the stage for a Dutch commercial hegemony that lasted roughly a century; they also shaped the character of Dutch society and culture at all levels and across the entire country. While the fine arts record the wealthy merchants who owned the Amsterdam townhouses and commissioned portraits from the famous painters, this elite society was really only the tip of the iceberg. Just as commerce permeated Dutch society and culture, so too did remarkably broad forms of cultural consumption. A famous generation of oil painters flourished to meet the demands of newly wealthy merchants eager to have themselves and their families appear in portraits (and the profusion of tropical dyes coming from overseas gave painters economical access to brilliant coloring agents). Scenes from everyday life were much prized. No less a figure than Rembrandt van Rijn, in "The Mill" (1645–48), painted the signature Dutch windmills. The consumption of oil paintings was, in comparative terms, exceptionally widespread. "The most modest shop-keeper had his collection of pictures, and hung them in every room," writes Paul Zumthor. A now-collectable Ruisdael landscape or Steen genre picture then cost about a quarter of the weekly wage of a Leiden textile worker.[35]

Wherever one looks—at the diverse stockholders of the two great overseas trading companies, the extensive *trekvaarten* network, the numerous owners of the trading ships, and even for a few years the distinctly down-market traders of tulips—Dutch commerce engaged the talents and wealth of an exceptionally wide swath of society. The depth and breadth of the changes that these activities represent lent a distinctly modern character to Dutch society, not only in the details of "modern" financial institutions and economic growth patterns, but in the pervasiveness of the effect that commerce and technology had on the society as a whole. By contrast, in England, as we will see in the following chapter, a distinctively industrial society would emerge.

HISTORIANS OF LATE have not treated the "industrial revolution" kindly. A generation ago the picture was quite different. The industrial revolution in Britain, according to Eric Hobsbawm's classic account, "initiated the characteristic modern phase of history, self-sustained economic growth by means of perpetual technological revolution and social transformation." It "transformed in the span of scarce two lifetimes the life of Western man, the nature of his society, and his relationship to the other peoples of the world," stated another influential account. For these historians, the industrial revolution, with its new ways of working and living, created an entirely new industrial society. So-called leading sectors were in the vanguard. Unprecedented growth in the cotton, iron, and coal industries during the decades surrounding 1800, culminating in the steam-powered factory system, powered a self-sustaining "take-off" in the British economy. The cotton industry provided the paradigm, experiencing a spate of output-boosting inventions in spinning and weaving that made the inventors Hargreaves, Arkwright, Crompton, and Cartwright famous (and Arkwright, at least, vastly wealthy). These inventions required factory-scale application, so the argument went, and factories required James Watt's steam engines. Increases in iron and coal output, for steam engines and cotton factories, made the factory synthesis into a juggernaut that changed the world. In this interpretation, Britain blazed a pioneering path to industrial society while the rest of the world followed behind in its wake.[1]

But today the industrial revolution no longer seems such a tidy package. Even if the changes in Britain's core industrial districts were especially dramatic and well documented, other regions within Britain and even other countries contributed to and participated in these changes. On closer inspection it seems that the British economy grew slowly and gradually, and now no one can identify a single take-off decade. Early factories in many countries were powered by water, animals, or humans—not steam—and even in coal-rich Britain water power outpaced steam power well into the nineteenth century. Through the 1830s the largest British steam-powered cotton-spinning mills, which employed 1,000 or more workers in eight-story factory buildings, depended on a legion of traditional handloom weavers working mostly at home to make the machine-spun yarn into useful cloth. Even in Britain only one industrial worker in ten ever saw the inside of a factory; most labored in smaller shops or in the so-called traditional sectors, such as building, wool, and leather. These other activities, not the singular steam-powered cotton factories, contributed most of the country's manufacturing value, and experienced substantial growth themselves. The new historical view is of "a more long-run, varied and complicated picture of the British path to Industrial Revolution."[2]

All the same, something fundamental was going on in Britain around 1800. The key industrial-era images—and such fundamental concepts as industry, class, and culture—snapped into focus during these decades. The term *industry* had earlier referred only to a human attribute in the sense of skill, perseverance, diligence. Adam Smith, in his *Wealth of Nations* (1776), was perhaps the first to use *industry* as a collective word for manufacturing and productive institutions. In the 1830s *industrialism* was coined to refer to the new social system. (The term *industrial revolution* passed into common English only in the 1880s, although French authors earlier in the century believed that France was undergoing a *révolution industrielle*, while a little-known German theorized about the *industriellen Umwälzung* in the 1840s.) *Class* also was an existing term, referring to school or college classes in traditional subjects. In reference to people, *lower class* came first, followed in the 1790s by *higher class* and *middle class; working class* emerged around 1815, *upper class* soon after. Even *culture*, which had earlier meant only "the tending of natural growth" as in *horticulture*, came in the early nineteenth century to mean "a general state or habit of the mind" or "the state of intellectual development in a society as a whole," and, toward the end of the century, "a whole way of life, material, intellectual and spiritual."[3] Writing during the 1850s, Karl Marx argued (originally in German) that the culture of the working class was a creation of industry; a generation earlier no one could have expressed this thought in English. This chapter examines the dynamics of industrial society in Britain, long hailed as the first industrial nation, through extended case studies of London, Manchester, and Sheffield. It then evaluates the notion of "industrial society" through comparisons with other industrial regions and countries.

The First Industrial City: London

In the older view of the industrial revolution, there was no need to look at London. "The capital cities would be present at the forthcoming industrial revolution, but in the role of spectators," wrote Fernand Braudel in his acclaimed *Capitalism and Material Life.* "Not London but Manchester, Birmingham, Leeds, Glasgow and innumerable small proletarian towns launched the new era." The industrial revolution was, in another influential image, "a storm that passed over London and broke elsewhere." London, supposedly, was stuck in a preindustrial age, with its "gentlemanly capitalists" not concerned to build up massive industries and striving only to enter the landholding aristocracy.[4] But the notion of an industrial London is worth a second and more careful look. Around 1800, when manufacturing employed one in three of London's workers, the city had more steam engines than any of the factory towns. In 1850 London had more manufacturing workers than the four largest factory towns in England put together. Chemical, furniture, brewing, printing, shoemaking, textile-finishing, precision-manufacturing, and heavy-engineering industries sprang up to the south and east of London's fashionable center, often in compact specialty districts, while just downstream shipbuilding, provisioning, and processing industries surrounded the Port of London.[5]

Not only was it the country's largest site of industry, London's insatiable hungers and unquenchable thirsts helped transform England from a rural-agricultural economy to an urban-industrial one. London's growth in these decades still astounds. In 1700 London, with a half-million residents, was the largest city in Europe (surpassing Paris) and ten times more populous than the next largest British town; of all the world's cities only Tokyo, perhaps, was larger. From 1800 to 1850 London *added* more residents (1.4 million) than the *total* 1850 populations of the country's dozen largest textile-factory towns, even though they had experienced rapid growth themselves. In 1850 London numbered 2.4 million residents. As Daniel Defoe observed, "All the people, and all the lands in

FIG. 3.1. City of London.

A view of the fashionable center of London, with the dome of St. Paul's Cathedral prominent beyond Southwark Bridge, seen from the gritty south bank of the Thames River. Cargoes arrived in the steamboats and larger vessels, then were offloaded onto shore by smaller boats called lighters, and finally moved on land with the aid of horse and human labor. Samuel Smiles, *Lives of the Engineers* (London, 1862), 2:190. Courtesy of Illinois Institute of Technology Special Collections.

England seem to be at work for, or employed by, or on account of this overgrown city."[6]

Already by 1700 London's escalating demand for food outpaced the production of its outlying market gardens. Soon enough, eggs, geese, sheep, cattle, grain, malt, herrings, turkeys, apples, and more were flowing into the city from the entire southern half of England. "Georgian London was said to consume each year 2,957,000 bushels of flour, 100,000 oxen, 700,000 sheep and lambs, 238,000 pigs, 115,000 bushels of oysters, 14,000,000 mackerel, 160,000 pounds of butter and 21,000 pounds of cheese."[7] The necessity of moving this mountain of food into London helped create a far-flung transport system of roads, rivers, and canals. By 1805 north London was linked by the Grand Junction canal to the farming centers and to the industrializing towns, such as Manchester, Birmingham, and Leeds.

Large ships sailed right into the center of London, bearing goods from the world's markets, just as they had done in the Dutch port cities (see chapter 2). Around 1700 London's crowded streetside wharves along the Thames River handled three-quarters of Britain's overseas trade. Crammed into the 488 yards between London Bridge and the Tower of London were sixteen special wharves (the so-called legal quays) plus the Customs House and Billingsgate fishmarket. (Fresh fish came from the Thames until pollution killed them off in the 1820s.) Additional wharves stood upstream from London Bridge, across river on the south bank, and downstream from the Tower. But the collection of customs duties required all shipments of goods, wherever they were first offloaded, to pass through one of those sixteen legal quays.

In response to the extreme congestion and the pervasive streetside pilfering that attended the handling of all this cargo, the city's merchants launched a £5 million dock-building campaign (see fig. 3.2). Beginning near the Tower, the new dockyard district eventually stretched twenty-five miles downstream. Built on artificial lakes carved from the ground, and connected to the Thames River by short canals, these dock-warehouse complexes began with the 600-ship West India Docks (opened in 1802) on the Isle of Dogs. Next came the London Docks at Wapping (1805), the East India Docks at Blackwall (1806), and the Surrey Docks in Rotherhithe (1809). Several years later came the St. Katherine's Docks (1828) next to the centrally located Tower.

The sprawling dock complex and its flow of cargoes created a vast commercial and shipping infrastructure staffed by a multitude of "agents, factors, brokers, insurers, bankers, negotiators, discounters, subscribers, contractors, remitters, ticket-mongers, stock-jobbers, and a great variety of other dealers in money." Through the mid-1860s London was also Britain's leading shipbuilding district. Sails, ropes, barrels, and pulleys were needed to outfit Britain's overseas fleet. On the south bank were the Royal Navy Yard at Deptford and the shipbuilders at Rotherhithe, who made the first iron steamship, the *Aaron Manby,* launched in 1822. At Millwall on the Isle of Dogs the prominent millwright William Fairbairn, a pioneer in iron construction, opened shop in 1833 and built many ships for the East India Company. Iron shipbuilding also flourished at nearby

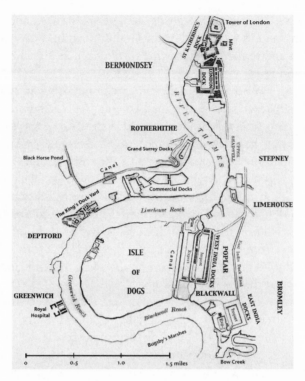

FIG. 3.2. Port of London.

Before this nineteenth-century dockyard complex was built, all goods entering London passed through the so-called legal quays, located immediately upstream from the Tower (*top of map*). The downriver dockyards
became a major site of shipbuilding, provisioning, and processing. Eventually the dock complexes extended far downstream beyond Bow Creek
(*bottom of map*). Charles Knight, *London* (London: Knight, 1842), 3:65.
Courtesy of Illinois Institute of Technology Special Collections.

Blackwall from the 1840s, culminating with the launch in 1857 of I. K.
Brunel's 600-foot-long *Great Eastern*. The port employed thousands. In
1851 there were 4,100 workers at the warehouses and docks themselves;
5,000 in shipbuilding, of whom nearly 3,300 were highly skilled shipwrights; 18,000 in ocean navigation and 5,700 more in inland navigation;
and no fewer than 1,214 government employees at the East India Company. Messengers and porters, at the docks and elsewhere in the commercial infrastructure, accounted for another 33,000 workers.[8]

Among the port's chief commodities was coal. Already by the 1760s Defoe noted "at London [we] see the prodigious fleet of ships which come constantly in with coals for this increasing city." Coal served for baking, brewing, iron forging, pottery firing, and glassmaking, among the heat-requiring trades, as well as for heating homes and offices. Adding to the coal smoke, noxious fumes issued from the "boiling" trades, which produced soap, candles, and glue and were the ultimate fate of countless animal carcasses. Travelers from the countryside entering the city perpetually denounced its coal-fouled environment. By 1830 the volume of coal shipped into London had reached 2 million tons, a 250 percent increase since Defoe's time; in the 1830s coal also began arriving by rail. The Coal Exchange opened in 1849, near the Billingsgate fishmarket. Two years later the coal sector in London alone employed more than 8,000 persons (1,702 in coal and gas manufacturing, 2,598 as coal merchants and dealers, and 4,023 as coal heavers, who included three rugged women in their ranks).[9]

Beer brewing affords a revealing window into industrial London while illustrating the links between industry and sanitation, consumption, and agriculture. In an age of filthy water, beer was virtually the only safe beverage, and Londoners drank it in huge quantities. In 1767 one dirt-poor clerk, said to be "even worse" off than a day laborer, nonetheless drank beer at each meal: "*Breakfast*—bread and cheese and small [weak] beer from the chandler's shop, *Dinner*—Chuck beef or scrag of mutton or sheep's trotters or pig's ear soused, cabbage or potatoes or parsnips, bread, and small beer with half a pint of porter; *Supper*—bread and cheese with radishes or cucumbers or onions, small beer and half a pint of porter."[10] In 1688, and perhaps as late as 1766, British farmers harvested more bushels of barley (mostly for brewing and distilling) than wheat for bread. Taxes on beer making generated one-quarter of Britain's tax revenue before 1750 and one-seventh as late as 1805. During these same decades, London's porter-brewing industry dramatically embodied the hallmarks of industrialization: vast increases in scale and capitalization, labor- and cost-saving technology innovation, and market concentration, as well as a profusion of specialized by-product and supply industries.[11] Reducing costs and increasing output—rather than enhancing quality, as in Dutch commerce—was the focus of technology in the industrial era.

The beer known as porter deserves full recognition as a prototypical industrial-age product alongside cotton, iron, and coal. Ales, as traditionally brewed, required light malts made from the finest barley, and ale brewers halted yeast fermentation when a little malt sugar remained. The best ales were light and clear, slightly sweet, and drunk soon after brewing. By contrast, porter, first brewed in 1722, was dark, thick, and bitter. (Its name derives from the dockworkers who were its earliest consumers.) Porter malts were dark roasted from cheap grades of barley, and no malt sugars remained when the brewers were done. Most important, porter was susceptible of industrial-scale production.

Brewers found that aging porter for a year or more mellowed its rough taste. Initially they simply stored their brews in 36-gallon wooden barrels, but with the increase in volumes, the practice eventually tied up a quarter or more of a firm's entire capital resources. In the 1730s, to trim these heavy costs, the larger brewers began building on-site vats capable of storing up to 1,500 barrels of the brew. Larger vats followed soon (fig. 3.3). In 1790 Richard Meux built a 10,000-barrel vat that was 60 feet in diameter and 23 feet high. "They claimed that 200 people had dined within it and a further 200, also inside, drank to its success." Five years later Meux erected a 20,000-barrel behemoth. The competition between brewers to build ever-larger vats waned after 1814, however, when a 7,600-barrel vat at the Horse Shoe Brewery burst open and flooded the neighborhood, killing eight persons "by drowning, injury, poisoning by the porter fumes or drunkenness."[12]

The immense size of these storage vats hints at the porter brewers' industrial scale of production. The annual production of the city's leading porter brewer was 50,000 barrels in 1748, 200,000 barrels in 1796, and 300,000 barrels in 1815; the brewing companies were Calvert, Whitbread, and Barclay, respectively. By 1815 a London porter brewer shipped more in a year than the largest London ale brewer did in a decade. Porter brewers had industrialized, while ale brewers had not. The twelve largest London brewers made virtually nothing but porter and the closely related stout. Between 1748 and 1815 they built factories that more than tripled their production of these "strong" beers (from 0.4 to 1.4 million barrels) expanding their share of the London market from 42 percent to 78 percent. During these decades of expanding output, the city's total number of brewers actually *fell* from 158 to 98.

Beer consumption figured as a significant item in urban household budgets, especially for the poor. If beer consumption is taken as a proxy for workers' standards of living, the early industrial economy did not lift real wage levels until after the tough times of the 1840s. National figures for per capita beer consumption (in England and Wales) stood at 33.9 gal-

FIG. 3.3. London Brewery Vat.

This vat for aging porter beer shows the scale of industrial enterprise. It held more than 100,000 gallons, saving the use of 3,000 smaller wooden barrels. Charles Knight, *London* (London: Knight, 1843), 4:13. Courtesy of Illinois Institute of Technology Special Collections.

lons (1800–1804), increased temporarily around 1830, fell to 31.6 gallons (1860–64), then rose to a much-discussed 40.5 gallons (1875–79). Gourvish and Wilson in their study of the British brewing industry calculate that during that last span of years the average British man each week put away 16 pints of beer.[13]

The porter brewers pioneered industrial scales of production and led the country in the capitalization of their enterprises. In 1790 the net capital valuation of Whitbread's White Hart Brewery stood at £271,000 while Truman's Black Eagle Brewery and Barclay-Perkins' Anchor Brewery were not far behind, at £225,000 and £135,000, respectively. These are extraordinary figures: at this time £20,000 was the capitalization of Birmingham's very largest manufacturers and, two decades later, of Manchester's largest spinning factory. And the biggest brewers got bigger. In 1828 Whit-bread's nine partners had invested capital of £440,000 while Barclay-Perkins' total capital was £756,000.[14]

These outstandingly high capitalization figures were a sign that porter brewers were early and enthusiastic converts to the industrial-era logic of labor-saving mechanization. Early on, porter brewers purchased horse-driven pumps to move beer throughout the brewery. The big London brewers, despite the heavy cost for maintaining workhorses in the city, mechanized malt grinding also using horse power. Barclay-Perkins employed just 100 workers to make 138,000 barrels of beer in 1796. In addition to its dozen well-paid brewers and technical managers, the company employed 4 stokers, 7 coopers, 9 yeastmen, 4 miller's men, 18 draymen, 10 stage men, 12 horse keepers, 30 spare men, and 7 drawers-off.[15] (See fig. 3.4.) In 1851 the census found just 2,616 London brewery workers.

While pumps and grinding mills represented a mechanical replacement for human labor, rotating steam engines retired workhorses from their ceaseless turnstile rounds. As James Watt's business partner observed in 1781, "the people in London, Manchester, and Birmingham are Steam Mill Mad." In 1784 brewers installed the first of London's new rotating Watt engines, a 4-horsepower model at Goodwyn's and a 10-horsepower one at Whitbread's. These installations came just two years after iron-master John Wilkinson installed Watt's very first rotating engine and a full two years before mechanization at the Albion Flour Mill, typically remembered as the site where Watt's rotating engine proved itself. At Whitbread's the steam engine was not only reliable and fuel-efficient but cost saving, retiring twenty-four horses. With the savings from just one year's care and feeding of its horses, at £40 per head, Whitbread's paid for the engine's entire installation cost (£1,000). Brewers indirectly fixed a key term of measurement born in the industrial era, since Watt had the "strong drayhorses of London breweries" in mind when he defined "horsepower" at 33,000 foot-pounds per minute. In 1790, Watt recorded, his firm made engines principally for brewers, distillers, cotton spinners, and iron men. Five years later only the cotton and coal industries had more steam engines than the brewing industry (which used twelve engines totaling 95 horsepower).[16]

The porter brewers' unprecedented scale, market domination, and immense capitalization, in addition to their quest for money-saving technical innovation, are signs of their *industrial* character. So is the rise of specialized by-product processing and supply industries, which created a

FIG. 3.4. Porter Brewery.

Inside of large brewery showing five men "mashing" water and malt with "oars" (one is held aloft at right). Boiling hot water, from the "coppers" at back, flowed to the "mash tuns" for mixing with malt. The resulting "wort" drained down to the "under-backs" beneath the floor level. Not visible is a horse-driven pump that lifted the wort, collected in the "jack-back" at far left front, back up to the coppers for a boiling with hops. The master brewer is center. Temple Henry Croker, *The Complete Dictionary of Arts and Sciences* (London: Wilson & Fell, 1764), vol. 1, plate 24.

radiating web of industrial activities around the centers of iron, coal, cotton, building—and beer. These ancillary industries have not received the attention they deserve, for they are key to understanding how and why industrial changes became self-sustaining and cumulative. Brewing again provides an apt example. Industrial-scale porter brewing fostered several ancillary industries. Spent grains from brewing fattened up cattle awaiting slaughter, while the rich dregs from beer barrels became fertilizers. Brewers' yeast spawned an entire specialized economy: London's gin and whiskey distillers usually bought their yeast directly from the brewers, while bakers dealt with specialized brewers' yeast dealers. One of these enterprising middlemen devised a patented process for drying and preserving yeast and, after contracting with six large brewers, in 1796 built a large factory to process yeast freshly skimmed from brewing vessels.

Specialized trades also sprang up for making beer barrels, copper brewing vessels, and clarifying agents. Beer engines, to pump beer up from a
public-house cellar for dispensing, were manufactured by at least three
firms located near Blackfriars Road.[17]

Having industrialized production, porter brewers sought influence
over beer consumption. (Other competitive means were not readily available, since for decades prices had been fixed by law or custom and the export trade was dominated by rival British brewers in Burton-on-Trent.)
Porter brewers' immediate object was the principal site of mass consumption: London's 5,000 public houses, or pubs. Traditionally brewers
had sold barrels of beer to the publican, making monthly deliveries and
taking payment on the spot, but the larger porter brewers began retaining ownership right until the beer splashed into the consumer's quart pot.
The publican then paid for only the beer he actually sold, while the brewer
insisted that the publican take beer deliveries from no one else. Thus
began the controversial practice of "tieing" a pub's trade. If left unpaid, a
publican's monthly balance might become an interest-bearing loan. If the
loan balance grew too large, the brewer would ask for the publican's building lease as collateral and if necessary take possession of a defaulted publican's establishment. Some brewers outright bought a public-house lease
and installed a dependent tenant. By the early nineteenth century perhaps
half of all London pubs were tied to brewers through exclusive deliveries,
financing, or leasing. Tying typified the porter industry. In 1810 Barclay's
had tied relationships with 58 percent of the 477 pubs it served, while
Truman's had tied 78 percent of 481 pubs, and Whitbread's 82 percent of
308 pubs.[18]

Porter brewers were big supporters of the Beer Act of 1830, which
scrapped the centuries-old regulations on the selling of beer. The act, in
an effort to save the poor from poisoning themselves with cheap distilled
gin, permitted virtually any property-tax payer, upon paying a modest fee,
to set up a "beer shop." Across Britain 40,000 willing taxpayers opened
such shops and per capita beer consumption shot up 25 percent. Not surprisingly the big porter brewers of London (the twelve largest now brewed
85 percent of the city's "strong" beer) cheered the Parliamentary act.
Among them, Charles Barclay best captured the industrial logic of porter
brewing and mass consumption. "Who are to supply these beer shops?
The persons who can sell the cheapest and the best, and we say we can sell
cheaper and better than others," he told his fellow Members of Parliament. "We are power-loom brewers, if I may so speak."[19]

We can complete our portrait of industrial London by stepping back to survey the city's multitude of trades and its building and engineering industries. From the first reliable figures, in the 1851 census, London's 333,000 *manufacturing workers* outnumbered Manchester's 303,000 *total residents.* Manufacturing accounted for a near-identical percentage of employment for women and men (33 and 34 percent, respectively), although London's women and men did not do the same jobs. Of the top ten sectors of manufacturing, only in boot- and shoemaking were there roughly equal numbers of men and women. Women worked mostly as boot- and shoemakers, milliners and seamstresses (making fancy hats and dresses). By contrast, men dominated the trades of tailoring, furniture making, baking, woodworking, printing and bookbinding, iron and steel, and nonferrous metals. Male-dominated shop cultures and prevailing conceptions of what type of work was women's "proper" place go some way in accounting for where women and men worked. Tailors, however, used stronger tactics to exclude women (according to labor reformer Francis Place): "where men have opposed the employment of women and children by not permitting their own family to work, or where work is such that women and children cannot perform it, their own wages are kept up to a point equal to the maintenance of a family. Tailors of London have not only *kept* up, but *forced* up their wages in this way, though theirs is an occupation better adapted to women than weaving."[20]

The building trade, although it was not enumerated as manufacturing employment, certainly became industrialized during these years. The incessant increase in London's population necessitated constant construction. Bricks were in such short supply in the 1760s "that the makers are tempted to mix the slop of the streets, ashes, scavenger's dirt and everything that will make the brick earth or clay go as far as possible." Bricks fresh out of the kiln came down the street piping hot. "The floor of a cart loaded with bricks took fire in Golden Lane Old Street and was consumed before the bricks could be unloaded," noted one account. By the 1820s such industrial-scale "merchant builders" as the Cubitt brothers emerged. In that decade Thomas and Lewis Cubitt employed more than 1,000 workers in building vast blocks in the fashionable Bloomsbury and Belgravia districts, while their brother William employed an additional 700. The Cubitts not only directly employed all the various building trades but also made their own bricks and operated a steam-powered sawmill. Another of the city's merchant builders, John Johnson, who described himself variously as a paver, stonemason, bricklayer, and carpenter, owned a granite quarry in Devon, a wharf along the Thames, and brickyards, gravel pits, and workshops around London. In the 1851 census London's building sector employed a total of 70,000 bricklayers, carpenters, masons, plasterers, plumbers, painters, and glaziers (including 129 female house decorators).[21]

A stock image of London in the middle of the nineteenth century has its streets filled with small shops, yet in *Days at the Factories* (1843) George Dodd described twenty-two establishments that "were conducted on a scale sufficiently large to involve something like 'factory' arrangements." At midcentury, six of the country's eleven largest printing works were in London, as were twenty-two of the fifty-two largest builders. London's twelve biggest brewers, as noted above, towered above the field. Official figures miscounted many larger firms and simply ignored the city's railways and gasworks, making a complete picture of industrial London difficult to construct. Official statistics also overlooked the 20,000 construction workers laboring on the London-Birmingham railroad between 1833 and 1838 as well as the untold thousands who in the next decade connected London by rail to the four points of the compass and erected five major railroad stations. The 1851 census found just half of the 334 engineering firms listed in the following year's *London Post Office Directory*. And while the census recorded just six London engineering firms employing 100 or more men, the *Northern Star* newspaper itemized nine such firms' locking out their workers during a strike in 1852. These were Maudslay & Field (800 workers), John Penn (700), Miller & Ravenhill (600), J. & A. Blyth (280), and five additional firms with between 100 and 200 workers.[22]

Despite the faulty census information on engineering firms, we can clearly see London's rightful place as England's first center of engineering by reviewing two impressive mechanical genealogies. Boulton & Watt, as we noted, installed two of their earliest rotating engines in London at prominent breweries while a third went to Albion Mills at Blackfriars, in which they had invested. Scottish engineer John Rennie came to London in their employ to supervise the construction of the flour mill (1784–88). The Albion Mills pioneered the use of steam for grinding flour as well as the use of iron for gearing and shafting. When the mill burned down in 1791, Rennie had already made his mark. He then set up his own engineering works in nearby Holland Street. Rennie built the Waterloo and Southwark bridges, and after his death his two sons completed the New London Bridge (1831). His elder son George specialized in building marine steam engines for the Admiralty and in 1840 designed the first propeller-driven naval vessel. At the time, Rennie's engineering works employed 400. In 1856 George Rennie, then heading the mechanical sciences branch of the prestigious British Association for the Advancement of Science, helped a young London inventor, Henry Bessemer, launch his revolutionary steelmaking process (see chapter 6).

Impressive as the Watt-Rennie-Bessemer lineage is, perhaps the most significant London engineering genealogy began with Joseph Bramah. Originally a cabinetmaker from Yorkshire, Bramah patented an "engine" for pumping beer up from a publican's cellar, an improved water closet, and a hydraulic press, and is best known for his precision locks. Henry Maudslay trained at the Woolwich Arsenal, a leading site for mechanical engineering, then worked with Bramah in London for eight years making machine tools and hydraulic presses. Maudslay helped realize the grandiose plans of Samuel Bentham, named inspector-general of naval works in 1795. Bentham was determined to alter the prevailing shipbuilding practices, in which, as a result of shipwrights' traditional craft privileges of taking payment-in-kind, "only a sixth of the timber entering Deptford Yard left it afloat." Maudslay's forty-three special machines completely mechanized the manufacture of ship's pulleys, essential in the day of sail, and were installed by Marc Brunel at the Portsmouth navy yard in 1810.[23]

By 1825 Maudslay and Bramah were among the London engineers hailed for their use of specialized machine tools to replace skilled handcraftsmanship. These "eminent engineers," observed a parliamentary report, "affirm that men and boys ... may be readily instructed in the making of Machines."[24] Maudslay invented screw cutters and planers and perfected the industrial lathe. He also taught the next generation of leading machine builders. Among them, Joseph Clement, Richard Roberts, Joseph Whitworth, and James Nasmyth can be credited with creating many of the basic machine tools and techniques used in shops today, including the horizontal shaper, steam hammer, two-directional planer, and a standard screw thread. (French and American mechanics were also pioneers in precision manufacturing, mechanization, and standardization, as discussed below.) Yet by the 1830s, while Clement struggled to build an early mechanical computer for Charles Babbage, a core group of Maudslay-trained engineers (Whitworth, Nasmyth, Roberts, Galloway) had departed London. Each of them set up machine-building shops in Manchester.

Shock City: Manchester

If immense size exemplified London, swift growth provides the conceptual key to Manchester. For many years Manchester served the region's many cotton-textile outworkers as a market town. In 1770 its population numbered 30,000. Then Manchester started doubling each generation, by 1851 reaching the figure of 300,000. In 1786 "only one chimney, that of Arkwright's spinning mill, was seen to rise above the town. Fifteen years later Manchester had about fifty spinning mills, most of them worked by steam." By 1841 Manchester and the surrounding Lancashire region had twenty-five huge cotton firms *each* employing more than 1,000 workers, 270 firms employing between 200 and 900 workers, and 680 smaller firms. "There are mighty energies slumbering in those masses," wrote one observer in the 1840s. "Had our ancestors witnessed the assemblage of such a multitude as is poured forth every evening from the mills of Union Street, magistrates would have assembled, special constables would have been sworn, the riot act read, the military called out, and most probably some fatal collision would have taken place."[25]

Legend has it that weavers fleeing the Spanish-occupied southern Netherlands (see chapter 2) founded the region's cotton trade. Daniel Defoe as early as 1727 wrote of Manchester, "the grand manufacture which has so much raised this town is that of cotton in all its varieties." At this time families working at home did the necessary cleaning, carding, spinning, and weaving. "Soon after I was able to walk I was employed in the cotton manufacture," wrote the eldest son of Samuel Crompton, the Lancashire spinning-machine inventor. The young Crompton helped out by stepping into a tub of soapy water and tramping on lumps of cotton. "My mother and my grandmother carded the cotton wool by hand, taking one of the dollops at a time, on the simple hand cards. When carded they were put aside in separate parcels for spinning."[26] Spinning by women and weaving (largely by men) were also done at home; many weavers built their own wooden looms. Manchester merchants imported raw cotton from the American South and distributed it to the region's families, then sent finished cloth to markets in London, Bristol or Liverpool. As early as the 1790s Britain was exporting the majority of its cotton textile output.

Manchester's proximity to the port city of Liverpool constituted only one of the region's natural advantages for cotton manufacture. The ocean bathed the Lancashire region with damp air that helped cotton fibers stick together, making them easier to spin into a continuous thread. (For the

FIG. 3.5. Manchester's Union Street, 1829.

"There are mighty energies slumbering in those massess," wrote one middle-class observer, somewhat nervously, of the thousands of workers who poured out of the massive eight-story factory buildings each evening. Period engraving from W. Cooke Taylor, *Notes of a Tour in the Manufacturing Districts of Lancashire* (London: Frank Cass, 1842).

same reason, mills later vented steam into the spinning room to supersaturate the air.) Coal from the South Lancashire field was readily at hand. Indeed, the Duke of Bridgewater built his pioneering Worsley Canal, opened in 1761, to connect his estate's coal mines to Manchester. Bridgewater's further improvements, including the cutting of underground canals directly to the coal face, halved the cost of coal in Manchester. The earliest steam engines burned so much coal that only coal mines could use them economically. Even the early Watt rotating engines were employed mainly where the cost of animal power was excessive (as in London) or where coal was cheap (as in Manchester).

One of the central puzzles concerning the industrial "revolution" is why it took so long to unfold. One clue is the need for the ancillary industries that link innovative sectors together and ramify change; another clue is to be found in how and why the steam-driven cotton factory took form. Even though the basic mechanical inventions in carding, spinning, and weaving of cotton, along with Watt's rotating engine, had all been made before 1790, it was not until around 1830 that Manchester had a unified cotton factory system and a recognizable modern factory sector. To comprehend this forty-year "delay," we must consider the gender shift that attended incremental innovations in spinning and weaving, along with the structural shift in Manchester's business community.

Spinning begins with a long cord of loose cotton fibers ("roving") and results in thread suitable for weaving or knitting into cloth. For centuries, spinning had been the province of women. A woman might spin thread her entire life, supporting herself and remaining unmarried, which gave rise to that meaning of the word *spinster*. Many women welcomed the spinning jenny, invented by James Hargreaves in 1764. The jenny was a small-scale hand-operated device that separated spinning into two distinct mechanical motions: an outward motion pulled and twisted the cotton roving, while an inward motion wound the twisted thread onto spindles. It was cheap to buy or build one's own jenny, and a woman might spin twenty or thirty times more thread on a home jenny than she could with a traditional single-spindle spinning wheel.[27]

Spinning took an industrial, factory-dominated form largely because of the work of Richard Arkwright, a barber and wigmaker from a Lancashire village. Arkwright's "water frame" resembled the Lewis-Paul spinning machines of the 1740s in that both operated continuously with rotating rollers to impart twist to the roving and a winding mechanism to roll up the completed thread. But while the Lewis-Paul machines used a single set of rollers, Arkwright used two sets of rollers, covered them with leather, and spaced them about the same distance apart as the length of individual cotton fibers. The second set spun slightly faster than the first, at once twisting and pulling the cotton fibers into thread.

Arkwright patented his water frame machine in 1769 and built small mills at Nottingham and Cromford using horse power and water power before opening Manchester's first steam mill in 1786. At his Manchester mill, which employed 600 workers, an atmospheric steam engine pumped water over a waterwheel to power the machinery. Arkwright's extraordinary profits led many writers ever since to assume that a technical imperative brought about the factory system: that machinery and the factory were one. Yet this view does not withstand close inspection. Indeed, the term "water" frame, which suggests that Arkwright's spinning technology required water power and hence factory scale, is entirely misleading. Like jennies, water frames were initially built in smaller sizes adaptable to home spinning. Early Arkwright machines were small, hand-cranked devices with just four spindles. The death blow to home spinning came when Arkwright restricted licenses for his water frame patent to mills with 1,000 or more spindles. (Having a relatively small number of large-scale water frame mills made it easier for Arkwright and his business partners to keep an eye on patent infringements.) Practically speaking, these large mills needed water or steam power to drive the heavy machines. Water frames could no longer be set up in a kitchen. Arkwright's mills spun only the coarsest grades of thread, but he and his partners employed poor women and paid them dismal salaries that left handsome profits for the owners. It was said that none of the jobs took longer than three weeks to learn. Arkwright's mills—with their low wages and skills, their high-volume production of lower-grade goods, and their extensive mechanization—embodied core features of the industrial era.

The cotton spinning industry, despite Arkwright's great wealth, was more the product of an inelegant and unpatented hybrid technology. In 1779 Samuel Crompton's first spinning "mule" grafted the twin rotating rollers of Arkwright's machine onto the back-and-forth movements of Hargreaves' jenny. Thereafter a host of anonymous mechanics tinkered with the mule to make it larger (with 300 or more spindles). Whereas small-scale jennies supported the home industry of female spinners, as did the first mules, the larger mules, still hand-powered, required the greater physical strength of male spinners. The shift to heavier mules, dominated by skilled male spinners, doomed the independent female spinner. By 1811 a nationwide count revealed that spindles on Crompton's mules totaled 4.6 million, as compared with Arkwright's water frames, which incorporated a total of 310,000 spindles, and Hargreaves' jennies, at just 156,000 spindles. "Within the space of one generation," wrote Ivy Pinchbeck, "what had been women's hereditary occupation was radically changed, and the only class of women spinners left were the unskilled workers in the new factories to house Arkwright's frames."[28]

This wave of spinning inventions did not create a factory system so much as split Manchester into two rival cotton industries, the "factory" (spinning) firms and the "warehouse" (weaving) firms. The town's sixty factory-spinning firms were led by two giant textile-machine builders, McConnel & Kennedy and Adam & George Murray. Each was valued at around £20,000 in 1812 and each employed more than 1,000 factory workers. Few spinning firms were on this large scale. Through 1815 it was surprisingly common for as many as ten separate spinning firms to share one factory building; at this time only one-quarter of the town's factory buildings were occupied by a single factory firm. Compared with the factory firms, the "warehouse" firms were even more numerous and smaller (the largest warehouse firm in 1812 was half the largest spinning company in valuation). Weaving firms bought spun yarn from the spinning firms, and "put out" the yarn to a vast network of handloom weavers, male and female, across the region. Weaving firms then collected the woven cloth and marketed it, often shipping to overseas markets via Liverpool.[29]

Through the 1820s substantial conflict and hostility existed in Manchester between these two rival branches of the cotton industry. Factory and warehouse firms clashed in strident debates on exports, tariffs, child labor, and factory inspections. The perennial yarn-export question provoked sharp conflict between them: whereas the spinning firms hoped to export their rising output, the warehouse firms naturally feared that exports would hike domestic prices for yarn and in turn favor foreign weavers. Warehouse firms committed to home-based weaving blasted factory-based spinning. In 1829 one warehouseman slammed spinning factories as "seminaries of vice" and, hinting at the unsettling gender shifts, contrasted spinning with the more beneficial occupations of weaving, bleaching and dyeing "which require manly exertions and are productive of health and independence."[30]

Something like an integrated factory system took shape in Manchester only around 1825. By then Manchester had a total of thirty-eight integrated spinning-and-weaving firms. These were mostly "new" firms, established since 1815, or warehouse-weaving firms that had bought or built spinning firms (thirteen of each). Also by 1825, in marked contrast from just a decade earlier, most factory firms (70 percent) wholly occupied a factory building. The emergence, around 1830, of powered "self-acting" mules and smooth-working power looms thus fit into and reinforced an existing factory-based industrial structure. A gender division became locked in place, too. "Mule spinning" was firmly men's work, as was all supervisory work, while spinning on the smaller "throstles" as well as factory weaving was preponderantly women's work. (Powered self-acting mules, no longer needing a man's strength, might have been tended by women but the strong male-dominated spinners union kept women out of this skilled and better-paid trade.)

In the 1841 governmental report Factory Returns from Lancashire, integrated firms had more employees and more installed horsepower than the set of nonintegrated fine spinning, coarse spinning, and power weaving firms *together*. Yet during the years from 1815 to 1841, the very largest firms (above 500 workers) fell from 44 percent to 32 percent of total factory employment in the region, while medium-sized firms (150–500 workers) more than filled the gap, growing to 56 percent of total factory employment. The largest firms of the earlier period were specialized behemoths—McConnel & Kennedy sent most of their huge yarn output to Glasgow weavers—and were unable to move quickly in rapidly changing markets. The *Manchester Gazette* in 1829 predicted Manchester's prominent role in overseas imperialism (discussed in chapter 4): "You can sell neither cloth or twist [yarn] at any price that will cover your expenses. . . . Your eyes are wandering over the map of the world for new markets."[31]

Ironically, just as Manchester gained notoriety as "Cottonopolis" the city's industrial base was changing. For decades, in fact, Manchester was not merely the region's largest textile-factory town but also its commercial center. The Manchester Exchange, a meeting place for merchants and manufacturers, opened in the very center of town in 1809. Manchester's cotton-textile sector grew rapidly during the early decades of the nineteenth century; in 1841, textile workers outnumbered workers in all other Manchester's manufacturing trades together. The size of the city's cotton industry peaked in 1851 at over 56,000 workers. Yet that year, for the first time, more Manchester workers were employed in finishing cloth, making machines, building houses, and miscellaneous manufacturing (together) than in simply making cloth. From this time on, growth would depend not on the city's but the region's cotton-textile mills.[32] In Manchester, the making of dyeing and bleaching agents grew up as specialized auxiliaries to the textile industry.

Yet, far and away the most important auxiliaries in the city were Manchester's machine builders. While the first generation of them had built textile machines and managed textile factories, the midcentury machine builders—the generation of London transplants—focused on designing, building, and selling machine tools. Richard Roberts (1789–1864) came from London to Manchester in 1816 and specialized in large industrial planers, along with gear-cutting, screw-cutting, and slotting machines. Roberts's first notable textile invention was a machine for making weavers' reeds, manufactured by the productive Roberts, Sharp & Company partnership. Roberts made his most famous invention, an automatic, self-acting version of Crompton's mule spinner, in the midst of a

cotton spinners' strike in 1824. During the strike a delegation of the region's cotton manufacturers came to Roberts and proposed, as the industrial writer Samuel Smiles related the story, that he "make the spinning-
mules run out and in at the proper speed by means of self-acting
machinery, and thus render them in some measure independent of the
more refractory class of their workmen."[33] After a four-month effort,
he hit upon the self-acting mule, patented in 1825 and improved in 1832
with a radial arm for "winding on" spun thread. Two years later Roberts,
Sharp & Company began building locomotives in large numbers; and in
1848, again in the midst of labor strife, Roberts devised a novel rivet-
punching machine that made possible the Britannia Bridge over the
Menai Straits.

Joseph Whitworth (1803–87) is the best known of the Maudslay pro-
tégés trained in London. In 1833 he went to Manchester to manufacture
a series of accuracy-enhancing inventions in machine designs and standard screw threads. His famous micrometer could measure a millionth-
part of an inch. Whitworth's machine tools, steel armor, and armaments
made him rich and famous. Oversize forgings were the specialty of James
Nasmyth (1808–90), whose huge steam-powered hammer has become an
icon of the industrial era. His Bridgewater Foundry, near George Stephenson's Liverpool & Manchester Railway and the Bridgewater Canal, employed as many as 1,500 workers in building steam engines, machines, and
locomotives. Nasmyth tirelessly promoted the superiority of machine
tools, especially self-acting ones, over skilled workmen. These machine
tools, he wrote, "never got drunk; their hands never shook from excess;
they were never absent from work; they did not strike for wages; they were
unfailing in their accuracy and regularity."[34]

Roberts's self-acting machines, invented at the behest of strike-bound
manufacturers, as well as Nasmyth's continual strife with his workmen
(he retired, as he told a friend, to be free from "this continually threaten-
ing trade union volcano") underscores that more was at issue than the
efficient production of goods. Manchester embodied the harsh side of the
industrial revolution. On 18 August 1819, the nerve-strung local militia
opened fire on a mass meeting of Chartists, working-class political reformers, at St. Peter's Field in Manchester, killing eleven Chartists and in-
juring hundreds more. In the charged aftermath of this so-called Peter-
loo Massacre, a leading London banker pointed to "the desire and policy
of men engaged in trade . . . to screw down the price of labour as low as
possible." The social strains caused by industrialism spilled out onto the

streets. "The smashing of machinery, the destruction of mills and other property, and the assaulting of 'blacklegs,' occurred with alarming frequency." The assaults on factories during the spinners' strike of 1812, the more organized machine wrecking in 1826, and the "plug riots" of 1842—in which disgruntled workers disabled factories' steam boilers—amounted to what one author calls "a campaign in a social war."[35]

Clearly, something alarming and unprecedented was happening. "Manchester is the chimney of the world," observed Major-General Sir Charles Napier, temporarily home from India in 1839. "What a place! the entrance to hell realized." Charles Dickens agreed with these sentiments. "What I have seen has disgusted me and astonished me beyond all measure," wrote Dickens to a friend, just after his first visit to Manchester in 1838. Dickens's classic industrial novel, *Hard Times,* describes the physical and moral degradation brought by industrialism, a theme also of Elizabeth Gaskell's *Mary Barton,* set in Manchester. So widely known were the stock images—"the forest of chimneys pouring forth volumes of steam and smoke, forming an inky canopy which seemed to embrace and involve the entire place"—that one American commentator published his "firsthand" observations of Manchester without bothering to visit England at all.[36]

Not all visitors were dismayed or disgusted. "A precious substance . . . , no dream but a reality, lies hidden in that noisome wrappage," observed Thomas Carlyle, typically remembered as an arch-conservative, after a visit in 1838. "Hast thou heard, with sound ears, the awakening of a Manchester, on Monday morning, at half-past five by the clock; the rushing-off of its thousand mills, like the boom of an Atlantic tide, ten-thousand times ten-thousand spools and spindles all set humming there,—it is perhaps, if thou knew it well, sublime as a Niagara, or more so." Benjamin Disraeli in *Coningsby* (1844) also drew attention to this "new world, pregnant with new ideas and suggestive of new trains of thought and feeling." He stated, "Rightly understood, Manchester is as great a human exploit as Athens."[37]

All this commotion attracted young Friedrich Engels to Manchester. Engels arrived in December 1842 and took detailed notes during his twenty-one-month stay. Engels unlike most literary visitors knew industrial textiles from direct experience. He had already apprenticed in his father's textile mill in Barmen (now Wuppertal), Germany, and sharpened his business skills at a textile-exporting firm. Just before leaving Germany, he met the editor of the *Rheinische Zeitung,* one Karl Marx, and promised him firsthand essays on English industry. For Engels, Manchester was ground zero for the industrial revolution (he wrote specifically of "industriellen Umwälzung"). His duties in Manchester at the Ermen and Engels mill, partly owned by his father, left plenty of time for observing the city's working-class districts. Engels deliberately disdained "the dinner parties, the champagne, and the port-wine of the middle-classes." Workers' doors were opened for him by a young Irish factory girl, Mary Burns, with whom he would live for the next twenty years.

Returning to Germany, Engels put the final touches to *The Condition of the Working Class in England* (1845).[38] It is a classic statement on industry. Engels saw Manchester as the "masterpiece" of the industrial revolution and the "mainspring of all the workers' movements." At Manchester, he wrote, "the essence of modern industry" was plainly in view: water and steam power replacing hand power, power looms and self-acting mules replacing the hand loom and spinning wheel, and the division of labor "pushed to its furthest limits." Accordingly, the "inevitable consequences of industrialisation," especially as they affected the working classes, were "most strikingly evident." Not only did Manchester reveal the degradation of workers brought by the introduction of steam power, machinery and the division of labor, but one could see there, as Engels put it, "the strenuous efforts of the proletariat to raise themselves" up (50).

Engels in his writings said little about the working conditions inside the textile mills besides noting, "I cannot recall having seen a single healthy and well-built girl in the throstle room of the mill in Manchester in which I worked" (185). His real object was to shock his readers with visceral portraits of the city's horrible living conditions: Tumble-down dwellings in one place were built to within six feet of "a narrow, coal-black stinking river full of filth and rubbish" fouled by runoff from tanneries, dye-works, bone mills, gas works, sewers, and privies. The poorest workers rented dirt-floor cellars that were below the river level and always damp. Pig breeders rented the courts of the apartment buildings, where garbage thrown from the windows fattened the pigs, and stinking privies polluted the air. Engels keeps count of the number of persons using a single privy (380 is tops). "Only industry," Engels concludes, "has made it possible for workers who have barely emerged from a state of serfdom to be again treated as chattels and not as human beings" (60–64).

The very worst of Manchester was Little Ireland, whose 4,000 inhabitants were hemmed in by the River Medlock, railroad tracks, cotton mills, a gasworks, and an iron foundry. "Heaps of refuse, offal and sickening filth are everywhere interspersed with pools of stagnant liquid. The atmosphere is polluted by the stench and is darkened by the thick smoke of a dozen factory chimneys. A horde of ragged women and children swarm about the streets and they are just as dirty as the pigs which wallow happily on the heaps of garbage and in the pools of filth" (71). Engels was himself shocked at the damp one-room cellars "in whose pestilential atmosphere from twelve to sixteen persons were crowded." Numerous instances in which a man shared a bed with both his wife and his adult sister-in-law, however, were too much for his (dare one say) middle-class sensibilities, as was the mixing of the sexes in common lodging houses, where "in every room five or seven beds are made up on the floor and human beings of both sexes are packed into them indiscriminately" resulting in "much conduct of an unnatural and revolting character" (77–78).

Chased out of Germany after he and Marx published the *Communist Manifesto* in 1848, Engels returned to Manchester, where he worked as senior clerk for the Ermen and Engels company, eventually becoming a full partner. For two decades (1850–69) Engels lived a split existence: capitalist by day and working-class radical by night. Marx during these years lived in London and, with financial support from Engels, spent his days at the British Museum writing *Das Kapital.* That profits from a capitalist textile firm so directly supported the greatest critic of capitalism lends a fine irony to the famous line in the *Communist Manifesto,* "What the bourgeoisie . . . produces, above all, is its own grave-diggers."[39] Marx, with no firsthand industrial experience of his own, took Engels' description of Manchester as the paradigm of capitalist industry. Neither of them noticed a quite different mode of industry forming in Sheffield.

Region for Steel: Sheffield

"Our journey between Manchester and Sheffield was not through a rich tract of country, but along a valley walled by bleak, ridgy hills, extending straight as a rampart, and across black moorlands, with here and there a plantation of trees," wrote novelist Nathaniel Hawthorne, who lived in nearby Liverpool during the mid-1850s. "The train stopped a minute or two, to allow the tickets to be taken, just before entering the Sheffield station, and thence I had a glimpse of the famous town of razors and pen knives, enveloped in a cloud of its own diffusing. My impressions of it are extremely vague and misty—or, rather, smoky—: for Sheffield seems to me smokier than Manchester, Liverpool, or Birmingham; smokier than all England besides, unless Newcastle be the exception. It might have been Pluto's own metropolis, shrouded in sulphurous vapour."[40]

Sheffield was internationally known as a center for high-quality steel and high-priced steel products. Sheffielders transformed Swedish bar iron into crucible steel, the only steel available in quantity before Henry Bessemer's steel experiments of the 1850s (see chapter 6). While a large number of Sheffield's craftsmen specialized in steelmaking, an even larger number worked in the numerous trades that cut, forged, and ground the bars of steel into pocket knives, cutlery, saws, and files. Sheffielders exported to the United States up to a third of their total production, finding a ready market there for steel traps, agricultural implements, hunting knives, and unfinished bars. Not Manchester-style factories but networks of skilled workers typified Sheffield's industry.

Like London's port and Manchester's moist air and coal, Sheffield's geography shaped its industry. Coal from the vast south Pennines field, scant miles away, stoked its furnaces. Sandstone for grinding wheels was hewed from local ground, while the fire-resistant "refractories" for clay pots and bricks came from nearby. Four sizable rivers, including the Sheaf, which named the town, drained into the River Don, forming 30 miles of river frontage where water power could propel industry. Some of the region's 115 water mill sites were built for milling grain, making paper, or grinding snuff, but most were used for grinding, forging, and rolling the region's characteristic steel products.

Already Britain's eighth largest town in 1775, Sheffield nearly tripled in size during the years from 1801 to 1851, when the population reached 135,000. The early growth was built on a flourishing cottage industry in the surrounding district. By the early seventeenth century its workshops hummed with the making of scissors, shears, files, nails, forks, and razors. In 1624 Sheffield's artisans formed the Cutlers Company to organize and protect their trade. Artisans looking for finer markets specialized in the making of small metal boxes for tobacco, snuff, and trinkets. In the 1740s, two famous inventions secured the region's technical advantage for nearly a century: crucible steel, examined below, and "Old Sheffield Plate." Sheffield plate married the luster of silver with the economy of copper by fusing silver foil onto copper dishes and utensils. Makers of saws, anvils, lancets, and household cutlery, easily the city's most famous early trade, were prominent well before 1800.

A chronicler of the city in 1824 found sixty-two identifiable trades. In an alphabetical list of them, the B's alone include bayonets, bellows, boil-

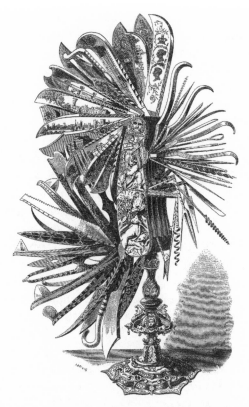

FIG. 3.6. Sheffield Cutlery at Crystal Palace, 1851.

Rodgers & Sons (Sheffield) sportsman's knife, 12 inches long, "containing eighty blades and other instruments . . . with gold inlaying, etching and engraving, representing various subjects, including views of the Exhibition Building, Windsor Castle, Osborne House, the Britannia Bridge, etc." Christopher Hobhouse, *1851 and the Crystal Palace* (London: John Murray, 1937), 94.

ers for steam, bone scales, brace bits, brass bolsters, brass in general, Britannia metal, butchers' steels, buttons, and button molds. The city's ten identifiable cutlery trades, led by the makers of table knives and pocket knives, employed more than 10,000. Some of the cutlery trade unions were tiny, such as the 50 souls devoted to plate-, spoon-, and fork-filing. In 1841, even though they were excluded from the apprenticeships necessary for skilled work, some women worked in the metal trades. Among them were 159 women cutlers, 158 scissors makers, 123 file makers, 42 fork makers, as well as numerous hafters and bone cutters. The skills afforded by such extreme specialization drew ever more workers (and ever more buyers) to the district. By 1841 Sheffield had 54 percent of Britain's file makers, 60 percent of its cutlers, and 80 percent of its saw makers.[41]

Sheffield's world-famous crucible steel, the second notable invention of the 1740s, owes its creation to a clockmaker named Benjamin Huntsman. In the early 1740s, in an effort to make suitable clock springs, he ran a coal-fired, brass-melting furnace hot enough to melt pieces of blister steel, iron that had been baked in charcoal for a week. Huntsman's success came in devising clay pots, or crucibles, that did not soften or crack under the extreme heat. In the early 1770s Huntsman built a steelmaking works of his own. By 1800 the Sheffield district comprised nine firms making blister steel, eleven firms melting the blister steel to cast ingots of crucible steel, and a number of separate water-powered mills hammering the cast ingots into rods and bars. With local, national, and overseas markets, Sheffield's crucible steel industry grew steadily. By the 1840s the Sheffield region's steel-making industry employed about 5,000 workers, who made 90 percent of Britain's steel output and around half of the world's.

It is crucial to understand that the factory system so important in Manchester was absent in Sheffield. "The manufacturers, for the most part," wrote one visitor in 1831, "are carried on in an unostentatious way, in small scattered workshops, and nowhere make the noise and bustle of a great ironworks." One famous razor-making works evidently confounded an American visitor, who, perhaps expecting a huge Manchester-style factory, observed that instead workmen were scattered across town. Knife blades were rough forged in one building, taken across the street for finishing, ground and polished at a third works "some distance off," and sent to a fourth building to be joined to handles, wrapped, packed, and finally dispatched to the warehouse-salesroom. Even the largest cutlery firms, such as Joseph Rodgers and George Wostenholm, each employing more than 500 workers by the 1850s, remained dependent on the region's characteristic outwork system. Cutlers were not really full-time employees, for they worked for more than one firm when it suited them. A cutler might work at a single kind of knife all his life. "Ask him to do anything else and there'd be trouble. He'd get moody, throw off his apron and get drunk for the rest of the day." A visitor to the Rodgers firm in 1844 observed that, whether large or small, "each class of manufacturers is so dependent on the others, and there is such a chain of links connecting them all, that we have found it convenient to speak of Sheffield as one huge workshop for steel goods."[42]

Joining the ranks of manufacturers in Sheffield did not require building, buying, or even renting a factory. It was possible to be "a full-blown manufacturer with nothing more than a stamp with your name on it and a tiny office or room in your house." All the skilled labor for making a knife or edge tool—forging, hardening, grinding, hafting, even making the identifying stamp—could be bought by visiting skilled workmen's shops up and down the street. An enterprising artisan might, by obtaining raw materials on credit and hiring his fellow artisans, become a "little mester," a step up from wage laborer. Some firms did nothing but coordinate such "hire-work" and market the finished goods, at home or overseas. These firms had the advantages of low capital, quick turnover, and the flexibility to "pick and choose to fit things in with whatever you were doing."[43]

For decades the Sheffield steel-making industry, too, was typified by small- and medium-sized enterprises linked into a regional network. Certain firms specialized in converting raw iron into blister steel, others in melting (or "refining") blister steel into cast crucible steel. Specialized tilting (hammering) and rolling mills formed cast ingots into useful bars and rods, even for the larger steel firms.[44] Beginning in the 1840s Sheffield steelmakers moved east of town to a new district along the Don River valley, with cheap land and access to the new railroad. There in the 1850s Jessop's Brightside works, with 10 converting furnaces and 120 crucible melting holes, briefly held honors as the country's largest steelmaker. Other steelmakers opening works in this district included Cammell's Cyclops works in 1846, Firth's Norfolk works in 1852, John Brown's Atlas works in 1855, and Vickers' River Don works in 1863.

These large firms, already producing a majority of the region's steel, grew even larger after the advent of Henry Bessemer's process (see chapter 6). Bessemer built a steel plant at Sheffield in 1858 with the purpose, as he put it, "not to work my process as a monopoly but simply to force the trade to adopt it by underselling them in their own market . . . while still retaining a very high rate of profit on all that was produced."[45] Bessemer in the early 1860s sold profitable licenses to John Brown, Charles Cammell, and Samuel Fox, who each constructed large-scale steel works. In 1864 John Brown's invested capital was nearly five times larger than the leading crucible-only firm (Vickers'), while Brown's workforce was more than three times larger. Brown's Atlas Works covered 21 acres. In the lat-

ter part of the nineteenth century these large steel mills and oversize forging shops symbolized a second generation of Sheffield's heavy industry.
A string of new alloy steels kept Sheffield's crucible steel industry at the
forefront of high-quality steelmaking well into the twentieth century.

The shift from water power to steam upended the region's industrial
geography, especially its power-dependent trades. Foremost among these
was the grinding trade, segmented into distinct branches for scissors,
forks, needles, razors, penknives, table blades, saws, files, and scythes.
Writes Sidney Pollard, who authored the definitive history of Sheffield's
industrial workers, "The grinders had been the last to leave the countryside, . . . where they worked along the rivers whose water-power they used,
'a law unto themselves,' with their own habits, customs, and traditions."[46]
Grinders hacked and turned rough-cut sandstone from the quarry into
wheels. Then, sitting astride the wheel, as on a horse saddle, the grinder
pressed the article to be smoothed or sharpened against the rotating
stone. The work was physically demanding and, as we will see, dangerous
beyond belief.

The application of steam power was the only "industrial" technology
that affected the Sheffield trades before the 1850s. First applied to a grinding "wheel" in 1786, use of steam power grew steadily, as table 3.1 shows.
(In Sheffield a "wheel" referred to a building containing a number of
workrooms, or "hulls," each of which had as many as six individual
"troughs" where a grinder worked his stone.) The application of steam
power, as Pollard describes it, "changed the way of life of grinders, who
became town-dwelling, full-time industrial workers instead of members
of a part-time rural industry. The application of steam to cutting, glazing
and drilling similarly drew cutlers, ivory cutters, hafters and others from
their homes or the small lean-to's, in which work had previously been carried on, into the town workshops provided with power." By 1854, when
Sheffield had 109 steam engines, the grinders consumed 58 percent of the
city's steam power, with cutlery and toolmakers accounting for most of
the rest.[47]

Even though grinding was centralized in town, the possibility of becoming a "little mester" persisted. The "little mesters" employed their fellow grinders in hope of becoming full-fledged manufacturers. In-workers
used a wheel owned by a manufacturer, and worked for set piece rates.
Out-workers might work for more than one manufacturer at a time, renting an individual trough at a public wheel whose owner supplied space

TABLE 3.1. Steam and Water Power in Sheffield Steel Grinding,
1770–1865

	Water Power		Steam Power	
	wheels	troughs	wheels	troughs
1770	133	896		
1794	83	1,415	3	320
1841	40		50	
1857	16*		80	
1865	32		132	

Source: Sidney Pollard, *A History of Labour in Sheffield* (Liverpool: Liverpool University Press, 1959), 53.
Note: Wheel = building containing multiple workrooms; *trough* = work station of one grinder and one grinding stone.
* Pollard noted that this figure was probably an underestimate, omitting the smaller wheels.

and power to his tenants for a weekly rent. An ambitious out-worker might still become a little mester by purchasing raw materials on his own account, arranging with a middleman to take the completed articles, and completing the work by renting several troughs and employing fellow grinders. Figures from one Sheffield wheel in 1824 indicate the continuous spectrum ranging from wage laborer to little mester: 35 grinders rented a single trough, 30 grinders rented between 1.5 and 3 troughs, while 1 grinder rented 4 troughs and another 6 troughs.

Steam grinding killed an alarming number of grinders. "Till steam-power was introduced in the trade, towards the end of the last century, the grinders' disease was scarcely known," observed one employment report from the mid-1860s. Previously, grinders had worked at water-driven wheels in the countryside, with substantial breaks for part-time farming or gardening. With the expansion of their work to a full-time, year-round occupation a shocking proportion of grinders fell sick and died from what we would today identify as silicosis and tuberculosis. The most dangerous grinding was done "dry" (without water on the wheel). Dry grinding removed more steel and allowed the article to be watched constantly, and it long remained the standard practice for grinding needles, forks, and razors.

The life expectancy for these dry grinders was terrifyingly short. During 1820–1840, a majority of the general population of the surrounding

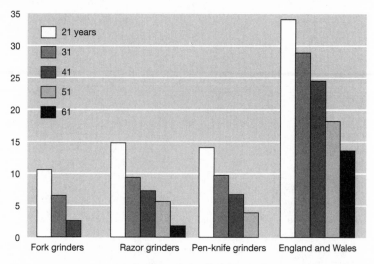

FIG. 3.7. Mortality of Sheffield Grinders, 1841.

Life expectancy (in years) of Sheffield grinders at different ages. Last group, "England and Wales," is the average population. For example, an average person in England or Wales at age 21 could expect to live 34 more years; at age 41 they could expect to live 25 more years. In sharp contrast, a Sheffield fork grinder at age 21 could expect to live 11 more years; if he reached age 41, just 2.5 more years. Data from Sidney Pollard, *A History of Labour in Sheffield* (Liverpool: Liverpool University Press, 1959), 328.

Midland counties (62 percent) could expect to live beyond age 50. However, a chilling 58 percent of fork and needle grinders were dead by age 30, while 75 percent of razor and file grinders were dead by age 40. Only the table blade, scythe, and saw grinders had any reasonable chance of living into their 50s. (Figure 3.7 shows the same dreadful picture, using calculations of life expectancy.) Fork grinding and razor grinding remained so dangerous that these grinders were excluded from other grinders' hulls and from the "sick clubs" which helped pay for medical care. It is a grim testimony of male grinders' determination to provide a gainful trade for their sons that they reportedly declined safer working conditions, reasoning that the longer grinders lived, the less work would be available for their sons. Grinders of saws, files, scythes, and table blades, for whom wet grinding was common, suffered less from lung diseases. By 1840 penknives and pocket blades were also wet ground. For all grinders, there was a significant danger that their wheels would blow apart while rotating. One grinder testified that in eighteen years no fewer than ten stones had burst under him.[48]

Steam not only directly killed many grinders, through dangerous working conditions, but also indirectly brought the deaths of many who crammed themselves and their families into the poorest central districts of industrial cities. In the 1840s sanitary reformers praised Sheffield for combating cholera epidemics, draining burial grounds, and cleaning streets. With a cleaner environment, the death rate for children under five years dropped from a desolate 32.7 percent in 1841 to 10.9 percent in 1851. But these modest sanitary improvements were simply overwhelmed by the city's central-area building boom in the 1850s. The conversion of many grinding, tilting, and rolling mills to steam power and their relocation from outlying areas packed the city's central area with factories. The construction of large steam-powered steel plants in the city's eastern district contributed to crowding, too. The few open areas in the city center were now covered with buildings, and many larger dwellings were converted to factories. "Sheffield, in all matters relating to sanitary appliances, is behind them all," wrote a qualified observer in 1861.

> The three rivers sluggishly flowing through the town are made the conduits of all imaginable filth, and at one spot . . . positively run blood. These rivers, that should water Sheffield so pleasantly, are polluted with dirt, dust, dung and carrion; the embankments are ragged and ruined; here and there overhung with privies; and often the site of ash and offal heaps—most desolate and sickening objects. No hope of health for people compelled to breathe so large an amount of putrefying refuse.[49]

Sheffield's dire sanitary conditions resembled those of London or Manchester for much the same reason: the city's densely packed population lacked clean water. Sheffield's water was piped from a reservoir a mile outside town, but it reached town unfiltered, clouded, and rusty. For the lucky 75 percent of houses in the borough (19,000 of 25,000) connected to the system in 1843, water could be drawn from standpipes in the yard for two hours on three days each week. For other times families had to store water in tubs or barrels or cisterns. The 25 percent of families without city water took their chances with wells, and the poorest sank wells in their building's walled-in courtyards, the location also of privies serving up to a dozen households. Even allowing for some exaggeration on the part of reformers, it seems clear that sanitary conditions in Sheffield were bad and got worse. In the mid-1860s there were credible reports that certain privies "did duty" for up to sixty persons. One correspondent counted twenty-eight rotting dog carcasses piled under a city bridge. Between the 1860s and 1890s the city's adult mortality rates (still above national averages) improved somewhat. Infant mortality did not. During these decades one in *six* Sheffield infants died before their first birthday.[50]

THE GEOGRAPHIES OF INDUSTRY surveyed in this chapter—multidimensional urban networks in London, factory systems in Manchester, and sector-specific regional networks in Sheffield—clinch the argument that there were many "paths" to the industrial revolution. No single process transformed Britain in lockstep fashion. London, driven by population growth, its port activities, and the imperatives of a capital city, industrialized along many dimensions at once, often in spatially compact districts and always tied together by ancillary industries such as those that grew up around the porter brewing industry. Brewing, shipbuilding, engineering, and building were among the chief innovative centers, but London would not have industrialized (and in turn would not have had such sustained population growth and market impact) absent its innumerable ancillary industries. The city showcased the country's industrial achievement at the Exhibition of 1851 in the Crystal Palace (fig. 3.8).

In sharp counterpoint to London's multidimensional growth is Manchester's single-industry cotton factory system. There, truly, cotton was king. As cotton textiles spread into the surrounding Lancashire district, Manchester traded in its cotton-textile factories for the ancillary industries of bleaching, printing, and dyeing. The region's demand for special cotton textile machines attracted the London machine builders, who moved to Manchester, bringing a crucial ancillary industry, and subsequently made machine tools, bridges, and locomotives. Sheffield, too, centered on a single sector, high-quality steel products. In a sense, the entire city constituted a spatially compact network of ancillary industries—"one huge workshop for steel goods"—providing all the specialized skills needed to make bars of steel into pocket knives, cutlery, saws, files, and other valuable items. Sheffield's network model of industry, and especially the persistence of the small-shop, "little mester," system, provided much greater scope for occupational mobility than Manchester's capital-intensive factories. Nor were Sheffielders time-bound preindustrial artisans, as attested by their innovations in silver plate and crucible steel, their adoption of Bessemer steel, and the city's far-reaching fame in top-quality specialty steels.

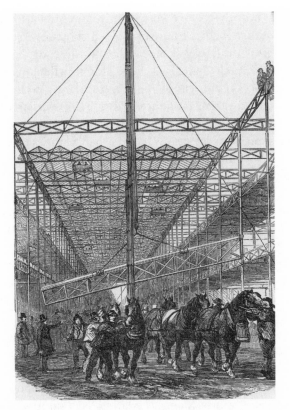

FIG. 3.8. Building the Crystal Palace, 1851.

The Crystal Palace, built for the 1851 Exposition in London, was the great
symbol of Britain's commercial and industrial might. At the top are the
glazing wagons workers used to place the 300,000 panes of glass, 12 by 49
inches square, that formed the roof. Christopher Hobhouse, *1851 and the
Crystal Palace* (London: John Murray, 1937), 48.

Yet the shocking filth and poverty evident in each of these three industrializing sites was fuel to the fires for critics of industrial society. It is
striking indeed that distinct paths to industrial revolution led to similarly
dire environmental outcomes. Here the centralizing power of steam appears fundamental. Workers in steam-driven occupations, whether in
London, Manchester, Sheffield, or the surrounding regions, were less
likely to be in the country, to eat fresh food, to drink clean water, and (especially if female) to be skilled and have reasonable wages.

Steam-driven factories drew into urban areas, where coal was easily delivered, not only the factory workers themselves but also the ancillary industries' workers and the host of needed shopkeepers and service-sector workers. In this way steam was responsible for cramming ever more residents into the industrial cities. This growth pattern continued as the greater thermal efficiencies available from larger steam engines gave profit-minded industrialists an incentive to build ever-larger and more efficient steam engines and correspondingly larger factories. Indeed, the evident social problems of steam-driven urban factories inspired Rudolf Diesel to invent an internal combustion engine that, he hoped, would return power, literally and figuratively, to decentralized, small-scale workshops that might be located in the countryside.

The industrializing cities were certainly striking to the senses and sensibilities, but of course they were not the whole picture. Indeed, the industrial revolution did not so much create a single homogenized "industrial society" during these decades as generate disturbing *differences* within British society. Visits to industrial urban slums shocked many commentators because they came with their sensibilities from town or rural life still very much in place. Traditional, rural ways of life persisted throughout the industrial period, often quite near cities. London's wealthy retreated to their rural estates within a day's horse ride, while the city's poor were stuck there. It is striking that even Friedrich Engels framed his critique of Manchester's squalid living conditions through a moralistic, middle-class set of values. Manchester attracted a storm of critical comment because it, and the other industrial centers, threw up a disturbing question about whether industry was dividing Britain into "two nations" split by irreparable differences. By comparison, working and living conditions in rural-based traditional industries, such as the woolen industry, passed largely without critical comment.

Looking farther afield, we can see that any complete view of the industrial revolution needs to incorporate not only the contributions of countries other than England but also their varied experiences of industrialization. Here I can give only the briefest sketch of the industrial revolution beyond Britain. To begin, the classic industrial technologies were typically the result of varied national contributions. The textile industries took form with continual interchange and technology transfer among England, France, the United States, and other countries. While Americans drew heavily on British textile machines, they built integrated textile factories arguably a decade earlier than did Manchester (with the pioneering Lowell-system mills in Waltham [1814–15] and Lowell [1821], Massachusetts). Similarly, one might see the distinctive "American system" of manufacturing as an ambitious attempt to replicate and mechanize the French achievement of manufacturing muskets, instituted as early as the 1780s, that featured interchangeable parts. Another distinctive American achievement, the steel industry, utterly depended on the transfer of production processes from Britain, France, Germany, and other countries.[51]

Different countries took several distinct paths to industrial revolution. Countries like Sweden and the United States were rich in wood and water power, and consequently they were largely spared the pressing problems of draining coal and mineral mines that had driven the development of steam engines in Britain. Svante Lindqvist provides a rough index of Swedish technological priorities in the late eighteenth century: out of 212 technical models held by the Royal Chamber of Models, there were numerous agriculture machines (43), wood-conserving fireplaces (29), mining machines (30), handicraft and textile machines (33), and various hydraulic works (35), but just *one* steam engine.[52] (Curiously, early attempts in both Sweden and America to transfer Newcomen steam engines from England ended in outright failures.) What is more, compared to Britain, coal-fired iron smelting came later to Sweden and the United States for the forceful reason that they had ready supplies of wood-based charcoal for fuel. Charcoal was used in making Sweden's famed bar iron, exported to Sheffield and around the world. Even more so than Britain, these countries depended on water power, which in turn, because water power sites needed to be spread out along rivers, lessened the rush to steam-and-coal-centered cities.

While variations in availability of raw materials obviously influenced a country's industrial path, cultural and political preferences mattered, too. France did not have a British-style industrial revolution but rather pursued artisan-based industries, which used a skilled and flexible labor force and were adapted to regional markets. Despite its population's being roughly twice that of Britain's, France had just one-third the number of large cities (three versus nine cities above 100,000). In the already urbanized Netherlands, there emerged a distinct mix of both craft-oriented and mass-production industries. In Germany, politically united only in the 1870s, industrialization was generally later to develop but featured rapid growth of industrial cities, such as those in the Ruhr, built on heavy industries of iron and coal (and soon enough, as we will see in chapter 5, on electricity and chemicals).

In chapters to come we'll see the differentiation of industrial technologies in the eras of systems, modernism, and warfare. Each of these eras was built on industrial technologies and yet led to their transformation. The next chapter follows up on Britain's industrialists, technologists, and government officials, whose eyes were "wandering over the map of the world for new markets" in a new, imperialist era for technology.

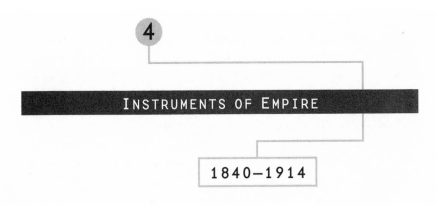

INSTRUMENTS OF EMPIRE

1840–1914

BRITISH TECHNOLOGY, propelled forward by the industrial revolution, had reached something of a plateau by the mid-nineteenth century. The display of Britain's mechanical marvels at London's Crystal Palace exposition in 1851 stirred the patriotic imagination, but now British industry faced a long march to stay abreast of rising competitors in Europe and North America. (And it was Germany and the United States that would spearhead a "second" industrial revolution in the decades to come.) At midcentury in Britain, and soon across much of the industrialized world, a new technological era took shape as colonial powers addressed the unparalleled problems of far-flung overseas empires. To a striking extent, inventors, engineers, traders, financiers, and government officials turned their attentions from blast furnaces and textile factories at home to steamships, telegraphs, and railway lines for the colonies. Imperialism altered these technologies, even as these technologies made possible the dramatic expansion of Western political and economic influence around the globe.

Britain was of course no newcomer to empire. In 1763, after the Seven Years' War, Britain had gained most of France's holdings in India and North America. In the next decade, however, American colonists rejected new taxes to pay for the heavy costs of this war and initiated their rebellion, which, in the view of one Englishman, would have failed if only Britain had possessed effective communications technology.[1] British rule in India was the next dilemma. The East India Company, a creature of Britain's mercantilist past, ran afoul of Britain's rising industrialists. Owing greatly to their relentless antimercantile lobbying, Parliament in

1813 stripped the East India Company of its monopoly of the lucrative Indian trade and, by rechartering the company in 1833, ended its control on the movements of private British traders within India and terminated its sole remaining commercial monopoly, that of trade with China. New players crowded in. The Peninsular & Oriental Steam Navigation Company, the legendary "P&O," began steamship service between England and India in 1840. The rise of "free trade" in the 1840s, also promoted by British industrialists keen to secure reliable sources of raw materials for their factories and cheap food for their factory workers, led to a wild scramble in the Far East. At least sixty British trading firms in China clamored for military assistance to uphold "free trade" there. The British government intervened, and this led to the first opium war (1840–42).

New technologies were critical to both the penetration phase of empire, in which the British deployed steam-powered gunboats and malaria-suppressing quinine to establish settlements inland beyond the coastal trading zones, and in the subsequent consolidation phase that stressed the maintenance and control of imperial outposts through a complex of public works.[2] Effective military technologies such as steam-powered gunboats, breechloading rifles, and later the fearsome rapid-firing machine guns helped the British extend their control over the Indian subcontinent and quell repeated uprisings among native populations. Even before the Indian Mutiny of 1857–58, which was a hard-fought battle against insurgent Indian troops who ranged across much of the northern and central regions of India and whose defeat cost the staggering sum of £40 million, there were major military campaigns nearly every decade (fig. 4.1). These included three Mahrattan wars (1775 to 1817), two Mysore wars in the 1780s and 1790s, the Gurkha war of 1814–15 in Nepal, two Anglo-Burmese wars in the 1820s and 1852, and the first of two opium wars in China. In the 1840s alone the British conducted a three-year military occupation to subdue and secure Afghanistan, which failed in 1842, a swift campaign crushing the Sinds in what is now Pakistan the following year, and the bloody Sikh wars over the Punjab in 1845 and 1848.

The tremendous cost of these military campaigns as well as the ongoing expenses for transporting, lodging, provisioning, and pensioning imperial officials simply ate up the profits of empire. We noted in chapter 3 that the East India Company put on the imperial payroll 1,200 workers in London alone. Recent research has demonstrated that, on balance, these collateral expenses of British empire completely absorbed the sizable profits of imperial trade. Imperialism in India did not generate

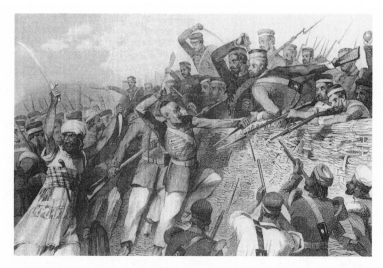

FIG. 4.1. Indian Mutiny of 1857–58.

"Attack of the Mutineers on the Redan Battery at Lucknow, July 30th 1857." British accounts inevitably stressed the "atrocities" committed by the rebel Indian soldiers, at Lucknow and across north-central India, but in his *Memories of the Mutiny* (London, 1894, pp. 273–80), Col. F. C. Maude detailed his own part in desecrating the bodies of executed Indian prisoners at Bussarat Gunj. Illustration from Charles Ball, *The History of the Indian Mutiny* (London: London Printing and Publishing, 1858–59), vol. 2, facing p. 9.

wealth. Rather it shifted wealth from taxpayers in India and Britain to wealthy traders, investors, military officers, and imperial officials, who became its principal beneficiaries.[3] This point is important to emphasize, because critics have long taken for granted that the imperatives of capitalism required imperialism (not least to dispose of surplus factory-made goods) and that the machinery of imperialism made money. Equally important, the wealth-*consuming* nature of imperial technologies sets off the imperial era from the earlier wealth-generating ones of commerce and industry. Imperial officials, and the visionary technology promoters they funded, spared no expense in developing and procuring those instruments of empire that promised to achieve rapid and comfortable transportation, quick and secure communication, and above all ample and effective military power.

Steam and Opium

Steam entered India innocently enough in 1817 when an 8-horsepower steam engine was brought to Calcutta in a short-lived attempt to dredge the Hooghly River. This plan was revived in 1822 when the East India Company bought the engine and again used it to clear a channel up the Hooghly to speed the passage of sailing vessels to Calcutta proper, some 50 miles inland from the ocean. At the time, Calcutta was the chief British port in India and the principal seat of its political power. The second city, Bombay, on India's western coast, was actually closer as the crow flies to London, but favorable winds made it quicker to sail to Calcutta via the African cape. The initial experiments with steam engines around Calcutta were, from a commercial point of view, rather disappointing. The steamship *Diana* worked on the Hooghly for a year as a passage boat while the more substantial *Enterprise*, a 140-foot wooden vessel with two 60-horsepower engines, a joint project of Maudslay & Field, the great London engineers, and Gordon & Company, was the first steam vessel to reach India under its own power, having steamed around the African cape in early 1825. The trip took a discouraging 113 days. The huge amount of fuel required by early steamers made them commercially viable only where abundant supplies of fuel were readily at hand, such as along the Mississippi River, or where their use secured some special advantage that covered their high operating costs.[4]

An early indication of the value of steamers in imperial India came in the first Anglo-Burmese war (1824–26). At first the war went badly for the British, who hoped to claim valuable tea-growing lands recently annexed by Burma. Britain's hope for a quick victory literally bogged down—in the narrow twisted channels of the lower Irrawaddy River. Britain's majestic sailing vessels were no match for the Burmese *prau,* a speedy wooden craft of around 90 feet in length propelled by up to seventy oarsmen and armed with heavy guns fixed to its bow. The British lost three-quarters of their force, mostly to disease, in the swamps of the Irrawaddy. The tide turned abruptly, however, when the British ordered up three steam vessels whose shallow draft and rapid maneuverability worked wonders. The *Enterprise* rapidly brought in reinforcements from Calcutta, while the *Pluto* and *Diana* directly aided the attack by towing British sailing ships into militarily advantageous positions. Another steam vessel that saw action later in the war was the *Irrawaddy,* arguably the first steam gunboat, with its complement of ten 9-pound guns and one swivel-mounted 12-pound gun on the forecastle. The defining image of this war was of the *Diana,* known to the Burmese as "fire devil," tirelessly running down the *praus* and their exhausted oarsmen. The king of Burma surrendered when the British force, assisted by the *Diana,* reached 400 miles upstream and threatened his capital.

Following this impressive performance in Burma, steamboat ventures proliferated in the next decade. The East India Company deployed steamers to tow sailing vessels between Calcutta and the ocean and dispatched the pioneering steamer up the Ganges River. (Accurately mapping the Ganges in the latter eighteenth century had been a necessary first step in transforming the vague territorial boundaries assumed by the company into a well-defined colonial state. To this end one could say that the first imperial technology deployed on the Ganges was James Rennell's detailed *Map of Hindoostan* [1782]. Rennell also published the equally valuable *Bengal Atlas* [1779].[5]) Regular steam service on the Ganges between Calcutta and upstream Allahabad began in 1834; the journey between these cities took between twenty and twenty-four days depending on the season. The river journey bore a high price. A cabin on a steamer between Calcutta and Allahabad cost 400 rupees, or £30, about half the cost of the entire journey from London to India and completely beyond the means of ordinary Indians. Freight rates from £6 to £20 per ton effectively limited cargoes to precious goods like silk and opium, in addition to the personal belongings of traveling officials and the necessary imperial supplies such as guns, medicines, stationery, official documents, and tax receipts. In the strange accounting of imperialism, however, even these whopping fares may have been a bargain for the East India Company, since in the latter 1820s it was paying a half-million rupees annually for hiring boats just to ferry European troops up and down the Ganges. Quicker river journeys also cut the generous traveling allowances paid to military officers. The economics of river transport were not only administrative ones. General Sir Charles Napier, who led the military campaign in the 1830s to open up for steam navigation the Indus River, India's second most important inland artery, pointed out direct commercial consequences. He wrote, "India should suck English manufacturers up her great rivers, and pour down these rivers her own varied products."[6]

Steam also promised to tighten up the imperial tie with London. Before the 1830s a letter traveled by way of a sailing vessel from London around the African cape and could take five to six months to arrive in India. And because no captain would venture into the Indian Ocean's sea-sonal monsoons, a reply might not arrive back in London for a full two years. Given these lengthy delays, India was not really in London's control. British residents in Madras urged, "Nothing will tend so materially to develop the resources of India . . . and to secure to the Crown . . . the integrity of its empire over India, as the rapid and continued intercourse between the two countries by means of steam." Merchants and colonial administrators in Bombay were eager to secure similar benefits. In the new age of steam Bombay's proximity to the Red Sea was a distinct advantage. From there the Mediterranean could be reached by a desert crossing between Suez and Alexandria. Efforts of the Bombay steam lobby resulted in the 1829 launch of the *Hugh Lindsay,* powered by twin 80-horsepower engines, which carried its first load of passengers and mail from Bombay to Suez in twenty-one days. Even adding the onward link to London, the *Hugh Lindsay* halved the time required for sending mail. So valuable was the Red Sea route that the British became far and away the largest users of the French-funded, Egyptian-built Suez Canal, opened in 1869. In 1875 Britain purchased the Egyptian ruler's entire share in the canal company for £4 million and in 1882 directly annexed Egypt, to maintain control over this vital imperial lifeline.

Already by 1840 steamboats in several ways had shown substantial promise for knitting together the various wayward strands of the British Empire. In the opium war of 1840–42 they proved their superiority in projecting raw imperial power. The opium wars, in the early 1840s and again in the late 1850s, were the direct result of China's frantic attempts to restrain "free trade" in opium. Opium was grown on the East India Company's lands in Bengal, auctioned in Calcutta, and then shipped by private traders to China. The large-scale trade in opium closed a yawning trade gap with the Celestial Empire, for whose tea Britain's citizens had developed an insatiable thirst. British exports of manufactured cotton to India completed the trade triangle. Opium, like all narcotics, is highly addictive. (One of the most chilling scenes in the Sherlock Holmes series is a visit to one of London's numerous opium dens, in "The Man with the Twisted Lip," where "through the gloom one could dimly catch a glimpse of bodies lying in strange fantastic poses . . . there glimmered little red circles of light, now bright, now faint, as the burning poison waxed or waned in the bowls of the metal pipes.") The opium war began when China took determined steps to ban the importation of the destructive substance, and the British government, acting on the demand of Britain's sixty trading firms with business in China, insisted on maintaining free trade in opium and dispatched a fleet to China to make good its demands.

Steamers played a decisive role in the opium war. The British fleet was able to do little more than harass coastal towns until the steamer *Nemesis* arrived in China in November 1840, after a grueling eight-month voyage around the African cape. The *Nemesis*, at 184 feet in length, was not merely the largest iron vessel of the time. It was, more to the point, built as a gunboat, with twin 60-horsepower engines, a shallow 5-foot draft, two swiveling 32-pound guns fore and aft, along with fifteen smaller cannon. The *Nemesis* was instrumental in the 1841 campaign that seized the major city of Canton. *Nemesis* sank or captured numerous Chinese war "junks" half its size, took possession of a 1,000-ton American-built commerce ship recently purchased by the Chinese, towed out of the way deadly oil-and-gunpowder "fire rafts" that the Chinese had created and lit, and attacked fortifications along the river passage up to Canton. The *Nemesis,* wrote its captain, "does the whole of the advanced work for the Expedition and what with towing transports, frigates, large junks, and carrying cargoes of provisions, troops and sailors, and repeatedly coming into contact with sunken junks—rocks, sand banks, and fishing stakes in these unknown waters, which we are obliged to navigate by night as well as by day, she must be the strongest of the strong to stand it."[7]

In 1842 the *Nemesis,* now leading a fleet comprising ten steamers, including its sister ship *Phlegethon,* eight sailing warships, and fifty smaller vessels, carried the campaign up the Yangtze River. At Woosung, site of the only serious battle of this campaign, *Nemesis* positioned an eighteen-gun warship, whose guns dispersed the Chinese fleet, including three human-powered paddle wheelers. The steamers promptly overtook them. The steamers were also essential in hauling the sailing vessels far up the river, over sandbars and mud, to take control of Chinkiang, at the junction of the Yangtze River and the Grand Canal. The Grand Canal was China's own imperial lifeline, linking the capital, Peking (as it was then spelled in English), in the north to the rice-growing districts in the south. The Chinese had little choice but to accept British terms. In 1869, in the aftermath of a second opium war, the Chinese Foreign Office pleaded with the British government to curtail the deadly trade:

> The Chinese merchant supplies your country with his goodly tea and silk, conferring thereby a benefit upon her; but the English merchant empoisons China with pestilent opium. Such conduct is unrighteous. Who can

justify it? What wonder if officials and people say that England is willfully working out China's ruin, and has no real friendly feeling for her? The wealth and generosity of England are spoken by all; she is anxious to prevent and anticipate all injury to her commercial interests. How is it, then, she can hesitate to remove an acknowledged evil? Indeed, it cannot be that England still holds to this evil business, earning the hatred of the officials and people of China, and making herself a reproach among the nations, because she would lose a little revenue were she to forfeit the cultivation of the poppy.[8]

Unfortunately for the Chinese people more than "a little revenue" was at play. Opium was a financial mainstay of the British Empire, accounting for one-seventh of the total revenues of British India in the nineteenth century. Recurrent attempts by humanitarian reformers in India and in Britain to eliminate the opium trade ran square into the sorry fact that British India was hooked on opium, too. While opium addiction was a severe problem in some districts of India, the imperial system depended on the flow of opium money. Annual net opium revenues—the export taxes on Malwa opium grown in western India added to the operating profits from growing opium in Bengal in the east, manufacturing it in government factories at Patna and Ghazipur, and exporting the product to China—were just over £3.5 million in 1907–8, with another £981,000 being added in the excise tax on opium consumption in India.[9] Finally, in 1909, Britain officially put an end to the odious enterprise.

Telegraphs and Public Works

In the industrializing countries of Western Europe and North America, telegraph systems grew up alongside railroads. Telegraph lines literally followed railway lines, since telegraph companies typically erected their poles in railroad right-of-ways. Telegraphs in these countries not only directed railroad traffic, a critical safety task because all railways had two-way traffic but most had only a single track; telegraphs also became the information pipelines between commercial centers, carrying price quotes, bids on commodity contracts, and all manner of market-moving news. In India, by contrast, the driving force behind the telegraph network was not commerce or industry but empire. As one Englishman phrased it, "the unity of feeling and of action which constitutes imperialism would scarcely have been possible without the telegraph."[10]

FIG. 4.2. Erecting the Indian Telegraph.

The telegraph network across India as well as between India and Europe depended on native Indian labor to erect, operate, and maintain the lines. Frederic John Goldsmid, *Telegraph and Travel* (London: Macmillan, 1874), frontispiece.

Telegraph lines were so important for imperial communication that in India they were built in advance of railway lines (fig. 4.2). The driving figure in this endeavor was Marquis of Dalhousie. As governor-general over India from 1848 to 1856 Dalhousie presided over an energetic campaign to bring Western ideas and Western technology to India. Dalhousie's territorial annexations in these years increased by almost half the size of British India and added substantially to the administrative complexity of governing it. The new possessions included the Punjab in the far northwest, the province of Pegu in Burma, and five native states including Satara, Sambalpur, Nagpur, Jhansi, and Oudh. The addition of Nagpur was especially welcomed by Lancashire cotton industrialists eager to secure alternative sources of raw cotton (to lessen their dependence on the American South as sectional strife loomed); colonial troops deployed to Nagpur helped fortify the continent-spanning road between Bombay and Calcutta. To help consolidate these far-flung holdings Dalhousie launched or completed a number of technological ventures, including the

Grand Trunk Road and the Ganges Canal, in addition to the railroad and wide-ranging Public Works Department discussed below. His first priority was the telegraph.

Dalhousie shaped India's telegraph network to fulfill the administrative and military imperatives of empire. The first, experimental line was built in two phases and ran from Calcutta to the mouth of the Hooghly River at Kedgeree. Events immediately proved its worth. News of the outbreak of the second Anglo-Burmese war arrived by ship at Kedgeree on 14 February 1852 and was telegraphed at once to Dalhousie at Calcutta. "If additional proof of its political value were required," Dalhousie wrote in the midst of war two months later, "it would be found in recent events when the existence of an electric telegraph would have gained for us days when even hours were precious instead of being dependent for the conveyance of a material portion of our orders upon the poor pace of a dak foot-runner."[11] In December Dalhousie outlined his nationwide plan. His top priority was "a telegraph line connecting Calcutta, Benaras, Allahabad, Agra, Amballa, Lahore and Peshawar" to link up all locations "in which the occurrence of political events was at all likely." After the line to Peshawar, at the Afghan border in the far west, came a line to Bombay, through which a line to London might be completed. Of lesser importance was a line to Madras, considered politically reliable. Moreover, to connect Calcutta with Madras Dalhousie planned not a direct telegraph line south down the eastern coastline but a much longer, indirect connection via Bombay. From Bombay to Madras this line passed through the military outposts at Poona, Bellary, Bangalore, and Arcot. The India Office in London quickly approved funds for the entire 3,150-mile network outlined in Dalhousie's plan.

Construction on the telegraph network began in November 1853, after a team of sixty installers was trained under the supervision of William O'Shaughnessy. O'Shaughnessy, a self-taught electrical expert and formerly deputy assay master of the Calcutta Mint, pushed the lines forward with breakneck speed. At first his installers placed a "flying line" of $5/16$-inch iron rod on uninsulated bamboo poles for immediate military use, later transferring it to insulated poles of stone, ironwood, or teak. Within five months the first trunk line, running the 800 miles from Calcutta to Agra, was opened; and by the end of 1854 the entire national backbone was complete, with links to Peshawar, Bombay, and Madras. Two years later, all the major military stations in India were interconnected by telegraph. Dalhousie, who aimed to mobilize 150,000 troops in one hour with the telegraph, had become acutely worried about increasing signs of military and civil discontent around him by the time he left India in 1856. ("Those who have travelled on an Indian line, or loitered at a Hindoo railway station, have seen the most persuasive missionary at work that ever preached in the East," wrote his biographer. "Thirty miles an hour is fatal to the slow deities of paganism."[12]) The wrenching cultural changes brought by his rapid-fire Westernization touched off a major rebellion.

The outbreak of the Indian Mutiny, on 10 May 1857, began a nerve-wracking "week of telegraphs." Earlier that spring native Indian troops near Calcutta had in several instances been openly mutinous, refusing to obey orders to load cartridges in the new Enfield rifle. Loading the rifle required soldiers to bite open cartridges that they believed to be coated with beef and pork fat, substances deeply offensive to Hindus and Muslims. (The Woolwich factory in England had indeed used beef fat, while the Dum-Dum factory in India had apparently not.) On 10 May native troops in Meerut, close to Delhi in the north central part of the country, seized control of that station, after eighty-five native members of a cavalry troop stationed there had been court-marshaled, publicly stripped of their uniforms, and imprisoned for refusing the suspect Enfield cartridges. The rebels marched on Delhi, took control of the city, and proclaimed a new Mogul emperor of India. By destroying the surrounding telegraph lines, the rebels cut off communication with the Punjab to the north, but not before a telegraph message had gone out on the morning of 12 May warning the British officers in the Punjab that certain native troops planned a rebellion there the following evening. Officers at Punjab quickly disarmed the native regiments before they got word of the uprising, which had been sent by foot runner. "Under Providence, the electric telegraph saved us," affirmed one British official in Punjab. Calcutta heard by telegraph of the fall of Delhi on 14 May and immediately dispatched messages requesting reinforcements for Delhi and Agra and inquiring about numerous potential trouble spots. "All is quiet here but affairs are critical," Calcutta heard on 16 May from Lucknow station. "Get every European you can from China, Ceylon, and elsewhere; also all the Goorkas from the hills; time is everything."[13]

Quick use of the telegraph saved not merely the British in Punjab but arguably the rest of British India as well. Most dramatic was that the telegraph made possible a massive troop movement targeted at the most serious sites of rebellion. British and loyal native troops returning from Persia were redirected to trouble spots in Calcutta, Delhi, and the Punjab;

loyal native troops from Madras moved to reinforce Calcutta; while British troops in Ceylon, Burma, and Singapore were called in as well. Officials in Calcutta coordinated steamers and sailing vessels and in short order resolved numerous logistical difficulties, all by telegraph. Time really was of the essence, since in June and July the mutiny spread across the northern and central provinces of India. But by then the deployment of numerous regiments loyal to the British prevented the Indian rebels from gaining ground. The promptness of the British responses astonished them. In the field campaigns that followed, the most famous use of telegraphs was in the Cawnpore-Lucknow "flying line" that aided the British troops in their assaults on the beleaguered Lucknow in November 1857 and March 1858. Isolated acts of rebellion continued until the capture of an important rebel leader in April 1859, and memories of atrocities on both sides poisoned trust between rulers and ruled for decades. One rebel on his way to execution pointed out the telegraph wire overhead as "the accursed string that strangles me."[14]

News of the Mutiny took forty days to arrive in London, traveling by steamers, camels, and European telegraphs. Consequently, imperial officials there were helpless bystanders as the conflict unfolded, was fought, and eventually ended. Insistent calls for action in the wake of this exasperating situation led to the inevitable government subsidies, but the first attempts to lay undersea telegraph cables or to use land lines across the Middle East proved expensive, slow, and unreliable. Messages relayed by non-English-speaking telegraph clerks might arrive a month after being sent and be totally unreadable. Not until 1870 was there a reliable telegraph connection between London and India (see fig. 4.3). The first line to open, a double land line running across Europe to Tehran, where it connected with a cable to Karachi via the Persian Gulf, was built by the German firm of Siemens and Halske, a leader in the second industrial revolution (see chapter 5). A second telegraph, also opened in 1870, went wholly by undersea cables from England to Bombay via the Atlantic, Mediterranean, and Red Sea. By 1873 telegraph messages between England and India took three hours. The British went on to lay undersea telegraph cables literally around the world, culminating with its famous "all red" route—named for the color of imperial possessions in the official maps—completed in 1902.[15]

By the time of the 1857 Mutiny, British rule in India had become dependent on telegraphs, steamships, roads, and irrigation works; soon to come was an expanded campaign of railway building prompted by the

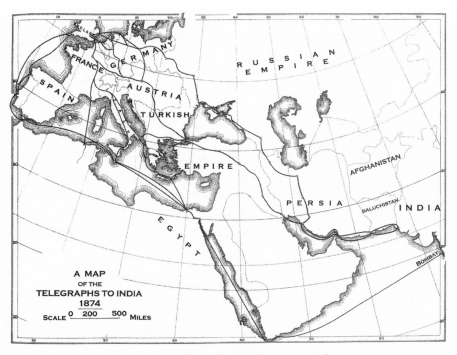

FIG. 4.3. Telegraph Lines between India
and Europe, 1874.

A high-technology imperial lifeline connecting Britain with India was established in 1870. Two overland telegraph lines ran from London, through
Europe and the Middle Eastern countries, while a third, all-underwater
line went through the Mediterranean Sea, the Suez Canal (opened in
1869), the Red Sea, and on to the Indian Ocean. The telegraph cut the
time needed to send a message between London and India from months
to hours. Frederic John Goldsmid, *Telegraph and Travel* (London: Macmillan, 1874), facing p. 325.

Mutiny itself. Yet, as hinted in the training of those sixty assistants needed
for the initial telegraph construction, the British also became dependent
on a cadre of technically trained native assistants to construct, operate,
and maintain the instruments of empire. Budgetary constraints made it
impossible to pay the high cost of importing British technicians for these
numerous lower-level positions. The colonial government in India had
no choice but to begin large-scale educational programs to train native
technicians. These pressures became acute in the decade following the

Mutiny because the colonial government embarked on a large and expensive program of roads, canals, and railroads designed to reinforce its rule. The East India Company was dissolved in the wake of the Mutiny, and the British government assumed direct rule through top-level officials in London and Calcutta.

During these same years, Lancashire industrialists were frantic to secure alternative supplies of cotton during the American Civil War. Their well-organized lobbying in this instance prevailed on the home government in London, which directed the colonial government in India to open up India's interior cotton-growing regions to nearby ports. A wide-ranging public works campaign might have led to balanced economic development, but the effect of British policy was to discourage the development of Indian industry. The prevailing view was neatly summarized by Thomas Bazley, president of the Manchester Chamber of Commerce and member of Parliament for Manchester: "The great interest of India was to be agricultural rather than manufacturing and mechanical."[16]

One can discern a decidedly nonmechanical slant in the structure of the Public Works Department itself, the technical education it presided over, and not least the public works projects that it helped construct. The Public Works Department (PWD), founded in 1854 to coordinate Dalhousie's numerous transportation and infrastructure projects, dominated state-sponsored technology in India. (Quite separately, wealthy Indian traders from Bombay revived the cotton textile industry that had flourished in the eighteenth century around that city, leading the Lancashire cotton lobby to redouble its effort in the 1870s to secure "free trade" in cotton by slashing import and export duties.) Most immediately the PWD set the agenda for technology in India through large construction efforts that included roads, canals, and irrigation projects, often—explicitly—with a view toward increasing exports of cotton or wheat.[17]

The character of the PWD's projects was no accident. The department reported to the British colonial officials in Calcutta, who could sometimes engage in creative interpretation of directives from the London-based secretary of state for India. But the policy was set in London, and there the officials responsible for India policy were receptive to domestic pressure groups such as the Lancashire textile industrialists. The secretary of state's office brimmed with letters, petitions, and all manner of insistent appeals from the Manchester Chamber of Commerce. In 1863, fearful of "the insurrection of cotton people," Charles Wood, secretary of state for India from 1859 to 1866, directed his colonial colleague in India to build "cotton roads" at great haste. "I cannot write too strongly on this point," he said. "The sensible cotton people say they acquit us of any serious neglect . . . but that we must make roads."

The first of two large projects taking shape in this political climate was a road-and-harbor project to link an inland region southeast of Bombay (Dharwar) to a new harbor site at Karwar, about 100 miles distant. Dharwar was of particular interest to the Manchester Cotton Company, a recently formed joint-stock company that aimed to ship new sources of raw Indian cotton to Lancashire, because the region grew the desirable long-staple cotton previously obtained from the American South. In October 1862 the British governor of Bombay, apprehensive that the complex project was being pushed so rapidly that proper engineering and cost estimates had not been made, nevertheless endorsed it: "The money value to India is very great, but its value to England cannot be told in money, and every thousand bales which we can get down to the sea coast before the season closes in June 1863 may not only save a score of weavers from starvation or crime but may play an important part in ensuring peace and prosperity to the manufacturing districts of more than one country in Europe."[18]

Even larger than the Dharwar-Karwar project, which cost a total of £225,000, was a grandiose plan to turn the 400-mile Godavari River into a major transportation artery linking the central Deccan plain with the eastern coast. The plan would send to the coast the cotton from Nagpur and Berar in central India, and this time the Cotton Supply Association was the principal Lancashire supporter. Work on this rock-studded, cholera-ridden river proved a vast money sink, however. By 1865, when the Lancashire lobby quietly gave up its campaign for Indian cotton and returned to peacetime American supplies, the Godavari scheme had cost £286,000 with little result. In July 1868 the first 200-mile stretch was opened, for limited traffic, and by the time the ill-conceived project was canceled in 1872 it had cost the grand sum of £750,000. As it turned out, the Lancashire lobby threw its support behind the Great Indian Peninsula Railway that connected cotton-rich Nagpur with the port of Bombay.

The Public Works Department's leading role also strongly stamped an imperial seal on technical education in India. The four principal engineering schools in India, founded between 1847 and 1866 at Roorkee, Calcutta, Madras, and Poona, had a heavy emphasis on civil engineering. The PWD was not only the source of many faculty at the engineering schools and of the examiners who approved their graduates but also far and away the leading employer of engineers in India. Indian courses were "unduly encumbered with subjects that are of little educational value for engineers, but which are possibly calculated to add to the immediate utility of the student in routine matters when he first goes on apprenticeship to the PWD," observed a witness before the Public Works Reorganization Committee in 1917. Another witness stated, "mechanical engineering has been greatly neglected."

The development of a well-rounded system of technical education in India was further hampered by the presence of the elite Royal Indian Engineering College, located, conveniently enough, at Cooper's Hill near London. It was founded in 1870 explicitly to prepare young, well-educated British gentlemen for supervisory engineering posts in India. Successful applicants had to pass a rigorous examination in mathematics and physical science; Latin, Greek, French, and German; the works of Shakespeare, Johnson, Scott, and Byron; and English history from 1688 to 1756. The college was permitted to enroll two "natives of India" each year "if there is room." At the PWD in 1886, Indians accounted for just 86 of 1,015 engineers, although they filled many of the lower ("upper subordinate" and "lower subordinate") grades. Indian prospects for technical education improved somewhat with the closing of the flagrantly discriminatory Cooper's Hill in 1903 and the founding of the native-directed Bengal Technical Institute (1906) and Indian Institute of Technology (1911). By the 1930s Indian students could gain degrees in electrical, mechanical, and metallurgical engineering in India.[19]

Railway Imperialism

Railroads in countries throughout Western Europe and North America were powerful agents of economic, political, and social change. Their immense capital requirements led to fundamental changes in the business structures of all those countries and in the financial markets that increasingly spanned them. Building and operating the railroads consumed large amounts of coal, iron, and steel, leading to rapid growth in heavy industries. Their ability to move goods cheaply led to the creation of national markets, as transportation costs became a much smaller consideration in how far away a factory's products might be profitably sold, while their ability to move troops rapidly strengthened the nation-states that possessed them. (See fig. 4.4.) In the 1860s the latter case was made by military victories by railroad-rich Prussia and the northern U.S. states.

But in the imperial arenas, the dynamics of empire subtly but per-

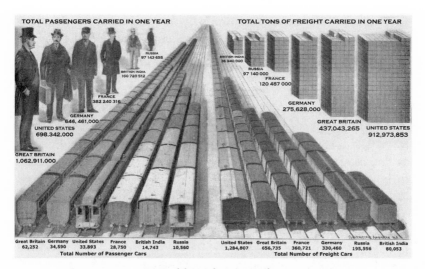

FIG. 4.4. World Leaders in Railways, 1899.

By the turn of the century, the Indian railway was the fifth largest in the world in passenger travel and sixth largest in freight. "Railways of the World Compared," *Scientific American* (23 December 1899): 401. Courtesy of Illinois Institute of Technology Special Collections.

ceptibly altered railways and the changes they brought. Where imperial officials were essential in arranging financing, their military and administrative priorities shaped the timing, pace, and routes of colonial railroads. Colonial railroads also reflected the economic priorities of bankers in London and other financial centers who floated huge loans for their construction; nearly always these bankers preferred open "free trade" markets to closed high-tariff ones, strong central colonial governments to divided regional ones, and easily collected import-and-export taxes. For all these reasons, railway imperialism typically led toward political centralization, economic concentration, and extractive development. This is not to say that railway imperialists always got what they wanted. For both the colonies and the metropoles, railroads were above all a way of conducting "politics by other means" often involving contests between local and global imperatives and powers. A survey of railway imperialism in India, North America, and South Africa concludes this chapter.

In India the political and military utilities of a wide-ranging railroad network were painfully obvious to the railroad promoters, since their thunder had been stolen by the prior construction of the telegraph network. Yet even as commands for troop movement could be sent down an iron wire at the speed of light, the troops themselves required a less ethereal mode of transport. Railways constituted political power in such a vast and politically unsteady place as colonial India. "It is not," wrote one railway economist in 1845, "with any hope of inspiring the company of British merchants trading to India with an expensive sympathy for the social and moral advancement of their millions of native subjects that we urge the formation of a well-considered means of railway communication,—but as a necessary means of giving strength, efficiency, and compactness to their political rule in those territories." Another writer saw a threat in the Russian Empire's sudden interest in building railroads. The Russian Tsar, too, "with no love for civilization has found it necessary to adopt the formation of railways in his vast and benighted territories." The Tsar had formed railroads "not with any consideration for the personal comfort of his subjects, but as a necessary means of strengthening his rule."[20]

Imperial priorities informed the two British engineers who planned the pioneering lines, Rowland M. Stephenson and John Chapman. Stephenson came from a family long associated with Indian commercial and political affairs and had no evident relation to the railroad pioneer George Stephenson. After becoming a civil engineer in the 1830s, Stephenson promoted various Indian steam ventures to investors in London. Having seen the success of the Peninsular & Oriental's steamship venture (he briefly served as secretary for one of its rivals and later for the P&O itself), Stephenson journeyed to Calcutta. Writing in the *Englishman* of Calcutta in 1844, Stephenson proposed a 5,000-mile network consisting of six major lines. "The first consideration is as a military measure for the better security with less outlay of the entire territory," he wrote of his plan. "The second is a commercial point of view, in which the chief object is to provide the means of conveyance to the nearest shipping ports of the rich and varied productions of the country, and to transmit back manufactured goods of Great Britain, salt, etc., in exchange." Developing native Indian industry, which his plan would simultaneously deprive of homegrown raw materials and overwhelm with British manufactured goods, was assuredly not among his goals. In 1845 Stephenson joined the newly formed East Indian Railway Company as managing director. Stephenson's plan was given official sanction by Dalhousie's 1853 "Minute" on railroads, which set railroad policy for decades. John Chapman was the chief technical figure of the Greater Indian Peninsula Railway, or GIP, formed also in 1845. The GIP's projected line, originating at the port of Bombay, climbed up the Western Ghats to the central Deccan plateau, a prime cotton-growing area. Three years later, Chapman observed that the Lancashire merchants thought of the GIP as "nothing more than an extension of their own line from Manchester to Liverpool."[21]

The first generation of Indian railways took form under a peculiar and still-controversial imperial arrangement. Under a precedent-setting 1849 contract with the East India Company (EIC, the statelike entity responsible for governing India until 1858), the pioneering East Indian Railway and the Greater Indian Peninsula Railway turned over to the EIC their entire paid-in capital. As the railroads planned routes that met the EIC's criteria, including specific routings, single- or twin-tracking, and various engineering standards, the EIC disbursed the "allowed" expenditures to the respective railroad. The EIC leased generous swaths of land to the railroads without cost for ninety-nine years. At any time through the ninety-eighth year, the railroads could turn over their companies to the state and demand full compensation; in the ninety-ninth year, somewhat too late, the state could claim the right to take over the roads without compensation.

The controversial point was the guaranteed return to investors. The EIC, which held the railroads' capital in Calcutta, promised interest payments to investors of 5 percent. Operating profits up to 5 percent went to the EIC to offset its guarantee payments, while any profits above 5 percent were split equally between the EIC and the railroad or, if the operating revenues had completely covered the EIC's 5 percent payments, with no backlog from previous years, the profits above 5 percent went entirely to the railroad company. The guaranteed interest payments rested ultimately on the EIC's ability to collect money from Indian taxpayers. The rub was that through 1870 the roads consistently made average annual profits of only around 3 percent and the EIC-backed investors were overwhelmingly British. (In 1868 less than 1 percent of shareholders were Indians, 397 out of 50,000—understandably enough, since shares traded only in London.) The scheme thus transferred money from Indian taxpayers to British investors. The finance minister of India, William Massie (1863–69), saw the problem clearly: "All the money came from the English capitalist, and so long as he was guaranteed 5 per cent on the revenues of India, it was immaterial to him whether the funds that he lent were thrown into the Hooghly or converted into bricks and mortar."[22]

In fact, however, since nearly all of the locomotives, most of the rails, and even some of the coal was imported from Britain, fully two-fifths of

FIG. 4.5. Gokteik Viaduct in Upper Burma.

The railway track is 825 feet above the level of the Chungzoune River, which flows through a tunnel beneath the bridge. Frederick Talbot, *The Railway Conquest of the World* (London, Heinemann, 1911), following p. 256. Courtesy of Illinois Institute of Technology Special Collections.

the money raised in Britain was spent in Britain. The Indian railroads' high expenses were the result not so much of flagrant corruption as of the subtle but again perceptible way in which the dynamics of empire shaped their form. Imperial considerations most obviously structured the financing and routing of the Indian railways. A vision of empire also inspired the vast, overbuilt railway terminals, such as Bombay's Victoria Station. One can furthermore see empire inscribed in the railroads' technical details, including their track gauges, bridge construction, and locomotive designs.

In 1846 the British Parliament passed the Act for Regulating the Gauge

FIG. 4.6. Bridging the Ganges River at Allahabad.

When building the Curzon Bridge to cross the mighty Ganges River at Allahabad in northern India, native laborers narrowed the Ganges at that point from the width of its hardened clay riverbanks (3 miles) to the width of its water channel (about 1¼ miles). The bridge's span was 3,000 feet. Frederick Talbot, *The Railway Conquest of the World* (London: Heinemann, 1911), following p. 254. Courtesy of Illinois Institute of Technology Special Collections.

of Railways, setting what would become the standard gauge in Europe and North America: 4 feet 8½ inches (or 1.435 meters). Nevertheless, the EIC's Court of Directors set the gauge of India's railroads at 5 feet 5 inches (or 1.676 meters) and furthermore decided in 1847 that all tunnels, bridges, and excavations must be made wide enough for double tracking. The mandate for a wide gauge and double tracking inflated construction costs in a number of ways: all bridge superstructures had to be extra wide, while all curves had to be extra broad. (Track construction up the Western Ghats, steep cliffs rising to the central Deccan plain, required a cumbrous arrangement of reversing stations, in addition to numerous tunnels and bridges to gain the needed elevation.) Some of the vast sums spent on bridges can be fairly traced to India's wide, deep, and at times fast-flowing rivers and their fearsome monsoon flooding (see figs. 4.5 and 4.6). Adding to the costs, however, was the British engineers' preference for expensive designs: no rough-and-ready timber trestles like those American railroad engineers were building. Instead, Indian bridges mostly conformed to British designs, built of wrought-iron trusses over masonry piers. Numbered bridge parts sent from Britain were riveted together into spans by Indian craftsmen and the spans placed by elephant-powered hoists. In 1862 the second largest bridge in the world was constructed this way across the Sone River near Delhi, and it cost the astounding sum of £430,000.

British-constructed Indian locomotives were also built for the ages and were correspondingly expensive. The most common class of locomotive in India, the Scindia, was built on a rigid frame with six sets of driving wheels (known as a 0-6-0 configuration) and featured copper fireboxes, copper or brass boiler tubes, forged valves, and inside cylinders with cranked axles. These were the Rolls-Royces of locomotives. By contrast, North American locomotives of this era had steel fireboxes and boiler tubes, cast-iron valves, and external cylinders, as well as leading "bogie" wheels that improved steering on sharp turns (4-4-0 or 2-8-0 configuration). Although India's railroad shops constructed approximately 700 locomotives before independence in 1947, the vast majority (80 percent) were built in Britain, supplemented by imports from America and Germany. During these years Indian railroads bought fully one-fifth of the British locomotive industry's total output.[23]

Railway construction under the first guarantee system picked up pace only after the Mutiny of 1857–58, when there were just 200 miles of rail line. In 1870 the Indian colonial government, reeling under the budget-breaking costs of the 5,000 miles of privately constructed but publicly financed railroads, embarked on a phase of state-built railroads featuring a narrow (meter-wide) gauge. The Indian colonial government built 2,200 miles of these roads at half the cost per mile of the guaranteed railroads. But in 1879 the secretary of state for India mandated that the Indian government only build railroads to the strategically sensitive northern (Afghan) frontier. Private companies, under a new guarantee scheme negotiated with the government in 1874, took up the sharp boom of railroad building across the rest of the country. At the turn of the century India had nearly 25,000 miles of railroad track. India's railway mileage surpassed Britain's in 1895, France's in 1900, and Germany's in 1920, by which point only the United States, USSR, and Canada had more mileage. Unfortunately for the Indian treasury, the roads built under the second guarantee scheme were also money pits (only the East Indian Railway's trunk line from Calcutta to Delhi consistently made money). Guarantee payments to the railroads between 1850 and 1900 totaled a whopping £50 million.[24] By the 1920s, Indian railroads, by then run-down for lack of investment, became a prime target of Indian nationalists agitating for the end of British rule.

Compared with India, railway imperialism in North America was a more complicated venture, not least because two imperial powers, Britain and the United States, had various claims on the continent. Railway building in the eastern half of the United States reflected commercial and industrial impulses. Merchants and traders in Baltimore and Philadelphia, for example, backed two huge railroad schemes to capture a share of the agricultural bounty that was flowing east via the Erie Canal to New York City. By 1860 a network of railroads from the Atlantic Ocean to the Mississippi River had created an industrial heartland that extended to Chicago. The United States, with 30,000 miles, had more than three times the railroad track of second-place Britain and nearly five times the mileage of third-place Germany.

Construction of the transcontinental railroads from the Mississippi to the Pacific Ocean during the next four decades (1860s–1900) boosted the country's railroad trackage to 260,000 miles. The defining governmental action—the Pacific Railroad Act of 1862—granted huge blocks of land to the transcontinental railroads for building their lines and for developing traffic for their lines by selling land to settlers. All but one of the transcontinental lines—including the Union Pacific, Central Pacific, Northern Pacific, Kansas Pacific, Southern Pacific, and Atchison, Topeka & Santa Fe railroads—were beneficiaries of these land grants. The Illinois Central had pioneered the land-grant, receiving 2.6 million acres in 1850 in subsidy for its line south to New Orleans.

The strange economics of empire also came into play in North America. The U.S. government faced a heavy financial burden in mobilizing and provisioning sufficient army units to safeguard settlers and railroad construction crews from the Native Americans who were determined not to give up their buffalo hunting grounds without a fight. In 1867 the *Omaha Weekly Herald* claimed that it cost the "large sum" of $500,000 for each Indian killed in the intermittent prairie battles. Railway promoters were quick to point out that the increased military mobility brought by the railroad, cutting the needed number of military units, would dramatically reduce the high cost of projecting power across these many sparsely settled miles. In this respect, the railroads in British India and the American West appear to have had more than a casual similarity. "The construction of the road virtually solved the Indian problem," stated one American railroad official two years prior to the massacre at Wounded Knee, South Dakota, in 1890.[25]

Citizens of British North America reacted with some alarm to these territorial developments. At the time, Canada was not a unified nation but a set of independent provinces. Yet not even the fear of being annexed to the United States, as was half of Mexico in the 1840s, united them. Merchants and traders along the St. Lawrence–Great Lakes canal system, a huge project of the 1840s, looked naturally to the shipping ports of the south; some even favored joining the United States. A railroad boom in the 1850s resulted in 2,000 miles of disconnected lines whose operating deficits emptied colonial Canadian treasuries and whose north-south orientation drained off Canadian products to the south. Would-be railway imperialists still lacked the necessary east-west lines that might bring economic and political cohesion to the provinces. Worse, during the Civil War the United States started a trade war with the Canadian provinces and made threats, deemed serious by many Canadians, to invade their western lands. An equally pressing problem involved the U.S. economic domination that might formally detach British Columbia from the British Empire and, on the other side of the continent, might informally control the maritime provinces.

A generous program of imperial railway subsidies was the glue that fixed the slippery provinces into place. In the 1860s a series of labyrinthine negotiations between Canadian colonial officials, British imperial officials, and London financiers arrived at this formula: London financiers would support large railway loans if the provinces were politically united; the independent provinces would agree to confederation if the Colonial Office sweetened the deal with government guarantees for railway construction (valuable not least for the patronage jobs railroads provided). So, the Colonial Office in London duly arranged government guarantees for the railway loans. Thus was the Dominion of Canada created in 1867 as a federation of Canada West and Canada East with the maritime provinces of New Brunswick and Nova Scotia. Railroads figured explicitly in the confederation agreement. The maritime provinces made their assent to the agreement conditional on the building of an intercolonial railroad in addition to the Halifax-Quebec railroad, already planned. Furthermore, expansion-minded citizens of Canada West received imperial guarantees for the intercolonial railway and promises of amiably settling the Hudson Bay Company's preemptive claim on western lands. By 1874 the *British* government had made guaranteed loans and grants totaling £8 million for the intercolonial lines and transcontinental Canadian Pacific Railway together. Of the Canadian Pacific the first premier of the Canadian confederation commented, "Until that road is built to British Columbia and the Pacific, this Dominion is a mere geographical expression, and not one great Dominion; until bound by the iron link, as we have bound Nova Scotia and New Brunswick with the Intercolonial Railway, we are not a Dominion in Fact."[26]

Railway imperialism in Mexico affords a glimpse of what might have happened to Canada absent the countervailing power of imperial Britain. The Mexican railroad system took shape under the long rule of the autocratic Porfirio Díaz (1876–1911). As early as the 1870s the Southern Pacific and Santa Fe railroads, then building extensive lines in the southwestern United States, began planning routes south into Mexico. Mexico at the time consisted of fourteen provinces whose disorganized finances left no hope of gaining external financing from London or other international money centers. With few options in sight, and a hope that railroads might bring "order and progress" to Mexico, Díaz gave concessions to the U.S. railroads to build five lines totaling 2,500 miles of track. Something like free-trade imperialism followed. In 1881 the U.S. secretary of state (James Blaine) informed the Mexican government that it would need to get rid of the "local complicated . . . tariff regulations which obtain between the different Mexican States themselves" and sign a reciprocal free-trade agreement with the United States. Díaz prevailed upon the Mexican Congress to ratify the trade pact, as the U.S. Senate had done, but the agreement did not go into effect owing to a quirk of U.S. domestic politics. U.S. railroad and mining promoters flooded south all the same. Mexico, as one railroad promoter effused in 1884, was "one magnificent but undeveloped mine—our India in commercial importance."[27]

Trade between Mexico and the United States increased sevenfold from 1880 to 1910. Total U.S. investment in Mexico soared to over $1 billion, an amount greater than all its other foreign investment combined and indeed greater than Mexico's own internal investment. Fully 62 percent of U.S. investment was in railroads; 24 percent was in mines.[28] By 1911 U.S. firms owned or controlled most of the 15,000 miles of railroad lines; three-quarters of mining and smelting concerns processing silver, zinc, lead, and copper; and more than half of oil lands, ranches, and plantations. Four great trunk lines shipped Mexican products north to the border. But the extractive mining boom brought wrenching social and political changes. The blatant U.S. domination of railroads inflamed the sensibilities of Mexican nationalists. The Díaz government nationalized two-thirds of the country's railroads in 1910, but the aging dictator was overthrown the next year. The legacy of railway imperialism in Mexico was not the "order and progress" that Díaz had aimed for but a confusing period of civil strife (1911–17) and a transportation system designed for an extractive economy.

In South Africa railroads at first had some of the centralizing and integrating effects that we have noted in India, Canada, the United States, and Mexico, but railway imperialism there ran headlong into a countervailing force, "railway republicanism." The result was not railroad-made confederation along Canadian lines, as many in Britain hoped, but a political fracturing of the region that ignited the second Anglo-Boer War (1899–1902). Southern Africa even before the railway age was divided into four distinct political units: two acknowledged colonies of Britain (the Cape Colony, at the southern-most tip of the continent, and Natal, up the eastern coast) and two Afrikaner republics (the inland Orange Free State, and the landlocked Transvaal republic) over which Britain from the mid-1880s claimed suzerainty. Britain understood this subcolonial status as a step to full integration into the British Empire, whereas the fiercely independent Afrikaners, descendants of seventeenth-century Dutch settlers who had trekked inland earlier in the nineteenth century to escape British control, saw it as one step from complete independence.

Imperial concerns mounted with the waves of British citizens brought to the region by the discovery of diamonds (1867) at Kimberley in eastern Orange Free State (fig. 4.7) and gold (1886) in the central Transvaal. One of the British newcomers was Cecil Rhodes who made his fortune in the Kimberley fields in the 1870s, formed the De Beers Mining Company in the 1880s (a successor of which still sets worldwide diamond prices), and secured a wide-ranging royal charter for his British South Africa Company in 1889. The story goes that Rhodes, his hands on a map of Africa, had gestured: "This is my dream, all English."[29] Rhodes, along with other centralizing imperialists, hoped to form southern Africa's many linguistic and ethnic groups into a single, unified colony dependent on Britain. Rhodes used the promise of building railroads with his chartered company to secure the political backing of the Cape Colony's Afrikaner majority. As the Cape's prime minister from 1890 to 1896 he led the railway-building campaign north across the Orange Free State and into the Transvaal, not least by arranging financing in London. In getting his rails

FIG. 4.7. Kimberley Diamond Mine, South Africa,
ca. 1880.

Southern Africa's premier diamond mine, with its 31-square-foot claims, was covered by hundreds of wire ropes connecting each claim to the unexcavated "reef" visible in the background. "Each claim was to all intents and purposes a separate mine." *Scientific American* (27 January 1900): 56. Courtesy of Illinois Institute of Technology Special Collections.

into the Transvaal Rhodes hoped to preempt that republic's plans to revive a defunct railway to the Portuguese port city of Lorenço Marques (now Mobutu, Mozambique). The early railway campaigns brought a degree of cooperation, through cross-traffic railroad agreements and a customs union, between the Cape Colony and Orange Free State, on the one hand, and the Natal colony and Transvaal on the other.

Ironically, railroads thwarted the imperial dream in South Africa. Rhodes found his match in the "railway republican" Paul Kruger, president of the Transvaal, and political leader of the region's Afrikaner Boers. Nature had dealt Kruger's Transvaal the supreme trump card: the massive Witwatersrand gold reef near Johannesburg, discovered in 1891 and around which would pivot much of the region's turbulent twentieth-century history. Mining the Rand's deep gold veins required heavy machinery and led to large-scale industrial development around Johannesburg. Not surprisingly, three of the region's four railroads had their terminals there.

Although Rhodes had hoped that his Cape Colony railroad's extension to the Transvaal would give him leverage over the inland republic's leader, quite the opposite happened. Kruger contested the imperialist's plan and gained the upper hand by appealing directly to London financiers himself. With a £3 million loan from the Rothschilds in 1892, Kruger completed the Transvaal's independent rail link to the port at Lorenço Marques.[30] From then on, Kruger could bestow as he saw fit the Rand's lucrative traffic among three railroads (his own, Natal's, and the Cape's). The tremendous fixed investment of these railroads, along with the light traffic elsewhere, gave whoever controlled the Rand traffic tremendous clout. Having failed to achieve anything like political union by railway imperialism, Rhodes tried it the old-fashioned way—militarily. In the Drifts Crisis of 1895, sparked by a dispute over railway freight charges around Johannesburg, the Cape Colony called for British military intervention against the Transvaal. With this crisis barely settled, Rhodes launched the Jameson Raid, an ill-conceived military invasion aimed at overthrowing Kruger.

By this time Britain may well have wished to wash its hands of the region's bitter disputes, but there remained the matter of protecting its investors' £28 million in colonial railway debts. In the face of Kruger's open hostility toward its citizens, Britain launched the Second Boer War (1899–1902), during which its army crushed the independence-minded Boer republics. The same Transvaal railway that carried gold out to the ocean also carried the British army in to the Rand, and carried Kruger away to exile. The region's railroad problem was a high priority for the British high commissioner charged with rebuilding the region after the war: "On the manner and spirit in which the peoples and Parliaments of South Africa handle this railway question depends the eventual answer to the larger question, whether South Africa is to contain one great and united people . . . or a congeries of separate and constantly quarreling little states," he wrote in 1905.[31] In his 1907 blueprint for federating the South African colonies, fully two-thirds of his 90,000-word text dealt with railroad issues. Railway union, he saw, was imperative, since independent provincial railroads would continually be instruments for sectional strife. The South Africa Act of 1909 created the regionwide South African Railways and Harbours Administration and helped unify the shaky Republic of South Africa, formed the next year. In 1914 the center of a unified railway administration was fittingly enough relocated to Johannesburg, the seat of railway power.

FIG. 4.8. Building the Cape Town to Cairo Railway.

Native laborers at work on Cecil Rhodes's grand imperial dream, a railroad that would connect the southern tip of Africa with a Mediterranean port. Here, workers set a continental record, laying 5¾ miles of track in ten hours. Frederick Talbot, *The Railway Conquest of the World* (London: Heinemann, 1911), following p. 144. Courtesy of Illinois Institute of Technology Special Collections.

Although his political career ended with the abortive Jameson Raid, Rhodes in his remaining years turned his considerable promotional abilities toward that most grandiose of all imperial schemes, the Cape Town to Cairo Railway, which aimed to stitch together a patchwork of mining and settlement ventures northward through Africa (fig. 4.8). Rhodes's British South Africa Company had no realistic chance of laying rails all the way to Egypt, not least because the scheme ran square into the ambitions of the Belgian Congo to control the rich Katanga copper belt in central Africa and of German Southeast Africa to dominate the territory from

FIG. 4.9. Spanning the Zambesi River at
Victoria Falls.

The 500-foot span over the Zambesi River just below the stunning Victoria Falls. Trains passed across the gorge 420 feet above low water. The
net beneath the bridge was "to catch falling tools and workmen." Frederick Talbot, *The Railway Conquest of the World* (London, 1911), following p. 144. Courtesy of Illinois Institute of Technology Special Collections.

there east to the Indian Ocean. Rhodes's scheme did hasten the European
settlement of the landlocked Rhodesias. The railroads just completed
through Southern Rhodesia (now Zimbabwe) prevented mass starvation
when *rinderpest* decimated the region's draft animals in the late 1890s,
while in 1911 Northern Rhodesia (now Zambia) was created through and
circumscribed by its mining and railroad activities (fig. 4.9). Perhaps the
farthest-reaching achievement of the Cape-to-Cairo scheme was in
durably linking the midcontinent copper belt with South Africa well into
the postcolonial decades.[32]

THE LEGACIES OF IMPERIALISM remain fraught with controversy in our
own postcolonial time. Indeed the arguments have sharpened with the
debate on globalization, since many of its critics denounce globalization
as little more than an updated imperialism (discussed in chapter 8). It is
difficult indeed to summon much sympathy for the historical actors who
believed in the "civilizing mission" of the imperialist era, who assumed

that selling opium or stringing telegraphs or building railroads would bring the unalloyed benefits of Western civilization to the natives of Asia, Africa, or South America. From our present-day sensibilities we see only too clearly the gunboats, rifle cartridges, and machine guns that these ventures entailed.[33] Both before and since the anticolonial independence movements of the late 1940s through 1960s, nationalists, especially in Asia and Africa, have condemned the schemes that brought imperialist-dominated "development" to their countries. Even today one can discern a shadow of the imperialist era in railroad maps of North America (look carefully at Canada, the western United States, and Mexico), in the prestige structure of technical education, and in the policy preferences of the orthodox development agencies in the United States and Europe.

Even for the dominant countries, imperialism was a venture with mixed economic results. We have mentioned that, on the best aggregate statistics, Britain as a whole did not make money with imperialism. While many traders and entrepreneurs, as well as the technological visionaries who tapped the imperial impulse, made their individual fortunes, the profits of the imperial economy did not outweigh the heavy expenses of sending imperial military forces overseas, maintaining the imperial bureaucracy, and funding the high-priced imperial technologies. We now have greater insight into why imperialism did not make money. Profit was simply not the point of imperial technologies: the expensive steam vessels, the gold-bricked locomotives, the double-tracked wide-gauge railways, the far-flung telegraph and cable networks.

In this respect, we can see that imperialism was not merely a continuation of the eras of commerce or industry; rather, to a significant extent, imperialism competed with and in some circumstances displaced industry as the primary focus of technologists. By creating a captive overseas market for British steamships, machine tools, locomotives, steel, and cotton textiles, imperialism insulated British industrialists in these sectors from upstart rivals and, in the long run, may even have hastened their decline in worldwide competitiveness.[34] Is it only a coincidence that Britain, a leader in the eras of industry and imperialism, was distinctly a follower behind Germany and the United States in the subsequent science-and-systems era? At the least, we must allow that the imperialist era had a distinct vision for social and economic development, dominated by the Western countries, and distinctive goals for technologies.

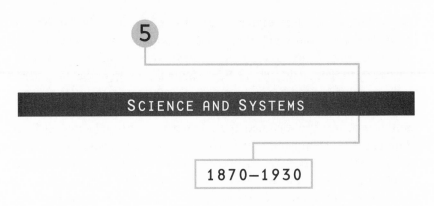

IN THE HALF-CENTURY AFTER 1870 a "second" industrial revolution, built from science-based technologies, altered not merely how goods were produced and consumed but also how industrial society evolved. The new industries included synthetic chemicals, electric light and power, refrigeration, and many others. By transforming curiosities of the laboratory into consumer products, through product innovation and energetic marketing schemes, science-based industry helped create a mass consumer society. A related development was the rise of corporate industry and its new relationships with research universities and government bureaus.

The economic imperatives of science and systems, along with the rise of corporate industry, decisively changed the character of industrial society. In surveying the strains and instabilities of the first industrial revolution, Karl Marx and Friedrich Engels prophesied that capitalism's contradictions would bring about its destruction, that (in their memorable phrase) the bourgeoisie would produce its own gravediggers. The recurrent economic panics—no fewer than nine major depressions between 1819 and 1929 in the United States, almost always linked to the British and European economies—lent force to their diagnosis that industrial capitalism was dangerous to society and doomed to failure. But the emergence of a technologically stabilized "organized capitalism" escaped their view. Technologists in this era, especially those hired in the science-based industries, increasingly focused on improving, stabilizing, and entrenching existing systems rather than inventing entirely new ones. Their system-stabilizing efforts—buttressed by patent laws, corporate ownership, industrial research laboratories, and engineering education—transformed industrial society. In these same decades *technology* took on its present-day meaning as a *set* of devices, a complex of industry, and an abstract society-changing force in itself.[1]

British industrialists largely failed in the science-based industries, even though British scientists discovered much of the relevant science. As chapter 3 recounted, London beer brewers pioneered factory scale, steam power, mechanization, and temperature control. "A brewery," wrote one industry expert in 1838, "may appropriately be termed a brewing chemical laboratory."[2] In electricity, it was the English chemist Humphrey Davy who first experimented with incandescent and arc lighting (1808) and the English physicist Michael Faraday who discovered electromagnetic induction and first built a dynamo, or direct-current generator (1831). In synthetic chemicals, the English chemist William Perkin first synthesized an artificial dye (1856). Yet technical and commercial leadership in each of these industries slipped from British hands.

The Business of Science

The first science-based industry started, improbably enough, with a sticky black tar derived from coal. By the time Charles Dickens set *Hard Times* (1854) in the fictional "Coketown," Britain was mining some 50 million tons of coal each year. Coke, a by-product of burning coal under certain conditions, had been used for decades as a high-grade industrial fuel for smelting iron and drying barley malt, while coal gas found ready employment in the lighting of factories (from the 1790s) and urban streets including those in London (1807) and Baltimore (1816). The baking of coal into valuable gas and coke in London alone employed 1,700 persons by midcentury. Sticky black coal tar was the unwanted by-product. One could say that the synthetic chemical industry was born as an ancillary industry.

Fittingly enough, it was in coal-fouled London that the possibilities of coal tar attracted the attention of an expatriate German chemist. August von Hofmann, surveying all that surplus coal tar, set his laboratory students at the Royal College the task of finding something useful in it. One student successfully made benzene and toluene but, tragically, burned himself to death when his still caught fire. William Perkin (1838–1907) was luckier in his mistakes. Aiming to work up coal tar into an antimalarial agent (quinine) Perkin instead produced one dark blob after another. One day in 1856, when washing his results with alcohol, the sink turned purple. Upon further testing he found that the purple crystals could turn cotton cloth purple, too. His purple dye became popular in Paris as *mauve,* and was the first in a famous family of aniline dyes.

Consequently, at age eighteen, Perkin quit the Royal College to set up a dye factory with his father and brother. Textile industrialists soon clamored for more dyes, in eye-catching colors and with fabric-bonding fastness. Dye factories sprang up in Manchester and in Lyons in France and Ludwigshaven in Germany. Like Perkin, their chemists found new dyes by an empirical, or cut-and-try, approach. Promising compounds, like the coal-tar derivative aniline, were treated with likely reagents, among which were arsenic compounds that caused major health problems.[3] French chemists achieved a red dye named *fuchsine* and a cluster of aniline blues, while Hofmann in London worked up a range of purples. The chemical structures of these early dyes were unknown at the time. It was German chemists—based in universities and with close ties to industry—who deciphered their chemical structures and set the stage for a science-based industry.

Perhaps distracted by the cotton shortages of the 1860s (see chapter 4), Britain's textile industrialists fumbled away a major ancillary industry. Between 1861 and 1867 German industrialists founded four of the five most important German chemical companies: Bayer, Hoechst, Badische Analin- & Soda-Fabrik (BASF), and Aktien-Gesellschaft für Anilinfabrik (AGFA). The fifth company, Casella, already in existence as a vegetable-dye dealer, entered the aniline dye field in 1867.[4] These firms, at first ancillary to the textile industry, soon dominated the worldwide synthetic chemicals industry. Even today consumers of aspirin, recording tape, and photographic film will readily recognize their names. (The Bayer company's American aspirin plant, built in 1903, was taken over during World War I by the U.S. government and afterwards sold to the Sterling Products Corporation, which has marketed Bayer-brand aspirin ever since.) In the 1860s German chemists who had gone abroad flocked back home. Hofmann in 1865 took up a chair of organic chemistry at the prestigious University of Berlin. Among his returning students were such up-and-coming leaders of the German chemical industry as Carl Martius, Otto Witt, and Heinrich Caro, who became, respectively, a founder of AGFA, chief chemist of Casella, and chief chemist and director of BASF. Losing so many talented German chemists was a severe blow to England's struggling dye industry.

Germany's science establishment, although vigorous, was not originally designed to support science-based industry. German state universities trained young men for the traditional professions, the government bureaucracy, and teaching in classical secondary schools. The universities emphasized "pure" research, leaving such practical topics as pharmacy, dentistry, agriculture, and technology to lower-prestige colleges. Technical colleges, intended originally to train engineers and lower-level technicians, gained the privilege of awarding doctorate degrees only in 1899. By 1905 there were ten such technical universities in Germany, located in Berlin, Munich, Karlsruhe, and other industrial centers. A third component to the German science system was a network of specialized state-sponsored research institutes. These included institutes for physics and technology (established in 1887), agriculture and forestry (1905), biology (1905), chemistry (1911), and coal (1912).

A major hire by a German university in the 1860s was August Kekulé. Kekulé, like Hofmann, had gone abroad, to Belgium, after his training with Justus Liebig at the University of Giessen. Kekulé created modern structural organic chemistry by theorizing that a carbon atom forms four single bonds and that benzene is a hexagonal ring. He propounded these theories in 1857 and 1865, respectively. Kekulé's structural insights contrast markedly with Perkin's empirical method. Instead of "tinkering with aniline," chemists with Kekulé's concepts were now "looking for the 'mother-substance,'" or skeleton, of the dyestuff's molecule.[5]

Kekulé's concepts helped chemists synthesize such naturally occurring dyes as alizarin and indigo, but the real payoff came with the creation of an entirely new field of chemistry. Preparing such commercial blockbusters as aniline yellow and Bismarck brown became the object of a fiercely competitive science race. Chemists in several countries worked furiously on a compound believed to be intermediate between those two colors. In 1877 Hofmann publicly announced the general structure of this dye and proposed a vast new class of "azo" dyes by pointing to the "coupling reaction" that produced their chemical structure (two benzene rings linked up by a pair of nitrogen atoms). Each of the twin benzene rings could be decorated with numerous chemical additions, and all such dyes had bright colors. BASF's Caro saw it as "an endless combination game." Exploiting the estimated 100 million possibilities would require a new type of scientific work, however, which he termed "scientific mass-labor" (*wissenschaftliche Massenarbeit*). Inventive activity in the industrial research laboratories set up to conduct this new type of chemical work was no longer, as it had been for Perkin, a matter of luck and tinkering. As Caro saw it, now invention was "construction bound to rules." The results were impressive.[6] By 1897 there were 4,000 German chemists *outside* the universities. Together with their colleagues in universities, they created the field of modern organic chemistry. The number of known structural formulas for carbon-containing compounds more than tripled (to 74,000) during the sixteen years before 1899 and would double again by 1910.[7]

The patent laws in Germany, by erecting a legal shield around the azo-dye field and preserving the scientific mass-labor, reinforced a system-stabilizing mode of innovation. Hofmann's public disclosure of the azo process in January 1877, six months before a major new patent law came into effect, was a practical argument in favor of public disclosure and patent term protection. From the viewpoint of the chemical companies the 1877 law had a severe shortcoming, however, in that the law explicitly did not cover new chemical substances. Chemical "inventions" could be patented only if they concerned "a particular process" for manufacturing such substances. It was doubtful that each of the individual dyes resulting from the innumerable azo-coupling reactions could be patented. In this climate, lawyers typically drafted azo-dye patents to cover a broad class of related substances, perhaps 200 or more, to strengthen the patent's legal standing. A revised patent law in 1887 required patentees to hand over physical samples of *each* of the numerous substances claimed in the patent. The 1887 law obviously hampered individual inventors and smaller firms, which rarely had the scientific mass-labor to prepare hundreds of samples, and consequently favored companies that did have such staffing.

Two years later, in the precedent-setting "Congo red" decision of 1889, Germany's highest court forcefully ratified the emerging paradigm of system stability. The case involved two chemical giants, AGFA and Bayer, which were suing a smaller rival for infringing on their patent. The smaller firm countered with a claim denying the validity of the Congo red patent itself. To support its claim, it called as an expert witness BASF's Caro, who testified that his laboratory assistant needed to see only the patent's *title* to prepare the dye. The court was swayed, however, by an impassioned speech by Bayer's Carl Duisberg, a rising star within the industry. If the court accepted Caro's line of reasoning, argued Duisberg, it would "deliver the death-blow to the whole chemistry of azo dyes and thereby [bring] most of the patents of [Caro's] own firm to the brink of the abyss."[8] Duisberg's argument carried the day. Under a freshly minted legal doctrine of "new technical effect," the court sided with AGFA and Bayer and upheld the dubious Congo red patent. The court fully granted that the patent lacked inventiveness, but it found that the substance did

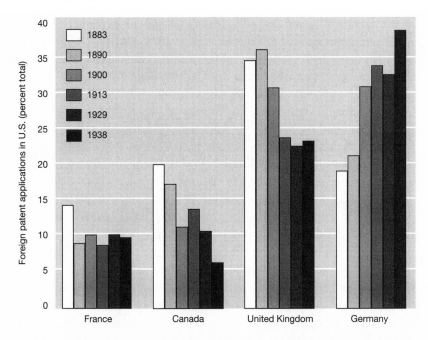

FIG. 5.1. Foreign Applications for U.S. Patents, 1883–1938.

After these four countries, no other received more than 4 percent of the patents granted during these years. Data from Keith Pavitt and Luc Soete, "International Differences in Economic Growth and the International Location of Innovation," in Herbert Giersch, ed., *Emerging Technologies* (Tübingen: Mohr, 1982), 109.

have significant commercial value, and that was enough. The new legal justification saved the scientific mass-labor, entrenching the system-stabilizing mode of innovation. "Mass production methods which dominate modern economic life have also penetrated experimental science," the chemist Emil Fischer stated in his Nobel Prize lecture in 1902. "Consequently the progress of science today is not so much determined by brilliant achievements of individual workers, but rather by the planned collaboration of many observers."[9] Duisberg put the same point more succinctly: "Nowhere any trace of a flash of genius."[10]

Patent statistics provide a ready measure of the German chemical industry's worldwide dominance. Germany, already holding 50 percent of the world market for dyes in the 1870s, achieved a full 88 percent at its peak in 1913. In 1907 Germany's only rivals in dyestuffs, England and

Switzerland, together managed to register just 35 dye patents while Germany registered 134—at the *British* patent office. At the German patent office this proportion was even more lopsided, 14 to 285.[11] Owing to its strong positions in chemicals and electricity, Germany's share of total foreign patents in the United States surpassed France's in 1883 and pulled ahead of Canada's in 1890 and England's by 1900. In 1938 Germany's U.S. patents equaled those of the other three countries combined. (See fig. 5.1.)

The synthetic-chemical empire led to the German chemical industry's direct involvement with the First World War and its appalling entanglement in the second. In World War I, popularly known as the chemist's war, chemists were directly involved in poison gas manufacture. Chemistry was also implicated in the trench warfare of the western front and the economic collapse of Germany that ended the war. The outbreak of war in the summer of 1914 cut off Germany from its imported sources of nitrates, required for fertilizers and explosives alike. Germany's on-hand stocks of nitrates were enough only to supply the army with explosives until June 1915. BASF kept the army in the field past this date with its high-pressure reactor vessels that chemically "fixed" atmospheric nitrogen (as ammonia), which could then be made into explosives or fertilizer. Of necessity the German army planned its offensives to match synthetic gunpowder production. For both sides, the mass production of explosives was a material precondition to keeping the machine guns firing, on which trench warfare grimly depended.

Chemistry also figured prominently in the attempt to break the stalemate of trench warfare. On 22 April 1915 German soldiers attacking the village of Ypres, in Flanders, opened 6,000 cylinders of chlorine (a familiar product BASF had supplied to textile bleachers for decades) and watched a yellow-green cloud scatter the French defenders. Soon chemical factories—on both sides—were turning out quantities of chlorine and other poisonous gases, as well as the blister-inducing mustard gas. As one poet-soldier recorded the experience, "Then smartly, poison hit us in the face. . . . / Dim, through the misty panes and heavy light, / As under a dark sea, I saw him drowning. / In all my dreams, before my helpless sight / He lunges at me, guttering, choking, drowning."[12]

Horrific as the gas clouds, gas shells, and gas mortars were—and gas caused at least 500,000 casualties—chemical warfare was not a breakthrough "winning weapon," because soldiers' defensive measures (such as the poet's "misty panes") were surprisingly effective, if cumbersome and uncomfortable. Only civilians, animals, and the Italian army faced poison gas unprotected. In just one instance, late in the conflict, did gas affect strategy. After its failed offensive in spring 1918, the German army effectively covered an orderly retreat with gas.[13] At the time, Germany still had plenty of poison gas; but it had run out of clothes, rubber, fuel, and food, a consequence of sending all fixed-nitrogen production into munitions while starving the agricultural sector.

The entanglement of the German chemical industry with the Third Reich also has much to do with the system-stabilizing innovation and the corporate and political forms needed for its perpetuation. By 1916, Germany's leading chemical companies, anticipating the end of the war, were already seeking a method of halting the ruinous competition that had plagued the industry before the war. That year the six leading concerns transformed their two trade associations (Dreibund and Dreiverband) into a "community of interest," together with two other independent firms. The eight companies pooled their profits and coordinated their activities in acquiring new factories, raising capital, and exchanging knowledge (fig. 5.2). Later, in the tumult of the 1920s, these eight companies in effect merged into one single corporation, I.G. Farben (1925–45). By 1930 the combine's 100 or so factories and mines employed about 120,000 people and accounted for 100 percent of German dyes, around 75 percent of its fixed nitrogen, nearly all its explosives, 90 percent of its mineral acids, 40 percent of its pharmaceuticals, and 30 percent of its rayon. Farben also had significant holdings in coal, banking, oil, and light metals. With all these heavy investments, Farben's executives felt they had little choice but to conform with Hitler's mad agenda after he seized power in 1933. Not Nazis themselves—one-fourth of the top-level supervisory board were Jews, until the Aryanization laws of 1938—they nevertheless became complicit in the murderous regime. During World War II Farben made synthetic explosives, synthetic fuels, and synthetic rubber for the National Socialist war effort. Many have never forgiven its provision to the Nazis of Zyklon B (the death-camp gas), its nerve-gas experiments on camp inmates, and its use of up to 35,000 slave laborers to build a synthetic rubber complex at Auschwitz.[14]

Flashes of Genius

The extreme routinization of innovation evident in the German dye laboratories with their "scientific mass-labor" was not so prominent in the field of electricity. In both Germany and the United States, a dynamic sector of electrical manufacturers and electric utilities featured corporate

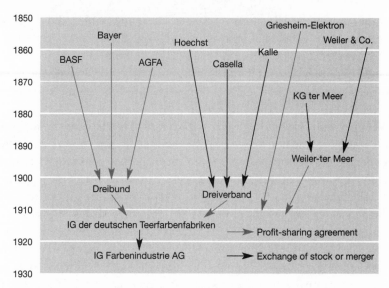

FIG. 5.2. Consolidation of the German Chemical
Industry, 1860–1925

Data from L. F. Haber, *The Chemical Industry during the Nineteenth Century* (Oxford: Clarendon Press, 1958), 128–36, 170–80; L. F. Haber, *The Chemical Industry 1900–1930* (Oxford: Clarendon Press, 1971), 121–28.

consolidation and research-driven patent strategies. In the United States the leading concerns were George Westinghouse's firm and General Electric, the successor to the Edison companies. In Germany, the leading firms were Siemens & Halske and German Edison, an offshoot of the American company.

The singular career of Thomas Edison aptly illustrates the subtle but profound difference separating system-originating inventions from system-stabilizing ones. Edison located his own inventive career at the fringe of large corporate concerns, including Western Union and General Electric. Edison's intensely competitive style of invention led to a lifetime total of 1,069 patents, among them an array of system-originating ones in electric lighting, phonographs, and motion pictures. His style of invention, however, ill-suited him to delivering the system-stabilizing inventions sought by General Electric. The company hired new talent and adopted an industrial-research model of innovation after 1900.

While Edison's boyhood was filled with farming, railroads, and newspapers, it was the telegraph industry that launched his technological career. A generation removed from the pioneering work by Samuel Morse and Moses Farmer, the telegraphers in Edison's time already confronted a nationwide system that had literally grown up alongside the railroads. Telegraph operators formed an early "virtual" community, tapping out tips, gossip, and professional news during slack times on the wire. Edison's first significant patents were a result of the telegraph industry's transition from an earlier city-to-city system, which Western Union had dominated, to an intraurban phase that demanded rapid on-site delivery of market-moving information. While Western Union's intercity lines might deliver such news from other cities to its central New York City office, a legion of bankers, lawyers, stock brokers, corporation executives, and newspaper editors wanted the news relayed instantly to their own offices. Responding to this opportunity, Edison moved to New York and made his mark as an inventor of printing telegraphs, automatic telegraphs, and stock tickers. (Edison's very first patent was for an automatic vote-recording machine, something the city's politicians of the time did not want.) As a telegraph inventor Edison's principal financial backers were Western Union, the telegraph combine, and Gold & Stock, a leading financial information provider.

Western Union continued to be a force in Edison's life even after he resigned from that firm in 1869 to pursue full-time invention. Working as an independent professional inventor saddled Edison with the responsibility of paying his own bills. For years, he struggled to balance inventive freedom with adequate funding. Manufacturing his telegraph equipment in Newark, New Jersey, Edison at one time oversaw as many as five factory establishments and 150 skilled workmen. His funding came from working on inventions deemed important by, among others, Western Union. In the 1870s Western Union engaged Edison's inventive talents to help it master the "patent intricacy" of duplex telegraphs and, as Edison dryly remarked, "as an insurance against other parties using them."[15] Duplex telegraphs, which sent two messages on one line, and quadruplex telegraphs, which sent four messages, were especially attractive to a large firm like Western Union seeking to wring additional efficiency from already established systems. Even though he resented the "small-brained capitalists" who commissioned this work, he still needed them to finance his inventive ventures.

In the spring of 1876 Edison completed his "invention factory" at Menlo Park, located in the New Jersey countryside midway between New

York and Philadelphia. There, as he famously promised, he would achieve "a minor invention every ten days and a big thing every six months or so."[16] Having devoted himself to full-time invention for seven years, Edison had achieved an average of forty patents each year, or one every nine days. He was granted 107 patents during 1882, his most productive year at Menlo Park, in the midst of the electric light campaign detailed below. An inventory of Menlo Park from fall 1878 itemizes a well-equipped electrical laboratory: galvanometers, static generators, Leyden jars, induction coils, a Rühmkorff coil capable of producing an 8-inch spark, condensers, Wheatstone bridges, and several types of specialized galvanometers. There was soon a well-stocked chemical laboratory, too.[17] At Menlo Park Edison directed a team of about twenty model makers, precision craftsmen, and skilled workmen, as well as a physicist and chemist. In 1877 Edison invented a tinfoil-cylinder "talking machine" (an early phonograph) and a carbon-diaphragm microphone for the telephone, invented a year earlier by Alexander Graham Bell. Publicizing the talking machine kept Edison in the public eye.

As Edison reviewed the technical literature he discovered that the outstanding problem in electrical lighting was that arc lights, popular for illuminating storefronts, were far too bright for indoor use, while incandescent bulbs were far too short-lived. Since the 1840s, inventors in several countries had patented twenty or more incandescent lamps. For achieving lower-intensity arc lighting, one alternative was placing three or four arc lights on a single circuit. Edison conceived of something similar for incandescent lights. His dramatic boast in September 1878 ("have struck a bonanza in electric light—infinite subdivision of light") was not an accurate technical claim. It was a commercial claim, fed to newspaper reporters, to generate investors' enthusiasm.

October 1878 was a busy month even for Edison. On the 5th he applied for a patent on a platinum-filament bulb, on the 15th he formed the Edison Electric Light Company to develop the invention and license his patents, and on the 20th he announced his plans for a complete lighting system in the *New York Sun*. In fact, absent practical incandescent bulbs, appropriate generators, and a system of distribution, Edison was months away from a complete system. Edison wrote that fall to an associate: "I have the right principle and am on the right track, but time, hard work and some good luck are necessary too. It has been just so in all of my inventions. The first step is an intuition, and comes with a burst, then difficulties arise—this thing gives out and [it is] then that 'Bugs'—as such little faults and difficulties are called—show themselves and months of intense watching, study and labor are requisite before commercial success or failure is certainly reached."[18]

Science arrived at Edison's laboratory in the person of Francis Upton, whom Edison hired in December 1878. Upton had a first-rate education from Phillips Andover Academy, Bowdoin College, and Princeton University and had recently returned from physics lectures by the legendary Helmholtz in Berlin. Edison wanted his electric lighting system to be cost competitive with gas lighting and knew that the direct-current system he envisioned was viable only in a densely populated urban center. Using Ohm's and Joule's laws of electricity allowed Upton and Edison to achieve these techno-economic goals. While Edison was already familiar with Ohm's law, Joule's law first appears in his notebooks just after Upton's arrival. Ohm's law states that electrical resistance equals voltage divided by current ($R = V/I$). Joule's law states that losses due to heat equal voltage times current, or (combining with Ohm's law) current-squared times resistance (heat $= V \times I$ or $I^2 \times R$).

Reducing the energy lost in transmitting direct-current electricity was the critical problem that these equations spotlighted. If Edison and his researchers reduced the electrical resistance of the copper wire by increasing its diameter, the costs for copper would skyrocket and doom the project financially. If they reduced the current, the bulbs would dim. But if they *raised* the voltage in the circuit, the current needed to deliver the same amount of electrical energy to the bulb would drop, thereby cutting the wasteful heat losses. Upton's calculations, then, indicated that Edison's system needed a light-bulb filament with high electrical resistance. Later, Joule's law also allowed Upton to make the counterintuitive suggestion of lowering the internal electrical resistance of the Edison system's dynamo. Other electricians followed the accepted practice of having their dynamo's internal resistance equal to the external circuit's resistance, and these dynamos converted into usable electricity only about 60 percent of the mechanical energy (supplied typically by a steam engine). In comparison, by 1880 Edison's low-resistance dynamos achieved 82 percent efficiency.[19]

On 4 November 1879, after two months of intensive experimenting on filaments, Edison filed his classic electric-bulb patent. Its claims to novelty included the "light-giving body of carbon wire or sheets," with high resistance and low surface area, as well as the filament's being enclosed in a "nearly perfect vacuum to prevent oxidation and injury to the conductor by the atmosphere."[20] These individual elements were not particularly novel. The big news was that Edison's bulb burned brightly for an extended period, it cost little to make, and was part of an economical operating system of electric lighting. No other bulbs had this set of qualities. When Edison tested his system in January 1881 he used a 16-candlepower bulb at 104 volts, with resistance of 114 ohms and current of 0.9 amps. The U.S. standard of 110 volts thus has its roots in Edison's precedent-setting early systems. (Similarly, the European standard of 220 volts can be traced to the decision in 1899 of the leading utility in Berlin [BEW] to cut its distribution costs by doubling the voltage it supplied to consumers, from 110 to 220 volts. BEW paid for the costs of converting consumers' appliances and motors to the higher voltage.)[21]

While Edison himself, the "Wizard of Menlo Park," cultivated newspaper reporters, it was Grosvenor P. Lowrey that steered Edison's lighting venture through the more complicated arena of New York City finance and politics. A leading corporate lawyer, Lowrey numbered Wells Fargo, the Baltimore & Ohio Railroad, and Western Union among his clients. He first met Edison in the mid-1860s in connection with telegraph-patent litigation for Western Union and became his attorney in 1877. After experiencing the dazzling reception given to a new system of arc lighting in Paris in 1878, Lowrey pressed Edison to focus on electric lighting. While many figures were clamoring for Edison to take up electric lighting, Lowrey arranged financing for his inventive effort from the Vanderbilt family, several Western Union officers, and Drexel, Morgan and Company. The Edison Electric Light Company, initially capitalized at $300,000, was Lowrey's creation.

So was a brilliant lobbying event that Lowrey staged on 20 December 1880 to help secure a city franchise. The city's go-ahead was necessary for the Edison company to dig up the city streets and install its lines underground. (The profusion of above-ground telegraph wires was already a major nuisance in the city's financial district, where Edison intended to build his system.) Edison did not clearly see the wisdom of winning over certain aldermen aligned with gas-lighting interests who rightly feared competition from his electrical system. Lowrey saw this plainly and devised an ingenious plan. He hired a special train that brought to Menlo Park eight New York aldermen, the superintendent of gas and lamps, as well as the parks and excise commissioners. They received a full two-hour tour of the Edison facilities followed by a lavish dinner, dramatically lit up by Edison lights. "For half an hour only the clatter of dishes and the popping of champagne corks could be heard, and then the wine began to

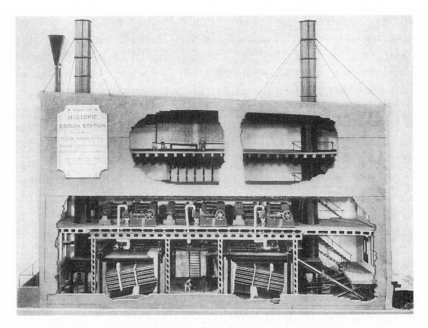

FIG. 5.3. Edison's Pearl Street Station.

This model of Thomas Edison's first central station (1882–94) was exhibited at the St. Louis Exposition in 1904. T. C. Martin and S. L. Coles, *The Story of Electricity* (New York: Marcy, 1919), 1:85.

work and the Aldermen, true to their political instincts, began to howl 'Speech, speech,'" wrote one bemused newspaper reporter.[22] Having put his guests in a receptive mood, Lowrey outlined his case on behalf of Edison. In short order, the franchise was approved.

Edison's first central electric station (fig. 5.3), at 257 Pearl Street, represented a synthesis of scientific insight, technical research, political talent, and perhaps a bit of the luck Edison had earlier acknowledged. In building the generating station and distribution system and installing the wiring, Edison sustained day-by-day management of construction. He simultaneously maintained a furious pace of invention. He applied for 60 patents in 1880, 89 in 1881, and a career-record 107 in 1882. On 4 September 1882, after a technician connected the steam-driven generator to the lighting circuit, Edison himself stood at 23 Wall Street—the offices of his financial backer Drexel & Morgan—and switched on the first 400 lights connected to the system. Within a month wiring was in place for 1,600 lamps; within a year that number topped 11,000. Even with his pioneer station completed, Edison faced a number of daunting technical and financial problems. The Edison utility did not charge for electricity during 1882, and it lost money throughout 1883. By 1886 Edison needed to create a broader customer base, but he simply could not raise the money to do so. A Drexel Morgan syndicate with sizable investments finally "[squeezed] out some money somewhere."[23] Fires from defective wiring were another serious problem. A fire on 2 January 1890 closed the Pearl Street station for eleven days. At J. P. Morgan's personal residence at 219 Madison Avenue an electrical fire damaged his library.[24]

By 1888, six years after opening of the pioneering Pearl Street station, Edison had established large companies or granted franchises in Philadelphia, New Orleans, Detroit, St. Paul, Chicago, and Brooklyn. In Europe central stations were operating in Milan and Berlin. By 1890 even the once-troubled New York utility reported impressive figures: compared with 1888, its customers had grown from 710 to 1,698; the number of 16-candlepower lamps had increased from 16,000 to 64,000; and the company's net earnings had grown from $116,000 to $229,000.[25] By this time Edison was at his newly constructed laboratory at West Orange working on a variety of inventions, mostly unrelated to electric light. While he retained his directorships in the Edison electric companies, he had withdrawn from the active management of them.

Battle of the Systems

The early dominance of Edison's lighting system may obscure the fact that it was a peculiar system devised for a specific environment. Edison had designed his direct-current system for the dense customer load in large urban areas. His DC system was not suited for the smaller cities and rural areas where most Americans still lived. (Only after 1920 would more than half of all Americans live in urban areas larger than 2,500 persons.) Meanwhile, the uses for electricity multiplied during the 1880s to include electric motors for street railways and factories, and there was further expansion in arc lighting. Inventors increasingly turned to alternating current electricity. By using transformers to step *up* the voltage for transmission (cutting I^2R heat losses) AC systems could send electricity over distances of dozens or even hundreds of miles, then step *down* the voltage for safe local distribution to consumers.

The Edison companies made step-by-step improvements in DC systems, but for the most part left the AC field to others. Some believe that Edison never understood alternating current. In fact, beginning in 1886

Edison experimented with AC electricity at his new West Orange laboratory. Edison sketched numerous AC generators, transformers, and distribution networks, and filed for at least four AC patents before concluding that it was economically unpromising. Edison was wary of the energy losses of transformers, the high capital costs of building large AC stations, and the difficulties of finding insulators that could safely handle 1,000 volts. "The use of the alternating current is unworthy of practical men," he wrote to one financial backer.[26]

Arc lighting for streets, AC incandescent systems for smaller towns, AC motors for factories, and the pell-mell world of street railways were among the lucrative fields that Edison's diagnosis overlooked. It was precisely these technologies, in addition to DC incandescent lighting for big cities, that Edison's principal rivals—the Thomson-Houston, Brush, and Westinghouse firms—succeeded at inventing, developing, and selling commercially. These several uses were brought together in the 1890s through the development of a "universal system." It was also during the 1890s that the companies founded by Edison and his principal rivals came increasingly to be steered by financiers. Whereas Edison had lived on the thrill of system-jarring competition, for these financiers such cutthroat competition brought instability to their systems and was undesirable—and avoidable. In the mid-1880s there were twenty firms competing in the field of electric lighting; by 1892, less than a decade later, after a period of rapid consolidation and mergers, there would be just two: Westinghouse and General Electric (see fig. 5.4).

In 1889 the several Edison firms and the Sprague Electric Railway & Motor Company were merged by Henry Villard, an investor in the early Edison companies, to form Edison General Electric. Its capital of $12 million came from German banks, Villard himself, and J. P. Morgan. Taking the title of president, Villard left the day-to-day management with Samuel Insull, formerly Edison's personal secretary and later, for many years, head of Chicago's Commonwealth Edison utility (1892–1934). Insull had the difficult job of coordinating three large production facilities—a huge machine works at Schenectady, New York, a lamp factory at Harrison, New Jersey, and a factory in New York City. Edison General Electric contracted with Edison to develop improved DC equipment at his West Orange laboratory. Insull even hoped that Edison would develop a full-fledged AC system, but Edison instead focused his inventive energies on improving his phonograph and purifying iron ore.

In 1878, when Edison was starting his electric lighting project at Menlo

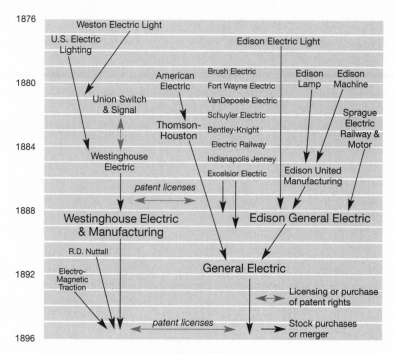

FIG. 5.4. Consolidation of the U.S. Electrical
Industry, 1876–1896.

Adapted from Arthur A. Bright, *The Electric Lamp Industry* (New York:
Macmillan, 1949), with data from W. B. Carlson, *Innovation as a Social
Process* (Cambridge: Cambridge University Press, 1991).

Park, Elihu Thomson was entering the field of arc lighting. Just six years
younger than Edison, Thomson (1853–1937) would distinguish himself as
the nation's third most prolific inventor, with a career total of 696 patents.
In the mid-1880s Thomson turned his inventive efforts on incandescent
lighting and AC systems. His other notable inventions include electric
welding, street railway components, improved transformers, watt meters,
and induction motors. These inventions were among the necessary technical components of the universal system of the 1890s.

To finance his inventions Thomson had struggled through a series of
ill-funded and unsatisfactory partnerships. The capital of American Electric Company ($0.5 million), formed to market Thomson's award-
winning arc lights, was a small fraction of the Edison concern ($3.8 mil-
lion). In 1882 Charles A. Coffin, a self-made Boston entrepreneur,

launched Thomson's career. While he had earlier averaged at best twenty patents a year, Thomson was soon applying for forty patents a year, fully equivalent to Edison's boast of an invention every ten days. Coffin and his investment syndicate understood that Thomson was most valuable to the company while inventing, and they encouraged Thomson to set up a special "Model Room" at the company's principal factory, at Lynn, Massachusetts.

Coffin and his investors also recognized that real success in the electrical field would require equal strengths in innovation, marketing, and manufacturing. A string of successfully marketed inventions propelled the firm into the big league of electrical manufacturers. Coffin also bought up no fewer than seven rivals, including the Brush firm, at a cost of $4 million. To raise such large amounts of money, Coffin turned to Lee, Higginson & Company, a Boston brokerage house that pioneered in the marketing of industrial securities. By 1892, the Thomson-Houston company was capitalized at $15 million, had 3,500 employees, and made annual profits of $1.5 million. By then, some 800 companies were using Thomson-Houston's arc and incandescent lighting equipment, while in the rapidly growing street-railway field an additional 204 companies were using its equipment.

The only serious rival to Edison General Electric and Thomson-Houston was the concern founded by George Westinghouse. Westinghouse (1846–1914) was already rich and famous from licensing his patents on railroad air brakes. In the mid-1880s he purchased key patents and hired the talent required to construct an AC lighting system. Westinghouse employed William Stanley as an in-house inventor; in 1886 Stanley perfected an AC transformer. That same year, Westinghouse installed its first AC lighting system, in Buffalo, New York. By 1889 Westinghouse had also developed AC industrial motors and street railways and concluded a patent-sharing agreement with Thomson-Houston. Under its terms, Westinghouse could sell Thomson's arc lighting, while Thomson-Houston could manufacture and sell its own AC systems without fear of the Westinghouse AC system patents.[27] By 1890, thanks to the popularity of its AC equipment, Westinghouse had sales of $4 million and total assets of $11 million.

The relentless competition of these three huge firms for customers, inventors, and patents played out in the race for contracts, in legal battles, and in a feeding frenzy to buy up smaller rivals. Rivals disappeared at a quick pace between 1888 and 1891, when Thomson-Houston acquired

7, Westinghouse 3, and Edison General Electric 2. By this strategy the big three put pesky rivals out of business, used their patents, and gained the services of their inventors. Competition could also take a more bizarre form. Edison General Electric, unable to compete technically with the other two big firms in AC systems, used its political connections to launch a lurid campaign to denounce AC as the "death current." The Edison firm backed the efforts of Harold Brown, once a salesman for Edison's electric pens and now an imaginative opponent of AC. With the quiet help of Edison's laboratory staff, Brown staged public spectacles in which he used AC current to electrocute dogs, calves, and horses. Typically, Brown would first apply a mild DC current, suggesting its relative safety, then dispatch the animal with a hefty jolt of the "deadly" AC. In 1890 he arranged for a 1,000-volt Westinghouse generator to be used, despite that company's opposition, in the first public electrocution at Sing Sing prison. The Edison publicity machine was then fired up against the dangers of the Westinghouse-supplied "executioner's current." In its campaign, Edison General Electric also lobbied state legislatures in New York, Ohio, and Virginia to limit the maximum voltage of electrical systems to 300, a legal move to scuttle the principal advantages of AC systems while preserving DC systems.[28]

Edison himself, smarting from the rough battle over patents during these years, derailed the first serious attempt at consolidating his concern with the Thomson-Houston firm. When he heard of the merger plans in 1889, he fumed, "My usefulness as an inventor would be gone. My services wouldn't be worth a penny. I can only invent under powerful incentive. No competition means no invention." But the financiers that ran both firms increasingly favored consolidation as a means to protect their substantial investments. J. P. Morgan, soon to reorganize a quarter of the nation's railways, expressed alarm at the electrical manufacturers' endless hunger for capital. "What we all want," wrote Charles Fairfield (a partner at Lee, Higginson) in 1891, "is the union of the large Electrical Companies." An attempt that year to merge Westinghouse with Thomson-Houston failed. George Westinghouse, while a brilliant technician and able factory manager, lacked the skills of a financier. "He irritates his rivals beyond endurance," Fairfield said.[29]

General Electric, a creation of the nation's leading financiers, took form in April 1892. After several months of negotiations between their respective bankers, the Thomson-Houston and Edison General Electric companies exchanged their stock for shares in the new company, capitalized at $50 million. This figure made it the second largest merger at the

FIG. 5.5. The Electric City.

Inspecting electric arc street lighting using a special electric vehicle. T. C.
Martin and S. L. Coles, *The Story of Electricity* (New York: Marcy, 1922), 2:32.

time. General Electric's board of directors comprised six bankers, two
men each from the Thomson and Edison companies, and a Morgan associate as chairman. (Thomas Edison was on the board only nominally;
Thomson outright declined a directorship in order to continue his technical work under GE auspices.) Coffin became GE's founding president.
Insull, offered the job of second vice-president in the new company, instead departed for Chicago, where he developed and installed one of the
early polyphase AC, or universal systems, and pioneered managerial and
financial techniques in building a regional utility empire.

Tenders of Technological Systems

Edison fought it, Thomson denied it, and Insull embraced it: a new pattern of technological change focused on stabilizing large-scale systems rather than inventing wholly new ones. In the most capital-intensive industries, including railroads, steel, chemicals, and electrical manufacturing, financiers like J. P. Morgan and Lee, Higginson in effect ended the ceaseless competition and fierce pace of freewheeling technological innovation. In doing so, they were the chief agents in giving birth to a stable organized capitalism. In the second industrial revolution, somewhat paradoxically, technological change became evolutionary. The arch-apostle of stability, U.S. Steel, extinguished at least two important innovations in steel making, the oversize H-beam mill of Henry Grey and its own experiments in the continuous rolling of thin sheets. (And in a delightful irony, the German chemical and steel industries both looked to U.S. Steel as a harbinger of what was modern and progressive in organizing industry, modeled themselves after its sprawling bureaucracy, and got much the same unwieldy result.) Exhausting battles with corporations contributed to the suicides of chemical-inventor Wallace Carothers and radio-inventor Edwin Armstrong. The nation's inventive efforts were increasingly channeled to expand and strengthen the existing technological systems.[30]

Besides financiers, the most important agents of industrial stability were scientists and engineers. Industrial scientists and science-based engineers stabilized the large systems by striving to fit into them and, most importantly, by solving technical problems deemed crucial to their orderly expansion. Neither of these professions existed in anything like their modern form as recently as 1870. Before then, engineers had been mostly either military or "civil" engineers who built fortifications, bridges, canals, and railways. During the second industrial revolution, engineering emerged as a profession. National societies were founded in the United States by mining engineers (1871), mechanical engineers (1880), electrical engineers (1884), and chemical engineers (1908). Industrial scientists too were created in these decades. In the German chemical industry, as we saw, the "scientific mass-labor" needed to synthesize and patent new dyes in the 1880s led to the first large-scale deployment of scientists in industry. Such a model of industrial research appeared somewhat later in the United States, with full-scale industrial research laboratories being organized by General Electric in 1901, by DuPont in 1902, and AT&T in 1911.

In 1900 there were hundreds of laboratories in U.S. industry but no "research and development" laboratories as we know them. Edison's laboratories at Menlo Park and West Orange had been extensions of Edison's personal drive for innovation more than established, on-going efforts. They foundered whenever Edison was physically absent, and they went dormant in the early 1890s when Edison left them to pursue an ill-fated iron-ore venture. From the 1870s, the railroad, steel, and heavy chemical industries had located laboratories on factory sites and sometimes in factory buildings, and their staffs were responsible for analyzing raw material and testing intermediate and finished products. By contrast, industrial research laboratories grew only in the largest companies, and mostly in the chemical and electrical sectors of science-based industry; additionally, these research laboratories were isolated—often physically— from production facilities.

The desire for new products and new processes, frequently touted as the chief inspiration for industrial research, is only part of the story. For example, DuPont invested heavily in industrial research beginning in the 1920s and loudly publicized its successes with such products as neoprene and nylon ("Better Things for Better Living through Chemistry" went its slogan). But acquiring competitors, rather than conducting research, remained DuPont's most important source of new technology for decades. A National Bureau of Economic Research study found that of twenty-five major product and process innovations at DuPont between 1920 and 1950, only ten resulted from its in-house research while fifteen resulted from acquiring outside inventions or companies.[31]

In 1901 General Electric saw its laboratory as a way to internalize innovation. Having suffered the expiration of its Edison-era patents on carbon-filament light bulbs, and facing determined competitors with new and promising bulb designs, GE founded its laboratory to head off future technological surprises. Initially GE hired sixteen chemists and physicists; their task, directed by Willis Whitney, was to build a better light bulb. In 1904 the laboratory produced a major improvement in the carbon filament, dubbed "General Electric Metalized." Yet even with the GEM bulbs, the company had to purchase patent rights from outside inventors; in 1907 it paid two German companies $350,000 for patent rights and in 1909 an additional $490,000 for the use of tungsten filaments. This expensive situation persisted until 1912 when, after years of concerted work, one of the laboratory's chemists patented an improved method for making long-lived tungsten filaments.

The success of fitting science into industry at GE must be largely credited to Willis Whitney. Whitney came to GE with a German doctorate in physical chemistry, a lucrative (patented) process of his own for recovering solvents, and an unsatisfactory experience teaching at MIT. He arranged for his laboratory's Ph.D. scientists to have plentiful equipment, ample intellectual leeway, and a stimulating environment that gave them journals, colloquia, and brisk discussion. Whitney's strong encouragement, especially for research work relevant to the company's commercial concerns, was legendary. Yet, like the German dye laboratories, the results were not the scientist's own. "Whatever invention results from his work becomes the property of the company," Whitney stated in 1909. "I believe that no other way is practicable."[32] Whitney took substantial pride in hiring Irving Langmuir, easily GE's most renowned scientist. Langmuir, after teaching at Stevens Institute of Technology, began research at GE in 1910 to perfect and patent a gas-filled light bulb. The so-called Mazda bulb brought GE stunning profits of $30 million a year. In 1928, thanks to its Mazda bulb sales, GE held 96 percent of the U.S. market for incandescent bulbs. Langmuir's follow-up research on the surface chemistry of such bulbs brought him a Nobel Prize in 1932.

Industrial research became a source of competitive advantage for the largest firms, including General Electric, AT&T, and General Motors. Companies that could mount well-funded research efforts gained technical, patent, and legal advantages over those that could not. Independent inventors, formerly the nation's leading source of new technology, either were squeezed out of promising market areas targeted by the large science-based firms or went to work for them solving problems of the companies' choosing. In the electrical industry, 82 percent of research personnel were employed by just one-quarter of the companies. By the 1930s, when General Electric and AT&T between them employed 40 percent of the entire membership of the American Physical Society, the industrial research model became a dominant mode for organizing innovation and employing scientists. By 1940 DuPont alone employed 2,500 chemists. At the time fully 80 percent of the nation's R&D funding ($250 million) came from industry.

The search for stability involved not only financiers, corporations, engineering societies, and research laboratories. The very content of engineering was at play, as we can see in the electrical engineering program at the Massachusetts Institute of Technology around 1900. Achieving stability had a precise technical meaning for electrical engineers in this period, since instabilities or transients in the growing regional electrical transmission systems were major obstacles to their orderly and safe development. The industrial orientation of electrical engineering at MIT from around 1900 into the 1930s contrasts markedly with its more scientific and military orientation during and after the Second World War.

Despite its sponsorship of Alexander Graham Bell's experiments on telephony in the mid-1870s, MIT did not found a four-year course in electrical engineering until 1882, and for years the program was mostly an applied offshoot of the physics faculty. In 1902, when a separate electrical engineering department was organized, Elihu Thomson joined the faculty as a special nonresident professor of applied electricity. From then on MIT would enjoy close relations with General Electric's massive Lynn factory, formerly the location of Thomson's Model Room, and would send students there for practical training and receive from GE electrical machinery to supply its laboratories. Thomas Edison and George Westinghouse also gave electrical machinery to the new program. Enrollments in the electrical engineering department grew quickly (it was briefly, in the mid-1890s, the institute's most popular program), and by the turn of the century electrical engineering had joined civil, mechanical, and mining engineering as MIT's dominant offerings.

A sharp departure from the department's origins in physics occurred under the leadership of Dugald Jackson, its chairman from 1907 to 1935. Jackson recognized that engineering was not merely applied science. As he put it in a joint address to the American Institute of Electrical Engineers and the Society for the Promotion of Engineering Education in 1903, the engineer was "competent to conceive, organize and direct extended industrial enterprises of a broadly varied character. Such a man . . . must have an extended, and even profound, knowledge of natural laws with their useful applications. Moreover, he must know men and the affairs of men—which is sociology; and he must be acquainted with business methods and the affairs of the business world."[33] Jackson upon arriving at MIT in February 1907 revamped the physics-laden curriculum in close consultation with a special departmental advisory committee. The imprint of business on the committee was unmistakable. Its members, in addition to Elihu Thomson (from General Electric), were Charles Edgar (president of Boston Edison), Hammond Hayes (chief engineer, AT&T), Louis Ferguson (vice-president, Chicago Edison), and Charles Scott (consulting engineer, Westinghouse). Harry Clifford competently handled courses in alternating current electricity and machinery, which Jackson had previously taught at Wisconsin. Jackson himself developed

FIG. 5.6. Electric Turbines and the "Wealth
of Nations."

Dugald Jackson revamped electrical engineering at MIT to engage the
technical and managerial problems of electric power systems, to tap (as
General Electric put it) "the wealth of nations generated in turbine
wheels." Pictured is Ohio Power Company's station at Philo, Ohio. *General Electric Review* (November 1929): cover. Courtesy of Illinois Institute
of Technology Special Collections.

a new fourth-year elective on managerial topics, "Organization and Administration of Public Service Companies" (i.e., electric utilities).

The research, consulting, and teaching activities during the Jackson
years drew MIT's electrical engineering offerings closer to the electrical
manufacturers and electrical utilities. Jackson consulted on several leading electrification projects—including the Conowingo hydroelectric project on the Susquehanna River in Maryland, the Fifteen Miles Falls project on the Connecticut River, the Great Northern Railway's electrification
of its Cascade Mountain Division, and the Lackawanna Railroad's electrification of its New Jersey suburban service. Even when Jackson taught
other subjects, his students reported, "they were learning a great deal

TABLE 5.1. MIT Electrical Engineering Curriculum for 1916

Subject Number	Material Covered
	Principles of Electrical Engineering Series
601	Fundamental concepts of elecctric and magnetic circuits
602	DC machinery
603	Electrostatics, variable and alternating currents
604	AC machinery
605	AC machinery, transmission
	Fourth-Year Electives
632	Transmission equipment
635	Industrial applications of electric power
637	Central stations
642	Electric railways
645–646	Principles of dynamo design
655	Illumination
658	Telephone engineering
	Elective Graduate Subjects
625	AC machinery
627	Power and telephone transmission
634	Organization and administration of public service companies
639	Power stations and distribution systems
643	Electric railways

Source: Wildes and Lindgren, *A Century of Electrical Engineering and Computer Science at MIT, 1882–1982* (Cambridge: MIT Press, 1985), pp. 58–59.
Note: Laboratories in Technical electrical measurement and Dynamo-electric machinery were required in the third and fourth years.

about public utility companies and their management."[34] In 1913 Jackson tapped Harold Pender, the author of two influential textbooks, to head the department's research division, funded by General Electric, Public Service Railway, Stone & Webster, and AT&T. In line with the department's research and consulting activities, the curriculum revision in 1916 "was predominantly aimed at the problems of electrical power systems" (see table 5.1). Illumination and telephony received some emphasis; the newer fields of radio and electronics did not.[35]

MIT's Technology Plan of 1920 had the effect of implanting Jackson's model of industrial consulting on the institute as a whole. On balance, the plan was not a well-thought-out initiative. In 1916 MIT had vacated its perennially cramped facilities in Boston's Back Bay and, with large gifts from Coleman du Pont and George Eastman, built a new complex of buildings on its present site along the Charles River. The move to Cambridge was expedited by an agreement made in 1914 to share the large bequest given to Harvard University by the shoe manufacturer Gordon McKay. In 1919, however, after a state court struck down the terms of the MIT-Harvard agreement and the McKay funds ceased, MIT was plunged into a financial crisis. Making matters worse, effective in 1921 the state legislature cut off its appropriation to MIT of $100,000 a year.

In response to this crisis George Eastman offered $4 million in Eastman Kodak stock—if MIT raised an equal sum by January 1920. Raising such a huge matching gift from individual donors seemed impossible, so president Richard Maclaurin turned to industry. In effect Maclaurin's Technology Plan cashed in on the reputation of MIT's faculty members, laboratories, and libraries, and offered them up for solving industrial problems. Each company receiving such access made a financial contribution to MIT. "Had MIT been allowed . . . to 'coast' on the McKay endowment, it would not have been forced to turn full-face to American industry and to become deeply engaged with its real problems of development."[36] While the Technology Plan staved off financial crisis, the type of narrowly conceived problems deemed important by industry was hardly a satisfactory basis for learning. In the early 1920s enrollments plunged in virtually all programs except for electrical engineering. In the 1930s MIT's president Karl Compton struggled to find other sources of funds, ones that did not so narrowly constrain the faculty's problem choices and publication options.

In Jackson's consulting, the MIT curriculum of 1916, and the Technology Plan, MIT's educational offerings and its consulting activities were responses to the problems of stabilizing large-scale technological systems. In Harold Hazen's work on the "network analyzer" during the 1920s and 1930s we can trace the same phenomenon in the very content of engineering research. One line of leading-edge engineering research during these years concerned the behavior and insulation of high voltages. Hazen's research responded to a more subtle problem, that of the instability of highly interconnected electrical systems, then emerging as local electric systems were wired together into regional systems.

The technical problems of interconnected systems were given sharp focus by the so-called Superpower and Giant Power regional electrification proposals of the 1920s. The problem was not so much that these systems would be geographically spread out, in a network stretching from Boston to Washington with lines running to western Pennsylvania or even beyond to the Midwest, and would feature transmission lines carrying up to 300,000 volts, roughly ten times the prevailing industry standard. The real problem was the system's interconnected nature, carrying multiple loads from multiple sources. Transients, or sharp fluctuations in voltage, were of particular concern. No utility would invest in such a mega-system just to have a transient bring it down (something that actually happened in 1965, causing a massive blackout in New York). In an effort to simulate these systems, engineers at General Electric and Westinghouse as early as 1919 built small-scale models using networks of generators and condensers.

Hazen's work on the "network analyzer" began with his 1924 bachelor's thesis under Vannevar Bush. Bush, a pioneer in analog computing, was working for Jackson's consulting firm studying the Pennsylvania-based Superpower scheme. Hazen's thesis, conducted jointly with Hugh Spencer, involved building a dining table–sized simulator capable of representing three generating stations, 200 miles of line, and six different loads. After graduation Hazen and Spencer joined GE's Schenectady plant and there worked for Robert Doherty, "who was much concerned with stability studies in power systems." When Hazen returned to MIT as Bush's research assistant in 1925, Bush asked him to simulate an entire urban power system. Simulating polyphase AC power systems involved not only generating multiple-phase currents (Hazen worked at 60 Hz) and shifting their phases to simulate motor loads but also devising highly accurate measuring equipment. For the latter, Hazen worked closely with GE's Lynn factory. By 1929 the measuring problems were solved and GE's Doherty approved the building of a full-scale network analyzer.

Built jointly by GE and MIT and physically located in the third-floor research laboratory in MIT's Building 10, the network analyzer was capable of simulating systems of great complexity. The analyzer was now an entire room full of equipment (fig. 5.7). It featured 8 phase-shifting transformers, 100 variable line resistors, 100 variable reactors, 32 static capacitors, and 40 load units. Electric utilities loved it. In its first ten years, its principal users were American Gas and Electric Service Corporation, General Electric, Canadian General Finance, Jackson & Moreland, Illinois Power and Light, Union Gas and Electric, the Tennessee Valley Authority, and half a dozen other large utility entities. During 1939 and 1940 a coalition of nine utilities, then studying a bold proposal for a nationwide electrical grid, paid for a massive upgrading and expansion of the facility. Used intensively by the electric utilities, the network analyzer became less

FIG. 5.7. MIT Network Analyzer.

Harold Hazen at the center of a signature artifact of the science and systems era. Electric utilities simulated complex electric power systems with the network analyzer, from the 1920s through its dismantling in 1953. Courtesy of Massachusetts Institute of Technology Museum.

and less relevant to MIT's educational and research activities. "While our board is still useful to power companies, it no longer serves as an adequate vehicle for creative first-rate graduate or staff research in the power area," stated a 1953 department memorandum recommending its termination. "One can say that it has had a glorious past." The analyzer was sold to Jackson & Moreland and shipped to Puerto Rico.[37] During Hazen's tenure as department head (1938–52), the rise of electronics and the press of military work dramatically changed electrical engineering, MIT itself, and the character of technology.

SYNTHETIC DYES, poison gases, DC light bulbs, AC systems, and analog computers such as Hazen's network analyzer constituted distinctive artifacts of the science-and-systems era. Broadening our view to include additional or different science-based industries would, I think, alter the particular stories but not change the overall patterns. (One can discern variations on the themes explored in this chapter in the pharmaceutical, automobile, steel, radio, and photographic industries.) The most important pattern was the underlying sociotechnical innovations of research laboratories, patent litigation, and the capital-intensive corporations of science-based industry. For the first time, industrial and university scientists participated equally with inventors, designers, and engineers in developing new technologies. Indeed, the R&D laboratory has become such a landmark that it is sometimes difficult for us to recall that new technologies can (and often are) a product of persons working elsewhere.

A decisive indicator that something distinct occurred in these decades was the decline of British industrial leadership and the rise of German and American primacy, especially in the science-based industries. A neat contrast can be made of the British cotton-textile industry that typified the first industrial revolution and the German synthetic dye industry and American electrical industry that together typified the second. German chemists like August von Hofmann and Heinrich Caro as well as American financiers like Charles Coffin and J. P. Morgan typified this era as much as did the technical artifacts listed above.

The presence of the financiers, corporations, chemists, and engineers produced a new mode of technical innovation and not coincidentally a new direction in social and cultural innovation. The system-stabilizing mode of technical innovation—"nowhere any trace of a flash of genius"—was actively sought by financiers, who had taken on massive financial liabilities in financing the large science-based industries, and by those industrial scientists like GE's Willis Whitney and engineers like MIT's Dugald Jackson who welcomed the opportunities it afforded. Edison, who distinctly preferred working at the edge of system-originating inventions, intuitively understood that his glory days as an inventor were over—even as his fame as a cultural icon grew apace. General Electric needed system-stabilizing industrial scientists, who might turn light bulbs into Nobel Prizes and profits, more than they needed a brilliant Edisonian inventor. The system-stabilizing innovations, with the heavyweights of industry and finance behind them, also created new mass-consumer markets for electricity, telephones, automobiles, household appliances, home furnishings, radios, and much else.

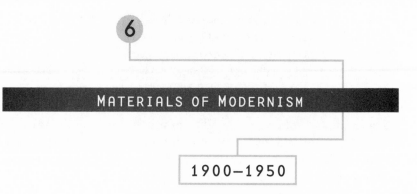

MATERIALS OF MODERNISM

1900–1950

"THE TRIUMPHANT PROGRESS of science makes profound changes in humanity inevitable, changes which are hacking an abyss between those docile slaves of past tradition and us free moderns, who are confident in the radiant splendor of our future."[1] It jars the ear today to hear this raw modernist language, a reminder that the determinist and disjunctive worldview of the modern movement—here voiced in a 1910 manifesto by a group of Italian Futurist painters—is something of a foreign memory for our own age. Yet in our age of skepticism about science and technology, it is important to understand how science and technology helped create a wide swath of modern culture. Modernism in art and architecture during the first half of the twentieth century can be best understood as a wide-ranging aesthetic movement, floated on the deeper currents of social and economic modernization driven by the science-and-systems technologies. Modernism's most persuasive promoters recognized the rhetorical force and practical effect of linking their vision to these wider socioeconomic changes. We shall see how these promoters took up a stance of "technological fundamentalism," which asserted the desirability and *necessity* of changing society and culture in the name of technology.

The modernists claimed the twentieth century. They selectively praised the builders of classical Greece and Rome and, closer at hand, the Crystal Palace in London, the Eiffel Tower in Paris, a number of massive grain silos and factories across North America, as well as railway carriages, steamships, power stations, and automobiles. Yet, modernists argued, cultural developments had failed to keep pace with the new materials and machine forms of the twentieth century. Bruno Taut, a determined

FIG. 6.1. Flat Roofs and Ribbon Windows.

Bruno Taut's modernist apartment block in Berlin's Neukölln district.
Bruno Taut, *Modern Architecture* (London: The Studio, 1929), 111.

booster of steel-and-glass buildings and a leading modern architect himself, condemned the traditional architect's "confused juggling with outward forms and styles" (fig. 6.1). Physically surrounded and artistically
suffocated by the monuments of imperial Vienna, Adolf Loos in his
polemical *Ornament and Crime* (1908) argued, "the evolution of culture
marches with the elimination of ornament from useful objects." Ornament, he wrote, was no longer integrated into our culture and was no
longer a valid expression of our culture.[2] An architecture attentive to the
spirit of the times, modernists argued, by interpreting the new aesthetic
possibilities in material form, would result in better schools, factories,
housing, and offices—indeed a better, modern society.

This chapter situates the development of aesthetic modernism in art
and architecture in the deeper currents of social, technological, and eco-

nomic modernization. It first tells a material history of modernism, stressing the impact of new factory-made materials—especially steel and glass—on discourse about what was "modern,"[3] then gives an intellectual and social history of modernism, especially in art and architecture, again focusing on modern materials but also stressing that concepts drawn from abstract art were at play. The account spotlights personal and intellectual interactions among three leading movements: the Futurists in Italy, de Stijl in the Netherlands, and the Bauhaus in Germany. Finally, the chapter evaluates the often ironic consequences of the modernist movement for building styles, household labor, and the rise of consumer society.

Materials for Modernism

The materials that modernists deemed expressive of the new era—steel, glass, and concrete—were not new. Glass was truly ancient, while concrete dated to Roman times. Steel was a mere 500 years old; for centuries skilled metalworkers in India, the Middle East, and Japan had hammered bars of high-quality steel (respectively called *wootz, Damascus,* and *tatara*) into razor-sharp daggers and fearsome swords. Beginning in the sixteenth century gunmakers in Turkey even coiled and forged Damascus steel bars into gun barrels. Europeans first made homegrown steel in the eighteenth century when the Englishman Benjamin Huntsman perfected his crucible steelmaking process (see chapter 3). Huntsman began with iron bars that he had baked in carbon until their surfaces had absorbed the small, but crucial amount of carbon that gave steel its desirable properties: it was tough, flexible on impact, and able to be hardened when quickly cooled from a high temperature. He then packed these bars—steel on the carbon-rich outside while plain iron inside—into closed clay crucibles, put them into a hot coal-fired oven that melted the metal, and finally cast ingots of steel. Huntsman's crucible steel was used in Sheffield and elsewhere for making cutlery, piano wire, scissors, scientific instruments, and umbrellas, but wider uses were limited by its high cost. Throughout the early industrial era, textile machines, factory framing, locomotives, and railroad tracks continued to be made of wrought iron or cast iron.

The first mass-produced steel in the world came from the experiments of another Englishman, Henry Bessemer. Bessemer, as we related in chapter 3, was a talented London inventor with a string of inventions already to his credit when he turned to iron and steel. A French artillery officer, worried because large cannons sometimes burst from the force within them, had challenged Bessemer to make a metal that could withstand the explosive force concentrated in a cannon barrel. Cannons could be forged from strips of wrought iron, but the process to make wrought iron required many hours of skilled labor, and these cannons' lengthwise seams sometimes split apart after repeated firings. Cannons were also cast whole from molten brass or cast iron, without seams, but brass was expensive and cast iron, while cheap, had problems of its own. Cast iron was brittle. Gunnery officers hated cast-iron cannons, since they might blow open without any warning. Bessemer set about to remedy this situation. What he needed was a way of producing a metal that was malleable like wrought iron but at low cost. In the 1850s he experimented with several ways of blowing air through liquid iron. If conditions were right, oxygen in the air combined with carbon in the iron, and the combustion that resulted made a towering white-hot blast. Again, as with Huntsman's crucibles, the hoped-for result was iron with just enough carbon to make it into steel.

Bessemer's dramatic process slashed the heavy costs to steelmakers for bushels of coal and hours of skilled work. For fuel, he simply used the carbon in the iron reacting with air (experienced ironmasters initially laughed at his 1856 paper claiming "The Manufacture of Malleable Iron and Steel without Fuel"), while his patented machinery displaced skilled labor. It turned out that Bessemer's process also could be made large— very large indeed. While his early converting vessel held around 800 pounds of metal, the converters he built in 1858 at his factory in Sheffield held more than ten times as much fluid metal, 5 tons. In time, 10- and even 25-ton converters were built in England, Germany, and the United States. (By comparison makers of crucible steel were limited to batches that could be hoisted out of the melting furnace by hand, around 120 pounds.)

The huge volume of Bessemer steel was a boon to high-quantity users like the railroads. In the 1870s, nearly 90 percent of all Bessemer steel in the United States was made into rails, and the transcontinental railroads that were built in the next two decades depended heavily on Bessemer steel. But American steelmakers, by focusing so single-mindedly on achieving large *volume* of production with the Bessemer process, failed to achieve satisfactory *quality*. In Chicago, one defective Bessemer beam from the Carnegie mills cracked neatly in two while being delivered by horsecart to the building site. Consequently, structural engineers effectively banned Bessemer steel from skyscrapers and bridges. In the 1890s the railroads themselves experienced dangerous cracks and splits in their Bessemer steel rails.

The successful structural use of steel was a result of European metallurgists' work to improve *quality* rather than maximize output. Finding
iron ores with chemical characteristics suitable for the Bessemer process
proved a difficult task in Europe. Since the original Bessemer process,
which used chemically acid converter linings, could not use the commonly available high-phosphorus iron, European steelmakers developed
the Thomas process. It used chemically basic linings in a Bessemer-like
converter to rid the steel of the phosphorus that caused it to be brittle.
Metallurgists soon found that open-hearth furnaces, too, could be lined
with the chemically basic firebricks. This trick allowed steelmakers on
both sides of the Atlantic to produce a reliable and cost-effective structural steel. Europeans had the Thomas process. Makers of structural steel
in the United States favored open-hearth furnaces. These required from
twelve to twenty-four hours to refine a batch of steel, so they were free
from the relentless production drive of Bessemer mills, where a blow
might take a scant ten minutes. From the 1890s on, architects on both
sides of the Atlantic had a workable structural steel.

Glass is by far the oldest of the "modern" materials. Early glass vases,
statues, cups, coins, and jewelry from Egypt and Syria are at least 5,000
years old. Phoenicians and later Romans brought glassmaking to the dominions of their empires, and from the Renaissance onward Venice was
renowned as a center for fine glassmaking. By the eighteenth century, Bohemia and Germany had become leading producers of window glass.
Glassmaking involved no complicated chemistry and no violent Bessemer blasts but only the careful melting of quartz sand with any desired
coloring substance, such as lead salts. The manufacture of both steel and
glass required extremely high temperatures; it is no coincidence that
Bessemer had worked on glass-melting furnaces before his steelmaking
experiments. But melting was only the start. Workers making glass needed
considerable strength and special skills for pressing or blowing the thick
mass of molten material into useful or decorative shapes. Initially, most
glass for windows was made by blowing a globe of glass then allowing it
to collapse flat on itself. In 1688 French glassmakers began casting and polishing large flat sheets, up to 84 by 50 inches, of "plate" glass. By the mid-
nineteenth century, the window-glass industry comprised four specialized trades. "Blowers" took iron pipes prepared by "gatherers" and created
cylinders of glass (often using a brass or iron form); then "cutters" and
"flatteners" split open the newly blown cylinders into flat sheets of window glass.[4]

Glass through most of the nineteenth century was in several ways similar to steel before Bessemer. It was an enormously useful material whose manufacture required much fuel and many hours of skilled labor and whose application was limited by its high cost. Beginning in the 1890s, however, a series of mechanical inventions transformed glassmaking into a highly mechanized, mass production industry. Belgian, English, French, and American glassmakers all played a role in this achievement. First coal-fired pots were replaced by gas-fired continuous-tank melting furnaces; then window-glass making was mechanized along "batch" lines; and finally plate-glass making was made into a wholly continuous process by Henry Ford's automobile engineers. (Broadly similar changes occurred in making glass containers and light bulbs.) By the 1920s the modernists, even more than they knew, had found in glass a material expressive of their fascination with machine production and continuous flow.

Window glass was mechanized early on. By 1880 Belgian window-glass makers were using so-called tank furnaces fired by artificial gas, and the first American installations of tank furnaces using natural gas soon followed. By 1895 fully 60 percent of American window glass was melted in tank furnaces. Around 1900 a continuous-tank factory in Pennsylvania, with three furnaces, required just seven workers to unload the raw materials, feed them into the hoppers, and stir and prepare the melted glass. Each of the three tank furnaces produced enough glass for ten skilled glass blowers and their numerous helpers, who gathered, blew, flattened, and cut the glass into window sheets. Melting window glass by the batch in pots persisted well into the twentieth century, however. While Pennsylvania accounted for two-fifths of America's total glass production, the discovery of cheap natural gas in Ohio, Indiana, Illinois, and Missouri led to the profusion in these states of small pot-melting furnaces.

The first window-glass blowing machine did not reduce the total number of glassmakers, but did dramatically alter the skills required. John Lubbers, a window-glass flattener in the employ of the American Window Glass Company—an 1899 merger of firms that accounted for 70 percent of the nation's window-glass capacity and nearly all its tank furnaces—experimented for seven years until 1903 when his batch process was able to compete successfully with hand-blown window glass. (French, Belgian, and other American inventors had in earlier decades experimented with rival schemes for mechanical blowing.) Lubbers' machine produced huge cylinders of glass, up to twice the diameter and five times as long as the hand-blown cylinders.

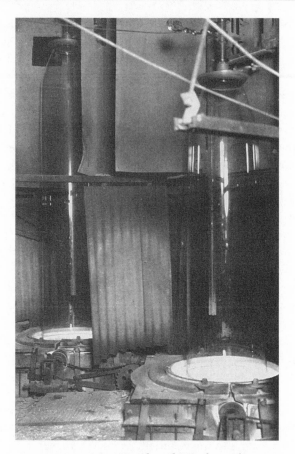

FIG. 6.2. Mass-Produced Window Glass.

Machine blowing cylinders of glass at Pilkington Brothers' St. Helens works in England. Cylinders were blown up to 40 feet in height before they were detached, cut into lengths, split open, and flattened into huge sheets of window glass. Raymond McGrath and A. C. Frost, *Glass in Architecture and Decoration* (London: Architectural Press, 1937), 61.

With its clanking and puffing, Lubbers' mechanical monster must have been something to watch (fig. 6.2). It performed a classic batch process. A hollow cast-iron cylinder, the "bait" to "catch" the molten glass, was lowered into a waiting vat of hot glass. After a moment, while glass solidified on the inside of the bait, two motors were started, one to slowly raise the bait with its cylinder of glass attached, and the other blowing air into the growing glass cylinder. When the glass cylinder reached its desired diameter, the blowing motor slowed down while the raising motor kept up its work, pulling out a tall cylinder of glass eventually 35 to 40 feet high. At the top of the cycle, the raising motor was briefly speeded up (to thin out the glass wall), then the huge cylinder of glass was cracked off and swung by crane onto a receiving area. There it was sliced into smaller cylinders, which were split and flattened into window sheets as before. This mechanical contraption required a dozen or more tenders, in addition to the cutters and flatteners. But the new mechanical technology obviously displaced the skilled glass *blowers* and, in the United States at least, wiped out their craft union. By 1905 American window-glass makers had installed 124 of these cylinder machines, and by 1915 a total of 284 cylinder machines (of several different types) accounted for two-thirds of U.S. window-glass production.

A continuous process more streamlined than Lubbers' batch machine helped propel the Libbey-Owens Sheet Glass Company to fame. Development of this process required nearly two decades, beginning with the early experiments of Irving W. Colburn, also working in Pennsylvania, and culminating in its commercial success in 1917, when Libbey built a six-unit factory in Charleston, West Virginia. In the patented Colburn machine, the bait was an iron rod, and it was dipped lengthwise into a shallow pan of hot glass. Once the glass had adhered to the rod, it was pulled by motor up from the vat and over a set of rollers, forming a flat sheet of glass directly. Water-cooled side rollers an inch or two above the molten glass kept the sheet at the proper width. When the sheet extended onto the nearby flattening table, a set of mechanical grip bars took over pulling the ever-lengthening sheet from the vat. From the flattening table the sheet passed directly into an annealing oven, through which it moved on 200 asbestos-covered rollers and was then cut, on a moveable cutting table, into suitable-size sheets. Several of Libbey's competitors imported from Belgium the Fourcault sheet-drawing machine, patented in 1902 but not commercialized until after the war. It drew glass straight up into a sheet, rather than horizontally over rollers, and reportedly saved even more labor; only cutters were still needed. Across the 1920s these semicontinuous cylinder and sheet machines together produced an ever-larger share of U.S. window glass. (Hand-blown window glass dropped from 34 to 2 percent of total production between 1919 and 1926.)

Plate glass was thicker than window glass and could safely be made into much larger sheets. Thick plate glass windows were of crucial importance in the tall office buildings going up in U.S. cities, because ample natural illumination was a primary goal of architects laying out office space in the era of hot, expensive incandescent lights. (American-made plate glass was used mostly for windows; high-quality European plate glass was required for mirrors.) Before 1900 American plate-glass making was wholly dependent on English techniques and machinery. Much labor and skill were required to cast the large plates of glass, weighing up to 1,400 pounds, while semiautomatic machines conducted the subsequent rounds of grinding and polishing to obtain a smooth surface. Industry lore has it that all polishing and grinding machinery was imported from England until an American manufacturer, engaging in a bit of industrial espionage, visited the Pilkington company's plant sometime before 1900 and, upon his return to the States, replicated that leading plant's machinery.

Plate-glass making between 1900 and 1920 underwent a series of evolutionary changes. Engineers made incremental improvements in processing the varied grades of sand used for grinding and on the rouge used for polishing, while electric motors were increasingly used to drive the huge (35-foot-diameter) round polishing tables. Factory designers sought to speed up the time-intensive grinding and polishing stages. Continuous-flow annealing ovens dramatically cut the time needed to ease the strains in the newly cast glass. A plate of glass might emerge in as little as three hours from a typical 300-foot-long sequence of five ovens. (By comparison batch-process annealing kilns required forty-eight hours just to cool down so that workmen could climb into the kiln and haul out the sheet by rope.) As a result of these mechanical developments, a large plate of glass that might have taken ten days to complete in 1890 could be finished in as little as thirty-six hours in 1923.

Developments after 1920 transformed plate-glass making into a wholly rather than partially continuous-production industry. In that year, while building his massively integrated River Rouge complex, Henry Ford assigned a team of engineers to work on glassmaking. It seems unclear whether Ford desired simply to raise the volume of production or whether he wanted them to focus on laminated safety glass, a technical challenge that established glassmakers had been unwilling to attempt. In any event, Ford's engineers hit on a continuous production process that became widely adopted by the entire plate-glass industry. In 1927 the Pittsburgh Plate Glass Company, a major manufacturer, owned five plate-glass factories and produced 50 percent of the nation's total, but automobile manufacturers owned eight plate-glass factories (including three of the four continuous-production plants) and their output accounted for 35 percent of the total. By 1929 fully half of American plate glass was manufactured by the continuous process.

In the Ford-style continuous-production scheme, glass flowed from a continuous melting tank (which had become the standard) in a stream onto an inclined plane that formed the glass into a flat sheet. A roller pressed the ever-moving sheet to proper thickness, then the hot glass sheet exited onto a moving table and into a continuous annealing oven. At the far end of the oven, the sheet was cut into desired sizes. Previously, fifty or more workers were needed for the casting stage; now just ten workers tended the entire process, from melting through annealing. Ford engineers also introduced assembly-line principles to the grinding and polishing stages, using a long, narrow continuous conveyer that successively ground and polished each sheet of glass. While the largest plates were still made by the traditional batch regime, the Ford-style continuous plate process worked best on smaller sizes, and soon these smaller plates were widely used for windows.

The appearance of a pane of glass in 1890 compared with one in 1925 was not so very different. And unit prices for window and plate glass were actually 30–50 percent higher in the mid-1920s than in the 1890s, owing to higher demand. What had changed most was the *amount* of window glass. Window-glass production in the U.S. grew threefold during these years, to 567 million square feet, while plate-glass production grew an astounding fifteenfold, to 148 thousand square feet. Just as with steel, the ability to produce a substance in large volumes led to its being "discovered" as a modern material.

Manifestos of Modernity

Modernism in architecture depended on the modern materials of mass-produced steel and glass. As early as 1929 the German architect Bruno Taut defined modernism as "flat roofs, huge sheets of glass, 'en tout cas' horizontal ribbon-rows of windows with pillars, which strike the eye as little as may be, by reason of black glass or dull paint, more sheets of concrete than are required for practical purposes, etc."[5] This modern style—as a distinctive style itself—was evident first at the Weissenhof housing exposition at Stuttgart in 1927 and was canonized officially by New York's Museum of Modern Art in 1932. It is recognizable instantly in the glass-box "corporate style" skyscrapers that went up in the 1950s and 1960s, but in the decades since it has become dull and hackneyed from mindless repetition. Critical to the development of the modern architectural style were the interactions among three groups: the Futurists in Italy, who gave modernism an enthusiastic technology-centered worldview; the members of de Stijl in the Netherlands, who articulated an aesthetic for modern materials; and the synthesis of theory and practice in the Bauhaus in Germany.

The Italian Futurists, a "close-knit fighting unit" led by Filippo Marinetti, found a modern aesthetic in the new world of automobiles, factories, and cities. Marinetti's wild enthusiasm for the machine found expression in a series of culture-defining "manifestos." In the years between 1910 and 1916, the Futurists' poets, painters, sculptors, and architects simply blasted a hole through the traditional views of art and architecture. Two of the group's most creative figures—Umberto Boccioni and Antonio Sant'Elia—were killed in World War I, but across the 1920s Marinetti brought their work to the attention of the nascent modern movement. The legacies of Futurism include Marinetti's insistence that modern materials were the foundation of modern culture: "I leave you with an explosive gift, this image that best completes our thought: 'Nothing is more beautiful than the steel of a house in construction.'"[6]

Marinetti returned from study in Paris to live and write in Milan, at the center of a region undergoing its own second industrial revolution. Textile production around Milan tripled between 1900 and 1912, iron and steel production likewise tripled, to 1,000,000 metric tons, while a world-class automobile industry sprang up with the establishment of Pirelli and Alfa Romeo and the great FIAT complex in nearby Turin. Automobiles are at the center of Marinetti's founding "Manifesto of Futurism," in which he launched the modernist vision of technology as a revolutionary cultural force. The 1909 manifesto begins with a set piece at his family's house in old Milan and a late-night discussion among friends. They had argued, he says, to the furthest limits of logic and covered sheets of paper with scrawls. In the middle of the night they felt alone, like proud beacons or forward sentries, curiously at one with "the stokers feeding the hellish fires of great ships . . . the red-hot bellies of locomotives." From the past, they heard the old canal, believed to be a work of Leonardo, "muttering its feeble prayers and the creaking bones of palaces dying above their damp green beards." Then, suddenly, beneath the windows the silence was broken by "the famished roar of automobiles."

Marinetti and his friends piled into three waiting automobiles and raced through the early-morning streets. Swerving around two bicyclists,

Marinetti flipped the car over and it landed in a ditch. Marinetti, dripping with "good factory muck," climbed out and proclaimed a manifesto to deliver Italy from "its foul gangrene of professors, archeologists, guides and antiquarians." Marinetti's images, arresting and enduring, forthrightly declared a modern aesthetic in the world of modern technology. "We affirm that the world's splendour has been enriched by a new beauty: the beauty of speed. A racing car whose hood is adorned with great pipes, like serpents of explosive breath—a roaring car that seems to ride on grapeshot—is more beautiful than the *Victory of Samothrace*."

> We will sing of great crowds excited by work, by pleasure, and by revolt; we will sing of the multicolored, polyphonic tides of revolution in the modern capitals; we will sing of the vibrant nightly fervour of arsenals and shipyards blazing with violent electric moons; greedy railway stations that devour smoke-plumed serpents; factories hung on clouds by the crooked lines of their smoke; bridges that leap the rivers like giant gymnasts, flashing in the sun with a glitter of knives; adventurous steamers that sniff the horizon; deep-chested locomotives pawing the tracks like enormous steel horses bridled by tubing; and the sleek flight of planes whose propellers chatter in the wind like banners and seem to cheer like an enthusiastic crowd.

The sculptor Boccioni phrased the cultural shift this way: "The era of the great mechanised individuals has begun, and all the rest is Paleontology."[7]

The first result of Marinetti's call for a techno-cultural revolution was a flurry of free verse with such titles as "L'Elettricità," "A un Aviatore," and "Il Canto della Città di Mannheim." A more significant result came with the paintings of Giacomo Balla and the sculpture of Umberto Boccioni. Their challenge was to deliver on the 1910 manifesto of Futurist painting, which had argued that living art must draw its life from the modern world: "Our forebears drew their artistic inspiration from a religious atmosphere which fed their souls; in the same way we must breathe in the tangible miracles of contemporary life—the iron network of speedy communications which envelops the earth, the transatlantic liners, the dreadnoughts, those marvelous flights which furrow our skies, the profound courage of our submarine navigators . . . the frenetic life of our great cities . . . ."[8]

Futurism was not about painting pictures of battleships or airplanes. Balla struggled to express in painting such abstract concepts as dynamism and elasticity, while Boccioni argued that sculptors must "destroy the pretended nobility . . . of bronze and marble" and instead use appropriate combinations of glass, cardboard, cement, iron, electric light, and other modern materials. No classical statues or nude models here. Futurist sculpture, such as Boccioni's *Unique Form of Continuity in Space* (1913), blended stylized human forms with the machine forms of the modern world. Modern objects, with their "marvelous mathematical and geometrical elements," Boccioni wrote, "will be embedded in the muscular lines of a body. We will see, for example, the wheel of a motor projecting from the armpit of a machinist, or the line of a table cutting through the head of a man who is reading, his book in turn subdividing his stomach with the spread fan of its sharp-edged pages."[9]

With their unbounded enthusiasm for modern technology, the Futurists understandably took a dim view of the traditional historical styles then popular. Indeed, in their opinion no proper architecture had existed since the eighteenth century, only the "senseless mixture of the different stylistic elements used to mask the skeletons of modern houses." Their architectural manifesto of 1914 hailed the "new beauty of cement and iron" and called on architects to respond constructively to "the multiplying of machinery, the constantly growing needs imposed by the speed of communications, the concentration of population, hygiene, and a hundred other phenomena of modern life." The Futurist concept of a house would embrace "all the resources of technology and science, generously satisfying all the demands of our habits and our spirit." In short, this would be "an architecture whose sole justification lies in the unique conditions of modern life and its aesthetic correspondence to our sensibilities."

Sant'Elia argued that modern architecture must break with tradition and make a fresh start: "Modern building materials and scientific concepts are absolutely incompatible with the discipline of historical styles, and are the main reason for the grotesque appearance of 'fashionable' buildings where the architect has tried to use the lightness and superb grace of the iron beam, the fragility of reinforced concrete, to render the heavy curve of the arch and the weight of marble." There was a new ideal of beauty, still emerging yet accessible to the masses. "We feel that we no longer belong to cathedrals, palaces and podiums. We are the men of the great hotels, the railway stations, the wide streets, colossal harbors, covered markets, luminous arcades, straight roads and beneficial demolitions."[10]

Sant'Elia summed up his ideas about modern materials and urban form in his great city-planning project, the famed Città Nuova. A futuristic vision of a hyperindustrialized Milan, Città Nuova was first exhibited in 1914 and continued to inspire modernist architects throughout the 1920s. Sant'Elia poured out a flood of modernistic images:

> We must invent and rebuild *ex novo* our Modern city like an immense and tumultuous shipyard, active, mobile and everywhere dynamic, and the modern building like a gigantic machine. Lifts must no longer hide away like solitary worms in the stairwells, but the stairs—now useless—must be abolished, and the lifts must swarm up the façades like serpents of glass and iron. The house of cement, iron, and glass, without carved or painted ornament, rich only in the inherent beauty of its lines and modelling, extraordinarily brutish in its mechanical simplicity, as big as need dictates, and not merely as zoning rules permit, must rise from the brink of a tumultuous abyss; the street which, itself, will no longer lie like a doormat at the level of the thresholds, but plunge storeys deep into the earth, gathering up the traffic of the metropolis connected for necessary transfers to metal cat-walks and high-speed conveyor belts.

We must create the new architecture, Sant'Elia proclaimed, "with strokes of genius, equipped only with a scientific and technological culture." Sant'Elia (possibly with some help from Marinetti, ever ready with verbal fireworks) finished up with his most widely quoted conclusion: "Things will endure less than us. Every generation must build its own city."[11]

Marinetti's provocative avant-garde stance, frank celebration of violence, and crypto-revolutionary polemics landed the Futurists squarely in the middle of postwar fascism. Violence was in the air, and Italy's liberal democracy was in tatters. More than a dozen groups, ranging from respectable university students to gun-toting street gangs, used *fascio* in their names. As "the new man," the presumed leader of this motley crew, Marinetti for a time rivaled even Mussolini, known chiefly as the editor of the Socialist Party's newspaper before the war, and at the time a stridently *anti*-socialist editor and journalist. For Marinetti, perhaps the high point (if one can call it that) came in April 1919, when he took a mob through the streets of Milan and wrecked the headquarters of the Socialist Party's newspaper, an event later known in the regime's legends as "the first victory of Fascism."

Marinetti and Mussolini saw eye-to-eye on war, violence, women, and airplanes, but not the established social and moral order. In 1920, Mussolini lurched rightward and achieved national political prominence through a new-found alliance with Italy's business and religious elites; two years later, following the infamous "march on Rome," he was sworn in as the Fascist prime minister of Italy.[12] While Mussolini solidified his grip on national power, Marinetti energetically took up the cause of international Futurism. At home, he signaled his political irreverence (and irrelevance) by calling for a revolution in Italian cooking. His *La cucina futurista* features such delicacies as "Car Crash"—the middle course of a "Dynamic Dinner"—consisting of "a hemisphere of pressed anchovies joined to a hemisphere of date puree, the whole wrapped up in a large, very thin slice of ham marinated in Marsala." The "Aeropoetic Futurist Dinner" is set in the cockpit of a Ford Trimotor flying at 3,000 meters, while the "Extremist Banquet" lasts two days. Pasta was banned in his Futurist cuisine.[13]

During the 1920s the Futurists slid off the stage of Italian politics but became a serious international artistic movement. The best-known Futurist work of architecture was an automobile factory complex built during 1914–26 for the Italian firm FIAT outside Turin. With its "daylight" (high and expansive) windows, its long "planar" (unbroken and unornamented) walls, and especially its dramatic roof—the site of a high-banked oval track where finished cars could be test-driven—it became a classic modernist icon and a mandatory waypoint on modernist pilgrimages. Marinetti worked tirelessly to bring Futurist concepts and images, especially the several manifestos and Sant'Elia's *Città Nuova*, to receptive audiences in the north of Europe. Before the war he proclaimed Futurist manifestos in Paris, London, Rotterdam, and Berlin, while a Futurist exhibition held in Paris subsequently traveled to no fewer than eleven European cities. One receptive audience, as early as 1912, was a group of German Expressionists in Berlin, many of whom were recruited to form the Bauhaus school of design in 1919, as we will see below. A second group that brought Futurist ideas into the larger avant-garde movement was the Dutch movement de Stijl (The Style), which also interacted with the Bauhaus, sharing students and staff.

De Stijl was a loosely interacting group of architects and painters; the name was also the title of their influential art magazine, published from 1917 until 1931. Sensing in de Stijl a kindred spirit, Marinetti in 1917 sent the Futurist architectural manifesto and a selection of Sant'Elia's drawings to Theo van Doesburg, the group's organizer and central figure. In response *De Stijl* published a warm appreciation ("the perfect management of this building taken as a whole, carried out in modern materials ... gives this work a freshness, a tautness and definiteness of expression") that secured Sant'Elia's international reputation. *De Stijl*'s far-reaching circulation made one particular drawing, through its multiple reproductions across Europe, the best known of all Sant'Elia's work.[14]

Chief among the de Stijl theorists was Piet Mondrian, a pioneer practitioner of abstract, nonfigurative painting. "The life of today's cultured person turns more and more away from nature; it is an increasingly abstract life," he announced. For de Stijl, the terms *nature* and *abstract* were on opposite sides of the "great divide" between tradition and modernity. Mondrian maintained that artists should recognize that there was an ultimate reality hiding behind everyday appearance and that artists should strive to see through the accidental qualities of surface appearance. He looked to the city as inspiration for the emerging modern style: "The genuinely Modern artist sees the metropolis as Abstract living converted into form; it is nearer to him than nature, and is more likely to stir in him the sense of beauty . . . that is why the metropolis is the place where the coming mathematical artistic temperament is being developed, the place where the new style will emerge." One like-minded modernist put the same point more simply: "After electricity, I lost interest in nature."[15]

Van Doesburg took Mondrian's notions about modern life one step further. In a famous lecture in 1922 called "The Will to Style: The New Form Expression of Life, Art and Technology," van Doesburg told audiences in Jena, Weimar, and Berlin: "Perhaps never before has the struggle between nature and spirit been expressed so clearly as in our time." Machinery, for van Doesburg, was among the progressive forces that promised to lift humans above the primitive state of nature and to foster cultural and spiritual development. The task of the artist was to derive a style—or universal collective manner of expression—that took into account the artistic consequences of modern science and technology:

> Concerning the cultural will to style, the machine comes to the fore. The machine represents the very essence of mental discipline. The attitude towards life and art which is called materialism regarded handiwork as the direct expression of the soul. The new concept of an art of the mind not only postulated the machine as a thing of beauty but also acknowledged immediately its endless opportunities for expression in art. A style which no longer aims to create individual paintings, ornaments or private houses but, rather, aims to study through team-work entire quarters of a town, skyscrapers and airports—as the economic situation prescribes—cannot be concerned with handicraft. This can be achieved only with the aid of the machine, because handicraft represents a distinctly individual

attitude which contemporary developments have surpassed. Handicraft debased *man* to the status of a machine; the correct use of the machine (to build up a culture) is the only path leading towards the opposite, social liberation.

Iron bridges, locomotives, automobiles, telescopes, airport hangars, funicular railways, and skyscrapers were among the sites van Doesburg identified where the new style was emerging.[16]

A more striking association of technology with a desired cultural change is difficult to imagine. This is the crucial shift: whereas the Futurists sang enthusiastic hymns to modern technology and the dynamic city, for de Stijl modern technology and the city were desirable because they were a *means* by which to achieve social liberation and "to build up a culture." This involved careful decisions ("the correct use of the machine"), not the Futurists' naïve embrace of automobiles or airplanes.

The buildings and theoretical writings of H. P. Berlage heavily influenced members of de Stijl. Architectural "style" in the modern age was for Berlage an elusive quality that an architect achieved by practicing "truth to materials" ("decoration and ornament are quite inessential") and creating spaces with proper geometrical proportions. Berlage's own Amsterdam Stock Exchange became a famous modernist building, but of equal importance was his early grasp and interpretation of Frank Lloyd Wright. Wright was virtually the only American architect noticed by the European modernists. Both Berlage's Stock Exchange (1902) and Wright's Larkin office building (1905) used a combination of brick and advanced structural techniques to create large open-air halls surrounded by galleries, in effect creating open space in the very middle of the buildings. Europeans did not learn much from Berlage about Wright's interest in the vernacular and nature worship. Instead, Berlage emphasized Wright's views on the technological inspiration of modern culture ("The machine is the normal tool of our civilization, give it work that it can do well; nothing is of greater importance"). In effect, Berlage selectively quoted Wright to support the technological framing of modernism sought by de Stijl: "The old structural forms, which up to the present time have been called architecture, are decayed. Their life went from them long ago and new conditions industrially, steel and concrete, and terra-cotta in particular, are prophesying a more plastic art." Berlage probably showed Le Corbusier, the prolific writer and influential architectural theorist, a Dutch-designed "modern villa at Bremen" that was Corbusier's first view of a modernist building.[17]

J. J. P. Oud was the leading practicing architect associated with de Stijl. Oud knew Wright's work and appreciated "the clarity of a higher reality" achieved by Sant'Elia, but his own practical experiences grounded his theory. His early work—including several houses, villas, shops, a block of workers' apartments (fig. 6.3), and a modernist vacation home—received such acclaim that in 1918, at the age of twenty-eight, he was named city architect for Rotterdam. In 1921 he wrote *On Modern Architecture and Its Architectonic Possibilities.* Oud clearly sensed and helped articulate the architectural possibilities of modern technology, but at the same time he avoided the vortex of technological utopianism: "I bow the knee to the wonders of technology, but I do not believe that a liner can be compared to the Parthenon [contra Futurists]. I long for a house that will satisfy my every demand for comfort, but a house is not for me a living-machine [contra Corbusier]." Disappointingly for him, "the art of building . . . acts as a drag on the necessary progress of life," Oud wrote; "the products of technological progress do not find immediate application in building, but are first scrutinized by the standards of the ruling aesthetic, and if, as usual, found to be in opposition to them, will have difficulty in maintaining themselves against the venerable weight of the architectural profession." To help architects embrace the new building materials, he articulated an aesthetic for plate glass, iron and steel, reinforced concrete, and machine-produced components.

> When iron came in, great hopes were entertained of a new architecture, but it fell aesthetically-speaking into the background through improper application. Because of its visible solidity—unlike plate-glass which is only solid to the touch—we have supposed its destination to be the creation of masses and planes, instead of reflecting that the characteristic feature of iron construction is that it offers the maximum of structural strength with the minimum of material. . . . Its architectural value therefore lies in the creation of voids, not solids, in contrast to mass-walling, not continuing it.

Plate glass at the time was usually employed in small panes joined by glazing bars, so that the window "optically continues the solidity of the wall over the openings as well," but Oud argued that glass should instead be used in the largest possible sheet with the smallest possible glazing

FIG. 6.3.   Dutch Modernism by J. J. P. Oud.

"Graceful development in the new tendency . . . modern architecture has definitely won through in Holland," was Bruno Taut's verdict after seeing these workmen's houses at Hook van Holland. Bruno Taut, *Modern Architecture* (London: The Studio, 1929), 91, 123.

bars. Reinforced concrete's tensile strength and smooth surface offered the possibility of "extensive horizontal spans and cantilevers" and finished surfaces of "a strict clean line" and "pure homogenous plane." In his conclusion, Oud called for an architecture "rationally based on the circumstances of life today." He not only called for a method of creation but also catalogued the proper qualities of modern materials. The new architecture's "ordained task will be, in perfect devotion to an almost impersonal method of technical creation, to shape organisms of clear form and proper proportions. In place of the natural attractions of uncultivated materials . . . it would unfold the stimulating qualities of sophisticated materials, the limpidity of glass, the shine and roundness of finishes, lustrous and shining colors, the glitter of steel, and so forth."[18]

The durable contribution of de Stijl, then, was not merely to assert, as the Futurists had done, that modern materials had artistic consequences, but to identify specific consequences and embed these in an overarching aesthetic theory. Architects now could associate factory-made building materials like steel, glass, and reinforced concrete with specific architectural forms, such as open spaces, extensive spans, and clean horizontal planes. Moreover, with the suggestion that architects devote themselves to an "impersonal method of technical creation," Oud took a fateful step by transforming the Futurists' flexible notion that every generation would have its own architecture into a fixed method of architectural design.

"Picasso, Jacobi, Chaplin, Eiffel, Freud, Stravinsky, Edison etc. all really belong to the Bauhaus," wrote one of the school's students in the 1920s. "Bauhaus is a progressive intellectual direction, an attitude of mind that could well be termed a religion."[19] These heady sentiments found practical expression in the Bauhaus, an advanced school for art and architecture active in Germany from 1919 to 1933. The Bauhaus was founded and grew during the country's fitful struggles to sustain democracy and the disastrous hyperinflation of 1921–23. Originally located in the capital city of Weimar, the Bauhaus relocated first to Dessau and finally to Berlin. After its break-up in 1933 its leading figures—Walter Gropius, Mies van der Rohe, Lazlo Moholy-Nagy—emigrated to the United States and took up distinguished teaching careers in Boston and Chicago. Gropius wrote of the Bauhaus, "Its responsibility is to educate men and women to understand the world in which they live, and to invent and create forms symbolizing that world."[20]

Such a visionary statement might be regarded as yet another wild-eyed manifesto, but by 1919 Gropius was among the progressive German architects, artists, and designers who had substantial experience with industrial design. The Bauhaus began as the fusion of two existing schools, an academy of fine arts and a Kunstgewerbe school, which brought together students of the fine and applied arts. While the academy's traditions stretched back into the past, the Kunstgewerbe school had been organized just fifteen years earlier in a wide-ranging campaign to raise the aesthetic awareness of German industry. Other initiatives of that time included the founding by Hermann Muthesius and Peter Behrens of the Deutscher Werkbund, an association of architects, designers, and industrialists that campaigned to convert German industry to the gospel of industrial design, as well as the employment of Behrens by the giant firm AEG (Allgemeine Electricitätsgesellschaft, a successor to German Edison). Behrens was in effect AEG's in-house style maven. Between 1907 and 1914, many of the major figures in architecture of the 1920s—including Mies van der Rohe, Bruno Taut, Le Corbusier, and Gropius himself—worked in Behrens' atelier. Indeed, many influential modernist buildings can be traced to this Werkbund-Behrens connection, including Behrens' own

FIG. 6.4. The First "Modern" Factory.

The Faguswerke factory (1911–13), designed by Adolf Meyer and Walter Gropius, made humble shoe lasts for shoemakers, but photographs of the building made history. Modernists praised the glass-enclosed corner stairwell as an open and unbounded vision of space. Bruno Taut, *Modern Architecture* (London: The Studio, 1929), 57.

factory buildings for AEG (1908–12), Gropius's Faguswerke (1911–13, fig. 6.4) and Werkbund Pavilion (1914), as well as Taut's exhibition pavilions for the steel and glass industries (1913–14), which dramatically displayed these modern materials.

The Bauhaus was unusually well positioned to synthesize and transform advanced concepts circulating in the 1920s about materials, space, and design. Many of its early staff members were drawn from Der Sturm, the Expressionist movement in Berlin that had provided an early audience for the Futurists. These avant-garde painters were especially receptive to the abstract machine-inspired Constructivist art emerging in the early Soviet Union, which the Russian El Lissitzky brought to the attention of Western Europeans through his association with de Stijl in the mid-1920s. Other leading members of de Stijl—including van Doesburg, Mondrian, and Oud—either lectured at the Bauhaus or published theoretical tracts in its influential book series. Van Doesburg's lectures at the Bauhaus (including his "Will to Style" quoted above) helped turn the

Bauhaus away from its early focus on Expressionism, oriental mysticism, and vegetarianism and toward an engagement with real-world problems and advanced artistic concepts.

Elementarism was an abstract concept drafted to the cause. Before the Bauhaus turned it into a distinctive architecture concept, *elementarism* had several diverse meanings. As early as 1915 the Russian abstract painter Kasimir Malevich pointed to the fundamental elements, or simple geometrical forms, that were basic units of his compositions; he later expanded his views in the Bauhaus book *Non-Objective World*. In 1917, another exemplar, Gerrit Rietveld, even before he associated with de Stijl, made the first of his famous chairs. They were not principally intended as places to sit comfortably. Rather, Rietveld separated and analyzed the functions of a chair—sitting, enclosing, supporting—and created a design in which each of the elements of his chairs, made of standard dimensional lumber, was visually separated from the other elements and held in a precise location in space. In 1924, van Doesburg pointed out that "the new architecture is *elementary*," in that it develops from the elements of construction understood in the most comprehensive sense, including function, mass, plane, time, space, light, color, and material. In 1925, self-consciously using the ideas of elementarism, Rietveld built the Schroeder house in Utrecht while van Doesburg coauthored a prize-winning plan for the reconstruction of the central district in Berlin.[21] As interpreted by the Bauhaus theorist Moholy-Nagy, himself an abstract painter, such "elements" were conceived of as the fundamental units of structure and space creation.

By the mid-1920s students at the Bauhaus were studying with some of the most original artists and architects in Europe. Students began with the six-month introductory course, the Vorkurs, then went on to a three-year formal apprenticeship in a particular craft (e.g., metalwork, pottery, weaving, woodwork) that resulted in a Journeyman's Diploma. Finally, students could elect a variable period of instruction in architecture or research leading to a Master's Diploma. By 1923 Gropius saw the school as preparing students for modern industry. "The Bauhaus believes the machine to be our modern medium of design and seeks to come to terms with it," he wrote. Training in a craft complemented the desirable "feeling for workmanship" always held by artists and prepared students for designing in mass-production industries.[22] The school's reorientation from mysticism to industry was expressed in its 1923 exposition, Art and Technology—A New Unity.

Von Material zu Architektur, by Lazlo Moholy-Nagy, a key theoretical reflection, was one of the key Bauhaus texts. In one sense, Moholy's title (*From Material to Architecture*) suggests the passage of students from the study of materials, properly theorized, to the study of architecture. This educational plan was given sharp expression in a circular diagram, attributed to Gropius, that showed students passing through the outer layer of the Vorkurs, then specializing in the study of a specific material, and at last taking up the study of architecture, which was at the core. Viewing construction materials as the medium of the modern world figured largely in Moholy-Nagy's career. Moholy-Nagy had come to Berlin in 1921, immersed himself in the avant-garde world of Der Sturm, de Stijl, and Russian abstraction. He was appointed to the Bauhaus in 1923 to oversee the metalworking shop; later in that year he also (with Josef Albers) took on the Vorkurs. One of his innovations was to transform the study of materials from an inspection of their inner nature to an objective physical assessment of their various properties. To test a material's strength or flexibility or workability or transparency, he devised a set of "tactile machines" that were used at the Bauhaus.

Ironies of Modernism

"We aim to create a clear, organic architecture whose inner logic will be radiant and naked," wrote Walter Gropius in *Idee und Aufbau* (1923). "We want an architecture adapted to our world of machines, radios and fast cars . . . with the increasing strength and solidity of the new materials— steel, concrete, glass—and with the new audacity of engineering, the ponderousness of the old methods of building is giving way to a new lightness and airiness."[23] At the time, Gropius was engaged in creating some modernist icons of his own. He had just completed his striking modernist entry for the Chicago Tribune Tower competition (1922). He would soon turn his architectural energies to designing the new Bauhaus buildings at Dessau, discussed below. The rush to proclaim a distinctive modern Bauhaus style provoked one critic to jest: "Tubular steel chairs: Bauhaus style. Lamp with nickel body and white glass shade: Bauhaus style. Wallpaper covered in cubes: Bauhaus style. Wall without pictures: Bauhaus style. Wall with pictures, no idea what it means: Bauhaus style. Printing with sans serif letters and bold rules: Bauhaus style. doing without capitals: bauhaus style."[24]

The efforts during the 1920s to proclaim a modern style in Germany occurred under unusual difficulties. The country's severe political turmoil, economic crisis, and street violence made the ferment in postwar Italy look calm in comparison. During 1923 the German economy collapsed under the strain of reparations payments imposed on it at the end of World War I. In January 1919 it took eight war-weakened German marks to purchase one U.S. dollar. Four years later it took 7,000 marks, and by December 1923 it would take the stupendous sum of 4.2 trillion marks to purchase one U.S. dollar. In February of that disastrous year, Gropius asked the government for 10 million marks (then equivalent to about $1,000) to help fund the art-and-technology exposition. By the end of that year's hyperinflation the sum of 10 million marks was worth less than one one-hundred-thousandth of a U.S. penny. The Bauhaus took up the surreal task of designing million-mark banknotes so housewives might buy bread without a wagon to transport the necessary bills. With the stabilization of the German currency in 1924 (the war debts were rescheduled) building projects began once again.

The lack of adequate housing especially plagued Germany's industrial cities, which had grown swiftly during the electricity and chemical booms of the second industrial revolution (see chapter 5). City governments in several parts of Germany began schemes to construct affordable workers' housing. Three cities—Dessau, Berlin, and Frankfurt—gave the German modernists their first opportunities for really large-scale building, and they have figured prominently in histories of modernism ever since. The mayor of industrial Dessau attracted the Bauhaus to his city, offering to fund the salaries of the staff and the construction of new school buildings, in exchange for assistance with his city's housing. For years Gropius had dreamed of rationalizing and industrializing the building process. Not only did he advocate the standardization of component parts, the use of capital-intensive special machinery, and the division of labor; he was also a close student of the labor and organizational methods of Henry Ford and efficiency engineer Frederick W. Taylor. And as we will see, such modernistic "rationalization" came with a similar vengeance to the household and the housewife who worked there.

At Dessau Gropius in effect offered the Bauhaus as an experimental laboratory for the housing industry. His commission from the city was to design and build 316 two-story houses, together with a four-story building for the local cooperative. Like Ford, Gropius specially planned the smooth flow of materials. Composite building blocks and reinforced-concrete beams were fabricated at the building site. The houses stood in rows, and rails laid between them carried in the building materials. Such labor-saving machines as concrete mixers, stone crushers, building-block makers, and reinforced-concrete beam fabricators created a factorylike environment. Like Taylor, Gropius had the planners write out detailed schedules and instructions for the building process. Individual workers performed the same tasks over and over on each of the standardized houses.[25]

The housing program in Berlin during the 1920s matched Dessau in innovative construction techniques and use of the modern style, but it dwarfed Dessau in scale. To deal with the capital city's housing shortage, Martin Wagner, soon to become Berlin's chief city architect, in 1924 founded a building society, Gemeinnützige Heimstätten-Spar-und-Bau A.G. (GEHAG). Wagner, an engineer and member of the Socialist Party, had previously formed cooperatives of building crafts workers; in turn, trade unions financed GEHAG, which grew to become one of Berlin's two largest building societies. GEHAG's twin aims were to develop economical building techniques and to build low-cost housing. In the five-year period 1925–29, the city's building societies together put up nearly 64,000 dwelling units, and around one-fifth of these were designed by modernist architects. (Private enterprise added 37,000 more units.) In preparation for this effort, Wagner visited the United States to examine industrialized building techniques while GEHAG's chief architect, Bruno Taut, studied garden cities in the Netherlands.

Berlin in the mid-1920s was something of a modernist mecca. In 1925 Taut began work on the Britz estate, which would encompass 1,480 GEHAG dwellings, the construction of which employed lifting and earth-moving equipment and rational division of labor. Taut used standardized forms to create a distinctive horseshoe-shaped block. Taut also designed large estates in the Berlin districts of Wedding, Reinickendorf, and Zehlendorf, the last including 1,600 dwellings in three- and four-story blocks as well as individual houses.[26] Gropius, along with four other architects, built the vast Siemens estate in Berlin for the employees of that electrical firm. These modernist dwellings all featured flat roofs, low rents, communal washhouses, and plenty of light, air, and open space.[27]

But neither Dessau nor Berlin matched the housing campaign of Frankfurt. There, in one of western Germany's largest cities, the building effort rehoused 9 percent of the entire population. Under the energetic direction of Ernst May, the city's official architect, no fewer than 15,174 dwelling units were completed between 1926 and 1930 (fig. 6.5). In 1925 May drew up a ten-year program that called for the city itself to build

FIG. 6.5. May Construction System in Frankfurt.

In the late 1920s Ernst May, the city architect for Frankfurt, oversaw the building of 15,000 dwelling units there. May's research team devised special construction techniques (e.g., the prefabricated "slabs" of concrete shown here) to hold down costs and speed up the building process. The result was a practical demonstration of mass-produced housing at reasonable cost. Bruno Taut, *Modern Architecture* (London: Studio, 1929), 114.

new housing and for building societies and foundations to do likewise following the city's plans and standards. May did much of the designing himself for the largest estates at the outskirts of the city. (Gropius did a block of 198 flats, while Mart Stam, whom Gropius had once asked to head the Bauhaus architecture effort, completed an 800-flat complex.) May's office established standards for the new building projects; these standards specified the size and placement of doors and windows, ground plans of different sizes, and the famous space-saving kitchens described below. May and his colleagues at the Municipal Building Department car-

ried out research into special building techniques. Cost savings were a
paramount concern; by 1927 a factory was turning out precast concrete
wall slabs that permitted the walls of a flat to be put up in less than a day
and a half.[28]

A bit south of Frankfurt, at Stuttgart, the emerging modern style had
its first highbrow showcase in 1927. Although the city commissioned only
sixty dwellings, the Weissenhoff housing exposition had a huge and last-
ing influence on the modern movement. Mies van der Rohe was designated as the director, and he invited fifteen of the best-known modern architects—including Oud and Stam from the Netherlands, Corbusier from
Paris, Josef Frank from Vienna, and many of the notable Germans, including Behrens, Poelzig, Taut, and Gropius. The model housing estate co-
incided with a major Werkbund exhibition and was on view for an entire
year before being occupied. When the exhibition opened, up to 20,000
people a day saw the new architecture. What is more, May brought numerous exhibition-goers to view his projects in nearby Frankfurt. As Mies
put it, the new architecture reflected "the struggle for a new way of habi-
tation, together with a rational use of new materials and new structures."[29]

The Stuttgart exposition of 1927 was the first salvo in a wide-ranging
campaign to frame a certain interpretation of modernism. It was to be rational, technological, and progressive; historical references and orna-
mentation were strictly forbidden. In 1932, the Museum of Modern Art in
New York gave top billing to its "International Style" show, which displayed
and canonized the preponderantly European works representing this
strain of modernist architecture. Homegrown American contributions to
the modern style included World's Fair expositions at Chicago (1933) and
New York (1939), especially the General Motors "World of Tomorrow"
pavilion, which linked science, rationalization, and progress through
technology.[30] The Congrès Internationaux d'Architecture Moderne,
known as CIAM (1928–56), with its noisy conferences and its edgy and
polemical "charters," also shaped the contours of the modern style; it did
so in such a pushy way that it gained a reputation as a kind of international modernist mafia. As we mentioned above, many leading modernists came to the United States after fleeing Hitler, who mandated "authentic German" styles in architecture and brutally suppressed the
left-wing social movements that had supported many of the modernists.[31]
The influential teaching of Bauhaus exiles Gropius, Moholy-Nagy, and
Mies van der Rohe in Boston and Chicago raised a generation of U.S.-
trained architects and designers who imbibed the modern movement di-

rectly from its masters. In the 1950s, in architecture at least, the International Style, or Modern Movement, became a well-entrenched orthodoxy.

While the public campaign to enshrine modernism in architecture is well known and one cannot overlook the thousands of modernist office buildings, apartments, hospitals, government buildings, and schools built worldwide in the decades since the 1920s, an equally influential set of developments brought modernism to the home. Here, modernism's rationalizing and scientizing impulses interacted in complex and interesting ways with established notions about the household and about women's roles as the principal homemakers. The "Frankfurt kitchen," designed in 1926 by Margarete Schütte-Lihotzky (Grete Lihotzky), became a classic and well-regarded modernist icon.

The outlines of Lihotzky's dramatic life story make her an irresistible heroic figure. Trained in Vienna as an architect, she was one of very few women to thrive in that male-dominated field. She worked energetically and creatively to bring workers' perspectives to her projects for housing, schools, and hospitals in Vienna (1921–25), Frankfurt (1926–29), Moscow (1930–37), and Istanbul (1938–40). Her Frankfurt kitchen designs became so well known that when chief city architect Ernst May moved to Moscow and she agreed to continue working with him, she did so only on the express condition that she *not* do any more kitchens. (In Moscow, she worked mostly on children's nurseries, clubs, and schools for the Soviet government.) In 1940 she returned to Austria to join the resistance movement fighting fascism, but within weeks she was arrested, narrowly escaped a death sentence, and spent the war in a Bavarian prison. After the war, she helped with the reconstruction of Vienna, was active in CIAM, and kept up her architectural practice, engaging in many international study trips, publications, and projects through the 1970s.[32]

Household reform in Germany during the 1920s, as Lihotzky discovered, was crowded with many diverse actors. The national government, while nominally socialist and formally committed to equal rights for women and men, enacted a policy of "female redomestication." This policy sought to encourage young women to embrace traditional women's roles of homemaker, mother, and supporter for her husband. The government hoped to end the "drudgery" of housework and to reconceive housework as a modern, scientific, and professional activity. Modernizing housework through the use of Tayloristic "scientific management" principles was precisely the point of Christine Frederick's *The New Housekeeping: Efficiency Studies in Home Management* (published in the United

States in 1913, translated into German in 1922 with the new title *Die rationelle Haushaltführung*) and was a central message of Elisabeth Lüders and Erna Meyer, both of whom were prolific authors on women's reform issues and who served as advisors to government and industry. In an essay published in a leading German engineering journal, Meyer maintained that "the household, exactly like the workshop and the factory, must be understood as a manufacturing enterprise."[33]

The German government agency charged with rationalizing workshops and factories also worked closely with several women's groups to rationalize the household. The Federation of German Women's Associations (Bund Deutscher Frauenvereine), with its 6,000-member clubs and one million members, was the country's largest association of women. With the federation's support the national government enacted compulsory home economics courses for girls and sponsored vocational secondary schools where women learned to be "professional" seamstresses, laundresses, and day-care attendants. Within the federation, the conservative Federal Union of German Housewives Associations (Reichverband Deutscher Hausfrauenvereine), originally founded to deal with the "servant problem," became a formal advisory body with special expertise on housewifery to the Reich Research Organization. The union's effort to modernize housekeeping resulted in numerous conferences, publications, and exhibitions, including one in Berlin in 1928 that featured a set of model kitchens.[34]

Lihotzhy's work on rational kitchen designs, then, emerged in the context of substantial governmental, industrial, and associational interest in the topic. Ernst May himself initiated a research program on "domestic culture" upon which to shape his Frankfurt housing designs. The program's investigations involved psychology, evaluations of materials and products, and scientific management principles; researchers studied such diverse areas as household products, consumer markets, appliances, and home economics classrooms. May's chosen products became an officially recommended line of household furnishings and were publicized in his journal, *The New Frankfurt*. Lihotzy developed her kitchen design using the principles of Frederick Taylor's time-and-motion studies (for example, she reduced the number of "unneeded" steps a housewife made within her kitchen and in taking food to the nearby eating room), as well as giving careful attention to materials.

Lihotzky's kitchen combined diverse colors and an effective design into a compact and photogenic whole (fig. 6.6). (By comparison a con-

FIG. 6.6. Lihotzky's Frankfurt Kitchen.

"This kitchen was not only designed to save time but also to create an attractive room in which it was pleasant to be," wrote Grete Lihotzky of her space-saving kitchen. Her compact design eliminated "unneeded" steps that made a housewife's work inefficient. More than 10,000 of these factory-built kitchen units were installed in Frankfurt's large-scale housing program. Peter Noever, ed., *Die Frankfurter Küche* (Berlin: Ernst & Sohn, n.d.), 45.

temporaneous kitchen designed by J. J. P. Oud and Erna Meyer for a Weissenhof house appears ugly, spare, and stark.) Summarizing Lihotzky's own description, one can tell that "some man" was not the designer. The gas cooker featured an enameled surface for easy cleaning and a "cooking box" (*Kochkiste*) where food that had been precooked in the morning could be left to stew all day, saving the working woman time and energy.

The flour drawer, made of oak, which contains tannic acid, kept worms out. Fully illuminated by an ample window, the work table, made of beech, featured an easily cleaned metal channel for vegetable waste. Cupboards for crockery were enclosed by glass windows and sealed against dust. Other features integrated into the compact plan of 1.9 by 3.44 meters were a fold-down ironing board, an insulated twin sink, a unit above the sink for drying and storing plates, and a moveable electric light. The kitchen fairly bristled with drawers. Most extant photographs of Lihotzky's Frankfurt kitchen are black-and-white, so they fail to reveal that Lihotzky featured color: "The combination of ultramarine blue wooden components (flies avoid the colour blue) with light grey–ochre tiles, the aluminum and white-metal parts together with the black, horizontal areas such as the flooring, work surfaces and cooker ensured that this kitchen was not only designed to save time but also to create an attractive room in which it was pleasant to be."[35]

During the peak years of the Frankfurt building campaign in the late 1920s, Lihotzky's kitchen was installed in 10,000 Frankfurt apartments. In fact, she had worked closely with the manufacturer, Georg Grumbach, to achieve an easily manufactured design. Grumbach's company assembled the kitchens as factory-built units and shipped them whole to the construction site, where they were lifted into place by cranes. Her kitchen also went into production in Sweden, after it was extensively praised in a Stockholm exhibition.

Looking at the modernist movement as a whole, then, a rich set of ironies pervades the history of aesthetic modernism and modern materials. While many of the key figures were, at least in the 1920s, left-wing activists committed to achieving better housing for workers, modernism in the 1950s became a corporate style associated with avowedly non-socialist IBM, Sears, and a bevy of cash-rich insurance corporations. Modernism as an overarching style professed itself to be a natural and inevitable development, reflecting the necessary logic of modern technological society; and yet the campaign to enthrone modernism was intensely proactive and political: in the 1920s it was furthered by sympathetic city housing projects, in the 1930s modernist architects were banned by the National Socialists, and in the 1940s onward into the Cold War years a certain interpretation of aesthetic modernism—carefully shorn of its socialist origins—was the object of intense promotional efforts by CIAM, the Museum of Modern Art, and other highbrow tastemakers. Finally, what can we make of Grete Lihotzky, an ever-committed communist, negotiating with a private manufacturer to mass-produce her kitchen designs? (The Grumbach factories also took orders directly from private homeowners who wanted the Frankfurt kitchen.) The Frankfurt housing developments themselves remained too expensive for the working-class families for whom they were designed. Instead, the Frankfurt apartments filled up with families of the middle class and of the best-paid skilled workers.

Designers and architects, motivated by a variety of impulses as was the entire modernism movement, actively took up the possibilities of mass-produced machine-age glass and steel. The materials—inexpensive, available in large quantities, and factory-manufactured—made it economically possible to dream of building housing for the masses, in a way that could not have been achieved with hand-cut stone or all-wood construction. The modern materials were also something of a nucleation point for technological fundamentalists asserting the imperative they felt to change culture in the name of technology. In examining how "technology changes culture" we see that social actors, often asserting a technological fundamentalism that resonates deeply in the culture, actively work to create aesthetic theories, exemplary artifacts, pertinent educational ventures, and broader social and political movements that embed their views in the wider society. When the techno-cultural actors fail, we largely forget them. If they succeed, we believe that technology itself has changed culture.

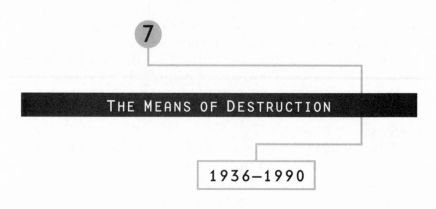

THE MEANS OF DESTRUCTION

1936–1990

No force in the twentieth century had a greater influence in defining and shaping technology than the military. In the earlier eras of modernism and systems, independent inventors, corporations, designers, and architects played leading roles in developing such culture-shaping technologies as electricity, synthetic dyes, skyscrapers, and household technologies, while governments and trading companies had dominated the era of imperialism. But not since the era of industry had a single force stamped a more indelible imprint on technology. One might imagine a world in 1890 without the technologies of empire, but it is difficult to envision the world in 1990 absent such military-spawned technologies as nuclear power, computer chips, artificial intelligence, and the Internet. Likewise, it was military-derived rockets that boosted the Apollo missions to the moon and military-deployed satellites that paid for the Space Shuttle missions. From the 1950s through the 1980s such stalwarts of the "high technology" (military-funded) economy as IBM, Boeing, Lockheed, Raytheon, General Dynamics, and MIT numbered among the leading U.S. military contractors. Lamenting the decline of classic profit-maximizing capitalism, industrial engineer Seymour Melman termed the new economic arrangement as contract-maximizing "Pentagon capitalism." During these years of two world wars and the Cold War, the technology priorities of the United States, the Soviet Union, and France, and to a lesser extent England, China, and Germany, were in varied ways oriented to the "means of destruction."

Merely the use of military technologies is not what distinguishes this era, of course. Fearsome attack chariots and huge cannon figured in

Renaissance-era warfare, while steam gunboats and Enfield rifles typified the imperialist era. What was new in the twentieth century was the pervasiveness of technological innovation and its centrality to military planning. This was true during the period of active world war between 1939 and 1945, as military officers, scientists, engineers, and technology-minded businessmen forged new relationships that led to the "military-industrial complex" (named by President Dwight Eisenhower, an ex-general himself), and also during the prolonged Cold War (1947–90) that followed. The great powers' universities, technology companies, government institutes, and military services committed themselves to finding and funding new technologies in the hope of gaining advantage on the battlefield or in the Cold War's convoluted diplomacy. Above all, for military officers no less than researchers, military technology funding was a way of advancing one's vision of the future—and often enough one's own career.

The swift pace of military-driven technical innovation did not come without cost. "The concentration of research on particular tasks greatly accelerated their achievement," writes Alan Milward of the Second World War, "but this was always at the expense of other lines of development." During the Cold War, "the increasing predominance of one patron, the military," writes Stuart Leslie, "indelibly imprint[ed] academic and industrial science with a distinct set of priorities" that set the agenda for decades. Such promising technologies as solar power, analog computers, and machinist-controlled computer machine tools languished when (for various reasons) the military backed rival technical options—nuclear power, digital computers, and computer controlled devices of many types—that consequently became the dominant designs in their fields.[1]

So pervasive was military funding for science and technology during these decades that perhaps only in our own (post–Cold War) era is a measured assessment possible. This is still no easy task, since for every instance of productive military-civilian "spin-off," such as Teflon, microchips, nuclear power, and system management, there is a budgetary black hole from which no useful device, military or civilian, ever came. Among these ill-considered schemes in the United States must number the little-publicized Lexington, Mohole, and Plowshare projects. These were lavishly funded schemes designed, respectively, to build nuclear-powered airplanes (1946–61), to drill through the earth's crust (1957–66), and to use atomic explosions in making huge earthen dams and harbors (1959–73). For Project Lexington the Air Force spent $1 billion—when a billion dollars really meant something—on the patently absurd nuclear plane. (The Soviet nuclear empire, even less constrained by common sense and safety concerns than its Western counterparts, spent untold sums on nuclear ships, nuclear explosions for earthmoving, and nuclear power.[2]) Assessing the technology-laden "puzzle palace" of the National Security Agency, whose budget and technologies remain veiled in secrecy, may never be possible. Its critics chalk off as a budgetary black hole the $35 billion Strategic Defense Initiative (1985–94), the latter-day "technological dream" popularly known as Star Wars.[3]

A War of Innovation

It may seem odd to distinguish between the two world wars, linked as they were by politics and economics, but in technology the First World War was not so much a war of innovation as one of mass production. The *existing* technologies of the time largely accounted for the outbreak, character, and outcome of this "war to end all wars." Telegraphs and railroads, no novelty at the time, figured prominently in the breakdown of diplomacy and outbreak of war in July–August 1914. Indeed, the fever pitch of telegraph negotiations, with their drumbeat of deadlines, simply overwhelmed diplomats on both sides. Telegraphs carried Austria's declaration of war on Serbia and later revealed to the United States Germany's designs on Mexico (in the intercepted Zimmerman telegram), while railroads carried German troops in a precise sequence of steps designed to defeat France and Russia. For German planners the problem was that the French held their ground at the Marne, just short of Paris. The two sides dug in. It would be a long war.

World War I's gruesome trench warfare—the great "sausage machine" that ground up young bodies—was the direct result of the machine gun and mass-produced explosives. Invented in 1885, the machine gun had been used by imperialist armies with devastating effect against lightly armed native armies. In the few scant hours of a confrontation in the Sudan during 1896, the Anglo-Egyptian army with machine guns slaughtered 20,000 charging Dervishes. When both sides possessed machine guns, however, each side could readily mow down the other's charging soldiers. Taking territory from an enemy defending itself with machine guns and barbed wire was practically impossible. The stalemate of trench warfare persisted just so long as both sides successfully mass produced gunpowder and munitions. Even the horrific poison gases used by both sides had little strategic effect (see chapter 5). Tanks were one novel technology that saw wartime action, and the British devised a war-winning campaign using tanks called Plan 1919. But the war ended in November 1918 before the plan could be tried. The defeat of Germany did not depend much on novel technologies such as Dreadnought-class battleships, radios, automobiles, or even aircraft. Germany lost the war when its industrial and agricultural economy collapsed.

As a madman bent on world domination, Adolf Hitler learned the right lessons from World War I. Wounded twice while serving in the Bavarian Army and partially blinded by poison gas, Hitler concluded that Germany would certainly lose another protracted war of production. Cut off from essential raw materials during the First World War, Germany, he resolved, would be self-sufficient for the second. Just three years after seizing power Hitler enacted the Four Year Plan (1936–40), his blueprint for self-sufficiency and rapid rearmament. Under his ideal model of autarky, Germany would produce everything it needed to fight a major war, including synthesizing gasoline and rubber from its plentiful domestic supplies of coal. Because of Germany's shortfalls of the copper and tin needed for munitions as well as the nickel and chromium needed for armor plate, that autarky would never be complete. As early as November 1937, Hitler grasped the point that Germany's early rearmament gave it only a short-lived technological and military advantage; the longer Germany delayed going to war, the more this advantage would fade. The inescapable conclusion, Hitler told his top advisors at the Hossbach conference, was that war must come soon. Meantime the German air force practiced bombing and strafing in Spain culminating with the Condor Legion's destruction of Guernica (1937).

Hitler's invasion of Poland in September 1939 revealed the tactic he had chosen to avoid trench warfare. His *blitzkrieg* or "lightning war" loosely resembled the British Plan 1919. A *blitzkrieg* began with massive aerial bombardment followed by a column of tanks, supported overhead by fighter aircraft, breaking through enemy lines and cutting off supply routes. Not merely a military tactic, *blitzkrieg* was more fundamentally a "strategic synthesis" that played to the strength of Germany's superior mobility technologies, especially aircraft and tanks, while avoiding the economic strain and social turmoil of a sustained mobilization. Back home Germany's industrial economy proved remarkably productive, using flexible general-purpose machine tools to alternate between military and civilian production. The intentional shortness of *blitzkrieg* attacks allowed for ample civilian production and substantial stockpiling of strategic raw materials between bouts of military action. Consequently the extreme wartime deprivation visited upon Japanese civilians at an early stage in the war—when the Japanese military took over the economy and directed all production toward military ends, literally starving the civilian economy—did not arrive in Germany until the last months of the war. Across Poland, France, Belgium, and far into Russia, until the Battle of Stalingrad in the winter of 1941–42, the German army was undefeated and appeared invincible. But the moment Germany was thrown back on the defensive, the war was lost. Hitler's strategic synthesis was undone.[4]

Fortunately, Hitler did not foresee all the lessons needed to win World War II. To begin with, the Germans never had the unified, efficient, single-minded "war machine" pictured in Allied wartime propaganda and never totally mobilized their economy, as the British did. Battles between rival factions within the Nazi party, not to mention creative "delays" by the non-Nazi civil service, often made a mockery of planning directives announced by Hitler and his advisors. For most of the war, German technologists matched their Allied counterparts in the innovation race. As military historian Martin van Creveld points out, both sides maintained a torrid pace of technological innovation that prevented either side from developing a war-winning weapon. While the Allies pioneered radar-based air defense, built the most powerful internal combustion aircraft engines, and developed the most sophisticated electronics systems, the Germans developed superior navigational aids for bombing, led the world in jet and rocket engines, and consistently had superior tanks, artillery, and machine guns.[5]

Germany's scientists and engineers kept pace with their Allied counterparts until April 1944. In that month the Nazis, facing an economic crisis stemming from the continual changes to factory technology required by vigorous innovation, effectively halted research-and-development work and froze technical designs. With this drastic move they hoped to produce vastly larger quantities of the existing munitions, aircraft, and weaponry: this was a return to mass production World War I style. This decision ended the contribution to the war of Germany's jet-powered aircraft. Only select research projects, with promise of achieving a military result and soon, were continued. One of these was the German atomic program.

The National Socialist atomic program, facing shortages of people, materials, and funding, was a long shot at the outset. German physicists Otto Hahn and Fritz Strassmann had in 1938 split the uranium nucleus.

Between 1940 and 1942 several German physicists, Werner Heisenberg most prominently, tried without success to sell an atom bomb project to the German Army and Reich Research Council. From 1942 onward Heisenberg and his colleagues had low-level funding that permitted them to work only on an atomic power reactor. Soon the country's economic turmoil and deteriorating military situation made it impossible to mount the industrial-scale effort necessary for a reactor, let alone to build an atomic bomb. The German atomic effort was further hampered by Hitler's anti-Semitic ravings, which had driven away Germany's Jewish scientists, among them many of the country's leading atomic physicists. Germany had neither the enriched uranium, the atomic physicists, nor the governmental resources to manufacture an atomic bomb. In early March 1945, with a cobbled-together "uranium machine" located in an underground bunker south of Berlin, and with his country crumbling around him, Werner Heisenberg still pursued a self-sustaining atomic chain reaction.[6]

"Turning the Whole Country into a Factory"

A revealing measure of just how much distance separated the German and Anglo-American atomic programs came on 6 August 1945. The war in Europe had wound down in early May when Soviet tanks entered Berlin. Quickly the Allies rounded up ten top-ranking German physicists, who were brought to a country estate in England. There, while the war continued in Asia, Allied intelligence agents closely questioned them about their atomic research. Mostly, the German physicists spoke with a haughty pride about their wartime achievements—until the evening when word arrived that the Americans had dropped an atom bomb on the Japanese city Hiroshima. The existence of an atom bomb was surprise enough. The Germans were simply astonished to learn that the explosion at Hiroshima had equaled 20,000 tons of high explosives, and that the Allied atom project had employed 125,000 people to construct the necessary factories while another 65,000 people worked in them to produce the bomb material. Their own effort, scarcely one-thousandth that size in personnel and immeasurably less in effect, paled in comparison. The German atom project had been nothing more than a laboratory exercise, and an unsuccessful one at that.[7]

If the First World War is known as the chemists' war owing to military use of synthetic explosives and poison gases, it was the Manhattan Project that denominated the Second World War as the physicists' war.

Physicists first began wartime atomic work at Columbia University, University of Chicago, and University of California-Berkeley. As early as December 1942 Enrico Fermi, at Chicago, had built a self-sustaining uranium reactor surpassing anything the Germans achieved during the war. Fermi's success paved the way for Robert Oppenheimer's Los Alamos laboratory. In the spring of 1943 physicists recruited by Oppenheimer began arriving at the lab's remote desert location in New Mexico. At its peak of activity, Los Alamos had 2,000 technical staff, among them eight Nobel Prize winners, who worked out the esoteric physics and subtle technology necessary to build a bomb. When the Los Alamos scientists, bright, articulate, and self-conscious about having made history, later told their stories, they in effect grabbed sole credit for the bomb project. Their tale was a wild exaggeration. In reality, Los Alamos served as the R&D center and assembly site for the bombs. The far greater part of the project was elsewhere, at two mammoth, top-secret factory complexes in Tennessee and Washington State. Indeed, these two factories were where most of those nearly 200,000 people worked and where virtually all of the project's whopping budget was spent (less than 4 percent of the project's plant and operating costs went to Los Alamos). Most important, it was the structure of these two factories, I think it can be fairly argued, that shaped the deployment of atomic weapons on Japan. One can also see, inscribed into the technical details of these factories, a certain vision for atomic weaponry that lasted far beyond the war.

In the physicists' fable, it was none other than the world's most famous physicist who convinced President Franklin Roosevelt to back a bomb project. In autumn 1939, with the European war much in their minds, Albert Einstein worked with Leo Szilard, one of the many émigré European physicists active in atomic research, setting down in a letter to the president the scientific possibilities and political dangers of atomic fission. Actually, it was Alexander Sachs, an economist, former New Deal staffer, and Roosevelt insider, carrying the physicists' letter, who made the pitch in person on 11 October 1939. Roosevelt saw immediately, before even reading the Einstein-Szilard letter, that a German atom project was menacing and that an Allied atom project needed action and soon. Roosevelt gave the go-ahead for a high-priority secret program. After several governmental committees considered its prospects, the project came to rest in the Office of Scientific Research and Development, or OSRD, a new government agency headed by the MIT engineer Vannevar Bush. Bush was charged with mobilizing the country's research and engineering talents in service to the war. From the first, then, the atom bomb project had the president's strong backing, an ample budget, smart and ambitious backers, and a stamp of "top secret."

The industrial-factory aspect of the Manhattan Project took shape in the months after June 1942. Until that point the atom project had been simply, if impressively, big science. It was funded and coordinated by Bush's OSRD in Washington; the agency directed research money to several universities around the country that had active atomic research. Earlier that year at the University of Chicago, physicist Arthur Holly Compton brought together several research groups, including Fermi's from Columbia, in an effort to centralize atomic research. The result was the misleadingly named Metallurgical Laboratory. That June, however, marked a decision point on whether the United States should expand the promising laboratory-scale work to a modest level of funding, aiming for an atomic reactor, or mount a massive, full-scale effort in pursuit of a bomb. Roosevelt, acting on Bush's recommendation, approved a large-scale project. A bomb it would be. Bush understood that this massive effort would overwhelm the capabilities of OSRD, whose mission was not industrial-scale production but conducting military-relevant research and developing weapon prototypes. Although the point is not frequently emphasized, it was entirely fitting that Roosevelt assigned the construction phase of the bomb project to the Army Corps of Engineers and that the Army assigned command over the Manhattan Engineering District to Brigadier General Leslie Groves, who had been the officer in charge of building the Pentagon complex.

Described as supremely egotistical, excessively blunt, and aggressively confident (after their first meeting Bush expressed doubt that the general had "sufficient tact for the job"), Groves went to work with dispatch.[8] In September, within days of taking command, he secured for the Manhattan Project a top-priority AAA procurement rating, purchased a 1,250-ton lot of high-grade uranium ore from the Belgian Congo, and initialed the Army's acquisition of 54,000 acres in eastern Tennessee. In December, shortly after Fermi's Chicago pile went critical, he secured Roosevelt's approval of $400 million for two huge factories: one for uranium separation at Oak Ridge, Tennessee, and one for plutonium production at Hanford, Washington. From then on, Groves' most substantial task was mobilizing the legion of contractors and subcontractors. The Army had already tapped Stone & Webster, the massive engineering, management, and financial consulting firm, to be the principal contractor. Groves him-

FIG. 7.1. Oak Ridge Uranium Factory.

The Oak Ridge complex in eastern Tennessee was the principal site for producing bomb-grade enriched uranium. The electromagnetic process, housed in the portion of the Oak Ridge complex shown here, promised a one-step separation of fissionable U-235 using atomic "racetracks" and huge magnets. Enriched uranium came also from Oak Ridge's equally massive gaseous diffusion plant and thermal diffusion plant. Vincent C. Jones, *Manhattan: The Army and the Atomic Bomb* (Washington, D.C.: Government Printing Office, 1985), 147.

self let subcontracts to divisions of Eastman Kodak, Westinghouse, General Electric, Chrysler, Allis-Chalmers, M. W. Kellogg, DuPont, and Union Carbide.

Groves needed all 54,000 acres of the Tennessee site, because Oak Ridge would be a mammoth undertaking. The crucial task at Oak Ridge was to produce enough enriched uranium, somewhere between 2 and 100 kilograms, no one knew precisely how much, to make a bomb (fig. 7.1). Common uranium had less than 1 percent of the fissionable 235-isotope. Since no one had ever isolated uranium-235 in such quantities, Groves quickly authorized large-scale work on two distinct processes, electromagnetic separation and gaseous diffusion, while a third process, thermal diffusion, was later drafted to the cause. Electromagnetic separation was the baby of Ernest Lawrence, already accomplished at big science, who offered up his Berkeley laboratory's atom smashers. The faster atomic particles could be propelled, everyone realized, the more interesting was the physics that resulted when opposing streams of particles collided. Lawrence's particular genius was in seeing that a charged stream of particles could be accelerated to unbelievably high energies by a simple trick. While precisely positioned magnets bent the stream of charged particles to follow an oval "racetrack," a carefully tuned oscillating electric field sped them up, faster and faster, each time they went around. It was a bit like pushing a child on a swing, a bit higher each pass. Armed with Lawrence's formidable skills at fund raising and machine building, his laboratory had, by 1932, accelerated hydrogen atoms to 1 million volts. By 1940, just after winning the Nobel Prize for the subatomic discoveries made by an impressive 16-million-volt machine, Lawrence lined up Rockefeller Foundation funds to build a really big accelerator in the Berkeley hills. Its 184-inch-diameter magnet was the world's largest. The cost of the machine was a heady $1.5 million.

Lawrence, ever the hard-driving optimist, promised Groves that his cyclotron principle could make a clean separation between the two uranium isotopes. Lawrence knew that his powerful magnets would bend the path of a charged particle, including a charged uranium atom. Passing through a constant magnetic field, he explained, the lighter, fissionable uranium-235 would be bent slightly more than the heavier, stable uranium-238, a manageable fraction of an inch. He invented a "calutron," a C-shaped device something like a cyclotron cut in half, to do the work of separating the isotopes. While supremely elegant in theory, the electromagnetic-separation process had in reality been effective only at the laboratory scale; in 1941, after a month of intensive effort using a smaller Berkeley machine, Lawrence had produced 100 *micro*grams of *partially* enriched uranium. A uranium bomb would need at least 100 million times more, perhaps 40 *kilo*grams of *highly* enriched uranium.

At Groves' direction Stone & Webster broke ground in February 1943 on the Tennessee complex. By August 20,000 construction workers were building the atomic racetracks and calutrons on a site that stretched over an area the size of twenty football fields. When all the necessary supporting infrastructure was completed, there would be 268 buildings. (On an inspection trip that spring, even Lawrence was dazzled by the size of the undertaking.) Obtaining enough copper wire for the innumerable magnet windings looked hopeless. Coming to the rescue, the Treasury De-

partment offered its stores of silver. Groves borrowed 13,540 tons of silver—a substance usually accounted for in ounces—valued at $300 mil-
lion, and used the precious metal in place of copper to wind magnets and
to build solid silver bus bars to conduct the heavy electric currents. All the
while, troublesome vacuum leaks, dangerous sparks, and innumerable
practical problems kept Lawrence and his assistants problem solving at a
frenetic pace. Facing intractable difficulties with the magnetic windings,
Groves shipped the entire first batch of 48 huge silver-wound magnets
back to the Allis-Chalmers factory in Milwaukee, idling the 4,800 Tennessee Eastman production workers for a full month. Production of the
precious uranium-235 resumed in January 1944.

The second uranium enrichment process tried at Oak Ridge required
even larger machines, although the science was not nearly so flashy.
Gaseous diffusion used small effects to achieve a big result. This method
of separating the two uranium isotopes exploited the established fact that
lighter molecules move more readily through a porous barrier than do
heavier ones. In this case, atoms of fissionable uranium-235 made into
uranium hexafluoride would move through a porous barrier a bit more
quickly than would the heavier uranium-238. Initial work by researchers
at Columbia University and Bell Telephone Laboratories suggested that a
nickel barrier would separate the isotopes effectively while withstanding
the highly corrosive uranium hexafluoride gas. Groves authorized the
Kellex Corporation, the prime contractor for this project, to build acres
of diffusion tubes at Oak Ridge, but the early results were somewhat disappointing. In autumn of 1943 one of Kellex's engineers, Clarence Johnson, hit upon the idea of sintering the nickel metal to form a barrier; sintering involved pressing powdered nickel into a semiporous block, and
it would require the company to handmake the new barriers, at staggering cost. Groves' willingness to scrap the existing nickel barriers, to junk
the Decatur, Illinois, factory that had been constructed to make them, and
to push on with the revamped gaseous diffusion plant convinced physicist Harold Urey, among others, that Groves was building for the ages.
Oak Ridge was to be not merely a temporary wartime factory but a permanent bomb-producing complex.

As the original goal of building a bomb by the end of 1944 slipped
past, with neither of the two processes having yielded the needed amounts
of enriched uranium, Groves brought a third process to Oak Ridge. Thermal diffusion was the simplest of the separation processes. Its chief
backer, Philip Abelson, had worked three years for the Navy—cut off from

the Army-led atom bomb project—on enriching uranium with an eye to making atomic reactors for submarines. In 1941 Abelson had perfected a method to make uranium hexafluoride, a liquid at more moderate temperatures than were used in the gaseous diffusion effort. He then lined up thermal diffusion tubes to enrich the liquid uranium. Trapped between hot and cold surfaces, the lighter uranium-235 drifted up while the heavier uranium-238 drifted down.

To exploit this separating effect Abelson built, at the Naval Research Laboratory, an experimental 36-foot-tall column into which the liquid uranium hexafluoride was piped between 400°F steam and 130°F water. Workers simply skimmed enriched uranium-235 off the top. In January 1944, with the Navy's go-ahead, Abelson began building a 100-column complex next to the steam plant of the Philadelphia Navy Yard. Abelson never promised the one-stop separation of Lawrence's cyclotrons; on the other hand, his columns worked reliably, yielding sizable quantities of slightly enriched uranium. In mid-June Groves contracted with the H. K. Ferguson engineering firm to build at Oak Ridge a complex consisting of twenty-one exact copies of the entire Philadelphia plant, or 2,100 columns in all. By 1945 the three processes were brought together: thermal diffusion provided an enriched input to gaseous diffusion and electromagnetic separation. Concerning the immense effort of making enough uranium-235 for a bomb, Niels Bohr told Edward Teller, promoter of the hydrogen bomb and another physicist laboring on the Manhattan Project, "I told you it couldn't be done without turning the whole country into a factory. You have done just that."[9]

The Oak Ridge complex with its network of designers, contractors, fabricators, and operators was just one division of this far-flung national factory. The site at Hanford, with 400,000 acres, was nearly eight times larger than Oak Ridge (fig. 7.2). Groves had chosen this remote site in south-central Washington State for several reasons. The Columbia River provided water for the atomic reactors, while electric power was readily available from the Grand Coulee and Bonneville hydroelectric dams. Most important of all, since it did not take an atomic physicist to understand that a large-scale plutonium factory would be a grave safety concern, Hanford was distant from major population centers. Conceptually, making plutonium was a snap. One needed only to bombard common uranium-238 (atomic number 92) with a stream of neutrons to make neptunium (atomic number 93). Neptunium thus made soon decayed into plutonium (atomic number 94). Berkeley physical chemist Glenn Seaborg

FIG. 7.2. Hanford Works Plutonium Factory.

The Hanford Works consisted of seven large complexes, separated for safety by 6 miles or more, built on 400,000 acres in south-central Washington State. The 100 B Pile, shown here, was one of three plutonium production reactors. The production pile has a single smokestack; the electricity plant has two smokestacks, and to the right beyond it is the pump house on the Columbia River. Vincent C. Jones, *Manhattan: The Army and the Atomic Bomb* (Washington, D.C.: Government Printing Office, 1985), 216.

had discovered plutonium in a targeted search for a fissionable element to replace enriched uranium in a bomb; he named it plutonium to continue the pattern of naming elements after the outermost planets, Uranus, Neptune, Pluto. After lengthy negotiations, Groves by December 1942 had convinced the DuPont company to become prime contractor for the Hanford works and take charge of designing, building, and operating a full-scale plutonium production and separation facility. A pilot plant at the Oak Ridge complex was also part of the plutonium project. At the time, DuPont had no experience in the atomic field, but it did have extensive experience in scaling up chemical plants. And separating plutonium from irradiated uranium entailed building one very large chemical-separation factory, involving a multistep bismuth-phosphate process.

At Hanford the most pressing problems were engineering ones. Whereas Seaborg's Berkeley laboratory had produced a few micrograms of plutonium, DuPont's engineers at Hanford needed to scale up production not merely a millionfold to produce a few grams of plutonium but a thousandfold beyond that to make the kilograms of plutonium believed necessary for a bomb. Intermediate between the laboratory and the factory was the pilot plant, where theoretical calculations could be tested and design data collected. Conducting research at the pilot-plant stage was usually a task for industrial scientists or engineers, but Chicago's academic Metallurgical Lab scientists had insisted that they direct the pilot-plant reactor. Groves, assenting reluctantly to their demand, directed them to conduct experiments to see if plutonium could be made in quantity. At the same time, and in advance of the pilot-plant data, DuPont engineers designed a full-scale plutonium facility at Hanford. Relations between the scientists and the engineers were strained from the start. The DuPont project manager quieted the Chicago scientists' criticism when he asked them to sign off on the thousands of blueprints. Only then did the scientists see the magnitude of the plutonium factory. It would consist of three huge uranium reactors to produce the plutonium and four gigantic chemical separation plants to isolate and purify the poisonous element. Construction at its peak in June 1944 employed 42,000 skilled welders, pipefitters, electricians, and carpenters.

By mid-September 1944 the Hanford plant, with its construction substantially complete, was ready for its first test run. The uranium pile went critical as expected—then petered out. After some hours the reactor could be restarted, but no chain reaction lasted for long. Morale plunged when it appeared that Hanford had turned out to be a vastly expensive failure. Some trouble-shooting revealed that the element xenon-135, a powerful neutron absorber with a brief half-life, was poisoning the reactor. When xenon-135 decayed, the reactor restarted but it could not run for long. Groves had ordered the Met Lab scientists to run their pilot-plant reactor at full power, and had they done so the xenon effect would have cropped up in their experiments. But they had not run at full power. Fortunately, the DuPont engineers had designed room for extra slugs of uranium (the Chicago scientists had denounced this apparent extravagance). With the additional uranium in place, 500 slugs added to the original 1,500, the increase in neutrons overwhelmed the xenon poisoning. The DuPont engineers' cautious design had saved the plant. It successfully went critical in mid-December.[10]

In the physicists' fable the next chapter is Los Alamos, where the best and brightest physicists created a workable atom bomb. The narrative climaxes with the incredible tension surrounding the Trinity test explosion on 16 July 1945, and the relief and wonderment at its success. The standard epilogue to the story acknowledges the actual atom bombing of Hiroshima on 6 August and sometimes even mentions the bombing of Nagasaki on the ninth. Various interpretations of the events exist. Our public memory has enshrined the most comforting version, President Truman's, that the atom bomb was dropped on Japan to end the Second World War, even though dissenting voices at the time made clear that this assessment received by no means universal confirmation. General Eisenhower told Secretary of the War Henry Stimson, "The Japanese were ready to surrender and it wasn't necessary to hit them with that awful thing."[11] Debate and discussion of these momentous events have continued down the years, and should continue. Here I would like to explore the machine-like momentum that had built up behind the atomic bomb project and proved irresistible to many. Winston Churchill, no foe of the bomb, wrote that "there never was a moment's discussion as to whether the bomb should be used or not."[12]

Many commentators, even Eisenhower and Churchill, miss the crucial point that the two bombs dropped on Japan were technologically quite distinct: the Hiroshima bomb used Oak Ridge's uranium while the Nagasaki bomb used Hanford's plutonium. The Hiroshima bomb, a simple gun-type device code-named Little Boy, worked by firing one subcritical mass of uranium-235 into another, creating a single larger critical mass; so confident were the designers that an atomic explosion would result that no test of this bomb was ever attempted. Hiroshima was therefore the site of the world's first uranium explosion. Entirely different was the Nagasaki bomb dropped three days later, code-named Fat Man, a plutonium device detonated by implosion. (Two supposedly subcritical masses of plutonium, such as were used in the uranium bomb, were deemed dangerously unstable.) Implosion relied on high explosives, properly detonated, to generate a spherical shock wave that compressed a hollow sphere of plutonium inward with great force and speed. When the bowling ball of plutonium was squeezed to grapefruit size, and a hazelnut-sized core released its initiating neutrons, the plutonium would go critical and explode. This was the bomb tested at Trinity in mid-July.

That spring, a committee of top-level army officers drafted a targeting plan for dropping atomic bombs on Japan. Hiroshima, Kokura, Niigata, and Nagasaki were the preferred targets. The order approving the plan was drafted by Leslie Groves and approved by Henry Stimson and George Marshall, respectively secretary of war and chairman of the joint chiefs of staff. Truman signed nothing; his approval can only be assumed. "I didn't have to have the President press the button on this affair," recalled Groves.[13] The Army air force planned to use whatever bombs it had at hand. The two bombs that were dropped were en route to the forward base at Tinian Island in the Pacific (fig. 7.3). A third bomb, still under final assembly at Los Alamos, would have been ready for shipping to Tinian by 12 or 13 August, but before it could be shipped out, Truman, rattled by the thought of "all those kids" who had been killed in Hiroshima, ordered a halt in the atomic bombing. He gave the order on 10 August, the same day the Japanese emperor's surrender terms reached Washington. The third bomb stayed at Los Alamos. While Truman and his advisors dickered over the precise terms of Japan's surrender, the fire bombing of Japan continued. General Hap Arnold ordered out 800 of his B-29s, escorted by 200 fighters, for a final all-out incendiary bombing of Honshu on 14 August. Even as the 12 million pounds of incendiaries and high explosives were falling, the emperor's acceptance of allied surrender terms was on its way to Washington. The war ended, officially, the next day.

The rush to drop the two atom bombs on Japan must be an enduring puzzle. President Truman's rationale, while comforting, does not account satisfactorily for a number of facts. As vice-president until FDR's death on 12 April Truman had not even been told of the bomb project, and, as Groves' comment above indicates, it is unclear how much of the decision was his anyway. An alternative view suggests that the bombs were dropped on Japan but aimed at the Soviet Union, then mobilizing its army for a long-promised invasion of Manchuria. In this view, propounded by physicist P. M. S. Blackett and historian Gar Alperovitz, the atom bombs did not so much end the Second World War as begin the Cold War. There is no doubt that the Allies wanted to keep the Soviets out of the Far East (the Red Army's presence in Eastern Europe was already a troublesome concern). Yet, with the Soviets moving troops on 8 August, the quickest end to the war would have been to accept Japan's proffered surrender on the tenth. (Possibly even earlier: on 11 July Japan's foreign minister directed a Japanese ambassador in Moscow to seek the Soviets' assistance in brokering a surrender agreement, a message that U.S. intelligence had intercepted and decoded.) And, in any event, these geopolitical considerations do not really explain the second bomb.

We know that concerns about budgetary and programmatic accountability pressed mercilessly on Groves. "If this weapon fizzles," Groves told his staff on Christmas Eve 1944, "each of you can look forward to a

FIG. 7.3. Nagasaki Medical College Hospital,
October 1945.

The buildings of Nagasaki Medical College Hospital, about 700 yards
from the hypocenter, just outside the zone of total destruction, withstood
the blast of the second bomb dropped on Japan (the plutonium one). The
interiors were burned out, and ruined; wooden structures in this area
were instantly destroyed. The first bomb (the uranium one) had destroyed the center of Hiroshima out to approximately 2,000 yards. Photograph by Hayashi Shigeo, used by permission of his wife Hayashi
Tsuneko. Published in Hiroshima-Nagasaki Publishing Committee,
Hiroshima-Nagasaki (Tokyo: Hiroshima-Nagasaki Publishing Committee, 1978), 220–21.

lifetime of testifying before congressional investigating committees."
Jimmy Byrnes—FDR's "assistant president" and Truman's secretary of
state—told Szilard, in the scientist's words, that "we had spent two billion
dollars on developing the bomb, and Congress would want to know what
we had got for the money spent."[14] The Manhattan Project's total cost of
$2 billion was 2 percent of the war-swollen 1945 U.S. federal budget, per-
haps $50 billion in current GDP terms.

One hesitates to put it this way, but the two bombs dropped on Japan
appear to have been "aimed" also at the U.S. Congress. After all, there were
two hugely expensive factories that needed justification. Indeed the pres-

sure to quickly drop the plutonium bomb *increased* after the uranium
bomb was dropped on Hiroshima. "With the success of the Hiroshima
weapon, the pressure to be ready with the much more complex implosion
device became excruciating," recalled one of Fat Man's assembly crew.[15]
The plutonium bomb's testing schedule was slashed to move its target
date up a full two days (to 9 August from 11 August). It does not seem accidental that Fat Man was dropped just before Truman's cut-off on the
tenth. Thus, in various measures, was Japan delivered a surrender-making
deathblow, while simultaneously the Soviets were shut out of the Far East
and the Manhattan Project's two massive atomic factories were justified
by the logic of war. Bohr's observation that the atomic project would
transform "the whole country into a factory," true enough in the obvious physical and organizational sense, may also be insightful in a moral
sense as well.

The *structure* of wartime atomic research had long-term consequences, even beyond the Hiroshima-Nagasaki debates, for the decades
that followed. The Army-dominated Manhattan Project had effectively
shut out the U.S. Navy. While Navy funds had developed the promising
thermal diffusion method of enriching uranium, and its pioneering trial
at the Philadelphia Navy Yard, the Navy was sidelined once this project
too was moved to Oak Ridge. The Army not only controlled the atom
bomb project during the war but also, as the result of recruiting the re-
doubtable Werhner von Braun and his German rocket scientists to its fa-

cility in Huntsville, Alabama, had the most advanced rocket project after the war. The newly organized Air Force meanwhile was laying its own claim to the atomic age with its long-range bombers.

These interservice blows shattered the Navy's prestige. While the Army and Air Force stepped confidently into the brave new atomic era, the Navy seemed stuck with its outmoded battleships and oversized aircraft carriers. A little-known Navy captain took up the proposal that the Navy develop its own atomic reactors for submarines. During the war Captain Hyman Rickover, who had a master's degree in electrical engineering and a solid record in managing electronics procurement, had been part of a low-level Navy delegation at Oak Ridge. Rickover, soon Admiral Rickover, would become the "father of the nuclear Navy."

By 1947 the hard-driving Rickover—something of a cross between the bureaucratically brilliant Groves and the technically talented Lawrence, with a double dose of their impatience and zeal—took charge of the Navy's nuclear power branch as well as the (nominally civilian) Atomic Energy Commission's naval reactors branch. He worked most closely and successfully with Westinghouse and Electric Boat. While Westinghouse worked on a novel pressurized light-water-cooled nuclear reactor, Electric Boat readied construction of the submarine itself at its Groton, Connecticut, yard. (Meanwhile a parallel project by General Electric to build a sodium-cooled reactor fell hopelessly behind schedule.) By early 1955, the Navy's first nuclear submarine, *Nautilus,* passed its initial sea trials and clocked unprecedented speeds. It also completed a submerged run of 1,300 miles to San Juan, Puerto Rico, ten times farther than the range of any existing submarine.

Nautilus, it turned out, was a precedent for more than just the U.S. Navy, which in time fully matched the other military branches with its nuclear-powered submarines capable of launching nuclear missiles. In 1953, while *Nautilus* was still in development, President Eisenhower announced his "Atoms for Peace" initiative. He aimed to dispel the heavy military cast of the country's nuclear program by stressing civilian applications. Electricity production topped his list. Eisenhower's insistence on speed and economy left the AEC little option but to draw heavily on the reactor options that were already on Rickover's drawing board. In fact, among his several duties, Rickover headed up the civilian reactor project. Working with Westinghouse once again, Rickover quickly scaled up the submarine reactor designs to produce a working civilian reactor. The reactor was installed at an AEC-owned site at Shippingport, Pennsylvania,

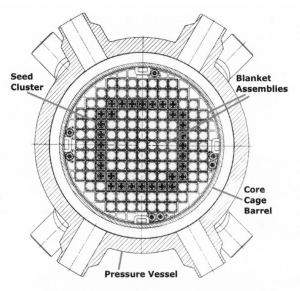

Seed Cluster

Blanket Assemblies

Core Cage Barrel

Pressure Vessel

FIG. 7.4. Shippingport Nuclear Reactor.

Commercial nuclear power in the U.S. grew up in the shadow of the U.S. Navy's work on submarine reactors. The pioneering Shippingport reactor's distinctive "seed and blanket" design was invented in 1953 at the Naval Reactors Branch of the Atomic Energy Commission. This cross-sectional diagram of the cylindrical reactor vessel shows the "seed" of highly enriched uranium (90 percent fissionable U-235) surrounded by "blanket" assemblies of natural uranium. Redrawn from Naval Reactors Branch, *The Shippingport Pressurized Water Reactor* (Washington, D.C.: Atomic Energy Commission, 1958), 6, 43, 61.

and its steam was piped over and sold to the Duquesne Light Company to spin its power turbines. Ten of the first twelve civilian reactors closely followed the Shippingport design, being knock-offs of submarine reactors.

The design decisions that made perfect sense in a military context, however, were not always well suited to a civilian, commercial context. Safety was a paramount concern in both domains, but the submarine reactor designers had not worried much about either installation or operating costs, which were of obvious importance to a profit-making company. Shippingport cost about five times what Westinghouse and General Electric considered commercially viable. Shippingport's "seed and blanket" submarine reactor (fig. 7.4) had been designed for maximum duration between refueling—of clear importance to submarines—but not for ease or speed or economy of maintenance. This mismatch between civilian and military imperatives further inflated the cost of the atomic power program.[16]

The enduring legacy of the Manhattan Project above and beyond its contribution to the atomic power effort was its creation of a nuclear weapons complex that framed years of bitter competition between the United States and the Soviet Union. The atomic fission bombs were small fry compared to the nuclear fusion bombs vigorously promoted by Edward Teller from mid-1942 onward. Where fission bombs split atoms, fusion bombs joined them. Fusion bombs, whose explosive yields were measured in multiple megatons of TNT, were a thousand-fold more powerful than the Hiroshima and Nagasaki bombs. (By comparison, the U.S. Army air force dropped approximately 3 megatons of explosives during the entire Second World War.) A cycle of recurrent technology gaps and arms races began with the atomic fission bombs (America's in 1945 and the Soviets' in 1949) and accelerated with the nuclear fusion bombs (in 1952 and 1953, respectively). The resulting nuclear arms race pitted the high technology industries of each superpower against the other in a campaign to invent, develop, produce, and deploy a succession of fusion bombs with ever-larger "throw weights" atop quicker and more accurate delivery vehicles.

In the nuclear age technology race, the advantage inclined to the offense. Each side's long-range bombers, intercontinental nuclear missiles, and finally multiwarhead MIRV missiles (multiple, independently targeted reentry vehicles) were truly fearsome weapons of mass destruction. The defensive technologies arrayed against them on both sides—the early-warning systems and the antiballistic missile, or ABM, systems—appeared hopelessly inadequate to the task. In fact, because its stupendous cost promised little safety, the United States shut down its $21 billion Safeguard ABM system located in North Dakota (1968–1978). The fantastical cost of remaining a player in the nuclear arms race strained the finances of both rivals. The cost from 1940 to 1986 of the U.S. nuclear arsenal is estimated at $5.5 trillion.[17] No one knows the fair dollar cost of the former Soviet Union's nuclear arsenal, but its currently crumbling state—nuclear technicians have in effect been told to find work elsewhere, while security over uranium and plutonium stocks is appallingly lax—constitutes arguably the foremost danger facing the planet today.

Command and Control: Solid-State Electronics

The Manhattan Project was far and away the best known of the wartime technology projects. Yet, together, the massive wartime efforts on radar, proximity fuzes, and solid-fuel rockets rivaled the atom bomb in cost. A definite rival to the bomb project in breadth of influence on technology was the $1.5 billion effort to develop high-frequency radar, a technical field in which British and American researchers worked closely, as they did in atomic physics. Centered at the Radiation Laboratory at the Massachusetts Institute of Technology, the radar project at its peak employed one in five of the country's physicists. The leading industrial contractors for the project included Bell Laboratories, Westinghouse, General Electric, Sylvania, and DuPont, as well as researchers at the University of Pennsylvania and Purdue University. Even as its radar aided the Allied war effort, the Rad Lab sowed the seeds for three classic elements of the Cold War military-industrial-university complex: digital electronic computing, high-performance solid-state electronics, and mission-oriented contract research. The U.S. military's influence expanded after the promise of digital computing and electronics opened up in the 1950s, just when the military services were enjoying open-ended research budgets and when the threats of the Cold War were placing new urgency on quick-acting command and control systems. Either by funding new technologies directly or by incorporating new electronics technologies in the huge projects focused on missile guidance, antimissile systems, nuclear weapons, and fire control, the military services imprinted a broad swath of technology with a distinct set of priorities. At the core was the military's pursuit of command and control—beginning with the bomber-alert systems of the 1950s through the electronic battlefield schemes tested in Vietnam down to the latter-day Strategic Defense Initiative.

At the Radiation Laboratory during World War II, researchers focused on three technical steps. Radar worked by, first, sending out a sharp and powerful burst of radio waves which bounced off objects in their path; receivers then detected the pulses of the returning signals; and finally the results were displayed on a video screen. While the first and third of these steps provided researchers with valuable experience in designing high-frequency electronics (which would be useful in building digital computers) the second step resulted in a great surge of research on semiconductors. Vacuum tubes were sensitive only to lower frequency signals, so when the radar project's leaders decided to concentrate on the microwave frequency (3,000 to 30,000 megahertz), they needed an electronic detector that could work in these very high frequencies. The radar researchers found that detecting devices, similar to the "cat's whisker" crystals used by ham-radio buffs (made with the semiconductors germanium and silicon) could handle the large power and high frequencies needed for radar. Much of the solid-state physics done during the war, then, focused on understanding these semiconductor materials and devising ways to purify them. "The significance of the wartime semiconductor developments in setting the stage for the invention of the transistor cannot be overemphasized," writes historian Charles Weiner.[18] Between 1941 and 1945 Bell Laboratories turned over three-fourths of its staff and facilities to some 1,500 military projects, including a large effort in radar research.

In July 1945, just as the war was ending, Bell launched a far-reaching research effort in solid-state physics. Bell's director of research, Mervin Kelly, authorized creation of a unit, the Solid State Department, to obtain "new knowledge that can be used in the development of completely new and improved components and apparatus elements of communications systems."[19] By January 1946, this department's semiconductor group had brought together the three physicists who would co-invent the transistor—William Shockley, Walter Brattain, and John Bardeen. Ironically, no one in this group had had direct experience in the radar project's semiconductor work (though other Bell researchers had). The group focused on germanium and silicon, familiar to scientists because of the wartime work with these materials. Across the next year and a half, the group's experimentalists carefully explored the surface properties of germanium— while the group's theorists among them struggled to explain the often-baffling experimental results. Unlike metals whose electrons carry electrical current, in silicon and germanium current seemed to be carried by something like a positively charged electron. With this insight the group built and demonstrated the first transistor in mid-December 1947.

Bell's transistor group explored several different devices. In December 1947 came the first transistor, a point-contact device that was descended from a ham-radio crystal. The junction transistor, invented two years later, was the type of transistor most widely used during the 1950s. It was a three-layered semiconductor sandwich in which the thin middle layer acted like a valve that controlled the flow of electrical current. The third device, a field-effect transistor, sketched in a laboratory notebook in April 1945, was first built only in the late 1950s. It exploited the electronic interactions between a plate of metal and a semiconductor. Beginning in the 1960s, such field-effect, or MOS (metal-oxide silicon), transistors formed the elements of integrated circuits. Their descendants are used in virtually all computer chips today.

Several well-known companies entered the new field by making junction transistors. Raytheon was manufacturing 10,000 junction transistors each month by March 1953, mainly for hearing aids. That same year Texas Instruments established its research laboratory, headed by a former Bell researcher, Gordon Teal. The first of the three major developments TI would contribute was a silicon transistor, which the military bought in large numbers because of its ability to operate at high temperatures and in high radiation fluxes (the Air Force's nuclear plane was the occasion for this design criterion). The second project resulted in the first pocket-sized, mass-market transistor radio, on the market by Christmas 1954 and selling for $49.95. The third TI project, purifying large amounts of silicon, was achieved in 1956. Silicon Valley also took form during these years. Bell's William Shockley moved back to his hometown of Palo Alto, California, in 1955. While a brilliant theoretical physicist, Shockley was a lousy manager, and his start-up company's employees were unhappy from the first. Eight of them left Shockley's company two years later to found Fairchild Semiconductor. Fairchild not only mastered the "planar" process for manufacturing transistors (described below), which in the 1960s allowed for mass-producing integrated circuits, but also served as the seedbed for Silicon Valley. By the early 1970s various Fairchild employees had moved on again to found at least forty-one companies in that region, including Intel, the firm responsible for commercializing the microprocessor—described as a "computer on a single chip."

While Bell Laboratories had funded the researchers who invented the transistor, the development of the device in the 1950s resulted from a close and sometimes strained partnership between Bell and the military services. In the transistor story, as in that of the Shippingport nuclear reactor, we see how the tension between military and commercial imperatives shaped the emergence of a technology that today is fundamental to our society. While Bell Laboratories saw the device as a replacement for mechanical relays in the telephone system, the military services became greatly interested in its potential for miniaturizing equipment, reducing power requirements, serving in high-speed data transmission and in computing. Bell Laboratories briefed the military services on the transistor a week before the invention's public unveiling on 30 June 1948. The military significance of the transistor was so large that the military serv-ices considered imposing a blanket secrecy ban on the technology, but, while many military applications were kept secret, no wide-ranging clas-sification was placed on the transistor field as a whole.

Indeed, instead of classifying transistors, the armed services assertively publicized military uses for them. As a specified task of its first military contract, signed in 1949, Bell Laboratories jointly sponsored with the mili-tary a five-day symposium in 1951. Attended by 300 representatives of uni-versities, industrial companies, and the armed services, this symposium advertised transistors' advantages—size, weight, and power require-ments—in military equipment and disseminated the circuit concepts needed to use the new devices. The symposium material formed a 792-page textbook. In April 1952, the Bell System sponsored a second week-long symposium for its patent licensees. For the hefty $25,000 en-trance fee, credited as advance payment on the transistor patents, the li-censees received a clear view of Bell's physics and fabrication technology. Each licensee brought home a two-volume textbook incorporating ma-terial from the first symposium. The two volumes, composing *Transistor Technology*, became known as the bible of the industry. They were origi-nally classified by the government as "restricted" but were declassified in 1953. Even four years later, an internal Bell report could describe these vol-umes as "the first and still only comprehensive detailed treatment of the complete material, technique and structure technology" of transistors and added that they "enabled all licensees to get into the military contracting business quickly and soundly."[20] A third volume in the textbook series *Transistor Technology* resulted from a Bell symposium held January 1956 to publicize its newly invented diffused base transistor. Diffused tran-sistors used oxide-masking and gaseous-diffusion to create very thin lay-ers of semiconductor that became transistors (the basis for the planar technique exploited by Fairchild). For several years Bell sold these high-performance diffused transistors only to the military services.

Beyond these publicity efforts, the military services directly financed research, development, and production technology. The military support at Bell for transistor research started small then rose to fully 50 percent of the company's total transistor funding for the years 1953 through 1955. While Bell's first military contract (1949–51) had focused on application and circuit studies, its second military contract (1951–58) specified that it was to devote services, facilities, and material to studies of military in-terest. Bell's military systems division, conducting large projects in antiballistic missile systems, specified the applications for which customized transistors were developed. From 1948 to 1957, the military funded 38 percent of Bell's transistor development expenses of $22.3 million.

In 1951 the Army Signal Corps was assigned the lead-agency responsibility for promoting military transistor development. In the late 1930s, the Signal Corps Engineering Laboratory at Fort Monmouth, New Jersey, had designed the portable radios used during the war. Just after the war, the Signal Corps had announced a major electronics development effort to apply some of the lessons learned during the war. Miniaturizing its equipment was a major objective of this effort, and the transistor fit the bill. In 1953, the Army Signal Corps underwrote the construction costs of a huge transistor plant at Laureldale, Pennsylvania, for Western Electric (the manufacturing arm of the Bell system). In addition to Western Electric, other companies benefiting from this mode of military support were General Electric, Raytheon, Radio Corporation of America, and Sylvania. Altogether the Army spent $13 million building pilot plants and production facilities.

The Army Signal Corps also steered the transistor field through its "engineering development" program, which carried prototypes to the point where they could be manufactured. For this costly process the Army let contracts from 1952 to 1964 totaling $50 million (averaging $4 million per year). Engineering development funds fostered the specific types of transistor technologies the military wanted. In 1956, when Bell released its diffusion process to industry, the Signal Corps let $15 million in engineering development contracts. The program, according to a Signal Corps historian, was to "make available to military users new devices capable of operating in the very high frequency (VHF) range which was of particular interest to the Signal Corps communications program."[21]

Bell Laboratories had not forgotten its telephone system, but its commercial applications of transistors were squeezed out by several large high-priority military projects. In 1954 Bell had planned a large effort in transistorizing the telephone network. Two large telephone projects alone, Rural Carrier and Line Concentrator, were supposed to receive 500,000 transistors in 1955 and a million the year after; all together Bell System projects were to use ten times the number of transistors as Bell's military projects. But these rosy forecasts fell victim to the Cold War's technology mobilization. In February 1955 the Army Ordnance Corps asked Bell to begin design work on the Nike II antiballistic missile system, a twenty-year-long effort that Bell historians describe as "the largest and most extensive program in depth and breadth of technology carried out by the Bell system for the military services."[22] Many of Bell's transistor development engineers, as well as much of its transistor production facilities, were redirected from telephone projects to a set of designated "preferred" military projects. These included Bell's own work on the Ballistic Missile Early Warning System (BMEWS) as well as the Nike-Hercules and Nike-Zeus missiles; work by other companies on the Atlas, Titan, and Polaris missiles; and several state-of-the-art military computer projects, including Stretch and Lightning. Given the press of this military work, Bell had little choice but to delay transistorizing its telephone network until the 1960s. It took two decades to realize Bell's 1945 plan for "completely new and improved components . . . [in] communications systems."

The integrated circuit was also to a large degree a military creation. The integrated circuit relied on the Bell-Fairchild planar technique for fabricating transistors described above. Normally production workers made a number of transistors on one side of a semiconductor wafer, then broke up the wafer and attached wire leads to the separate devices. In 1959 both Jack Kilby at Texas Instruments and Robert Noyce at Fairchild instead attached the transistors together on a single wafer. Military applications consumed nearly all the early integrated circuits (see table 7.1). Texas Instruments' first integrated circuit was a custom circuit for the military; in October 1961 the firm delivered a small computer to the Air Force. In 1962 Texas Instruments received a large military contract to build a family of twenty-two special circuits for the Minuteman missile program. Fairchild capitalized on a large contract to supply integrated circuits to NASA. By 1965 it emerged as the leader among the twenty-five firms manufacturing the new devices. Although Silicon Valley's latter-day promoters inevitably stress its origin in civilian electronics, the importance of the military market to many of the pioneering firms is clear.

Across the 1950s and 1960s, then, the military not only accelerated development in solid-state electronics but also gave structure to the industry, in part by encouraging a wide dissemination of (certain types of) transistor technology and also by helping set industrywide standards. The focus on military applications encouraged the development of several variants of transistors, chosen because their performance characteristics suited military applications and despite their higher cost. Some transistor technologies initially of use only to the military subsequently found wide application (silicon transistors, the diffusion-planar process) while other military-specific technologies came to dead ends (gold-bonded

TABLE 7.1. U.S. Integrated Circuit Production and Prices, 1962–68

Year	Total production ($ million)	Average price per unit ($)	Military production as percent of total production
1962	4	50.00	100
1963	16	31.60	94
1964	41	18.50	85
1965	79	8.33	72
1966	148	5.05	53
1967	228	3.32	43
1968	312	2.33	37

Source: John Tilton, *International Diffusion of Technology: The Case of Semiconductors* (Washington, D.C.: Brookings Institution, 1971), p. 91.

diodes and several schemes for miniaturization that were rendered obsolete by integrated circuits). The large demand generated by the guided-missile and space projects of the 1950s and 1960s caused a shortage of transistor engineers in Bell Laboratories' commercial (telephone) projects. These competing demands probably delayed the large-scale application of transistors to the telephone system at least a half-dozen years (from 1955 to the early 1960s).

Command and Control: Digital Computing

Compared to solid-state electronics, a field in which the principal military influence came in the demands of top-priority applications, military sponsorship in the field of digital computing came more directly, in the funding of research and engineering work. Most of the pioneers in digital computing either worked on military-funded contracts with the goal to develop digital computers or sold their first machines to the military services. Code-breaking, artillery range-finding, nuclear weapons designing, aircraft and missile controlling, and antimissile warning were among the leading military projects that shaped digital computing in its formative years, from the 1940s through the 1960s. Kenneth Flamm in his comprehensive study of the computer industry found that no fewer than seventeen of twenty-five major developments in computer technology during these years—including hardware components like transistors and integrated circuits as well as design elements like stored program code, index register, interrupt mechanism, graphics display, and virtual memory—not only received government research and development funding but also were first sold to the military services, the National Security Agency, or the AEC's Livermore weapons-design lab. In addition, micro-programming, floating-point hardware, and data channels benefited from one or the other mode of government support. Many specific developments in memory technology, computer hardware, and computer software (both time-sharing and batch operating systems) were also heavily backed by government R&D funds or direct sales, often through the network of military contractors that were using computers for missile and aviation projects.[23]

Most of the legendary names in early computing did significant work with military sponsorship in one form or another. The fundamental "stored program" concept emerged from work done during and immediately after the World War II on code-breaking, artillery range-finding, and nuclear weapons designing. At Harvard, Howard Aiken's Mark I relay computer, a joint project with IBM before the war, became a Navy-funded project during the war. At the University of Pennsylvania's Moore School, J. Presper Eckert gained experience with memory techniques and high-speed electronics through a subcontract from MIT's Radiation Laboratory. Soon he and John W. Mauchly built their ENIAC and EDVAC machines with funds from the Army Ordnance Department, which had employed 200 women as "computers" laboring at the task of computing artillery range tables by hand. Eckert and Mauchly, struggling to commercialize their work after the war, paid their bills by building a digital computer to control the Air Force's Snark missile through a subcontract with Northrop Aircraft. At the Institute for Advanced Study (IAS) at Princeton, John von Neumann started a major computer development project. During the war, von Neumann had worked with Los Alamos physicists doing tedious hand calculations for the atomic bomb designs. Immediately after the war, he collaborated with RCA (much as Aiken had done with IBM) on a joint Army and Navy–funded project to build a general-purpose digital computer, about which he published several path-breaking papers. Von Neumann arranged that the first major calculation made on the institute's machine was computer modeling for the thermonuclear bomb. Copies of this machine were built for government use at six designated sites: Aberdeen Proving Ground in Maryland, Los Alamos, Argonne outside Chicago, Oak Ridge, RAND (an Air Force think tank), and an Army machine (code-named ILLIAC) at the University of Illinois.[24]

The Harvard, Pennsylvania, and IAS computer projects are justly remembered for developing the basic concepts of digital computing. Project Whirlwind at MIT, by comparison, was notable for developing computer hardware—particularly, fast magnetic-core memory—and notorious for consuming an enormous amount of money in the process. It also showed that computers could do more than calculate equations, providing a paradigm of command and control. Whirlwind initially took shape during the war as a Navy-sponsored project to build a flight trainer. At MIT's Servomechanisms Laboratory, Jay Forrester used a successor of Vannevar Bush's analog differential analyzer (see chapter 5) in an attempt to do real-time airplane flight simulation. Forrester, an advanced graduate student at the time, used an array of electromechanical devices to solve numerous differential equations that modeled an airplane's flight. But in October 1945, at an MIT conference that discussed the Moore School's ENIAC and EDVAC machines, Forrester was bitten by the digital bug. Within months he and the MIT researchers dropped their analog project and persuaded the new Office of Naval Research to fund a digital computer, a thinly disguised general-purpose computer carrying a price tag of $2.4 million which they said might be used for flight simulations. Forrester's ambitious and expensive plans broke the bank, even though the ONR was at the time the principal research branch for the entire U.S. military (separate research offices for the Navy, Army, and Air Force, as well as the National Science Foundation, were all still on the horizon). Whirlwind's annual funding requests were five times larger than those for the ONR's other major computer effort, von Neumann's IAS project.

"Make no small plans, for they have no power to stir the hearts of men," architect Daniel Burnham had once advised. Forrester certainly subscribed to this philosophy. Faced with mounting skepticism from the Navy that it would ever see a working flight simulator, Forrester articulated an all-embracing vision of Whirlwind as a "Universal Computer" at the heart of a "coordinated CIC" (Combat Information Center) featuring "automatic defensive" capabilities as required in "rocket and guided missile warfare." Computers, he prophesied, would facilitate military research on such "dynamic systems" as aircraft control, radar tracking and fire control, guided missile stability, torpedoes, servomechanisms, and surface ships.[25] Forrester wanted Whirlwind to become another megaproject like the Radiation Laboratory or Manhattan Project. The Navy blanched at the prospect and dramatically scaled back its Whirlwind funding, although the ONR continued substantial funding across the 1950s for MIT's Center for Machine Computation. It was the Air Force,

FIG. 7.5. Whirlwind Computer Control Center.

The Whirlwind computer in 1950, at a moment of transition. Gone by this time was the original Navy-funded plan to use Whirlwind as a flight trainer. This image shows Forrester's vision of Whirlwind, then funded by the Air Force, as a command-and-control center. An array of computers built to the Whirlwind design formed the core of the bomber defense system known as SAGE (Semi-Automated Ground Environment). Left to right are Stephen Dodd, Jr., Jay Forrester, Robert Everett, and Ramona Ferenz. Courtesy of Massachusetts Institute of Technology Museum.

increasingly concerned with defense against Soviet bombers, that took up Forrester's grand vision for command and control (fig. 7.5). In 1950, with Whirlwind amply funded at just under $1 million that year—the bulk of the money now coming from the Air Force—MIT began a series of Air Force–commissioned studies of air defense. These advocacy studies would culminate in the fantastically complex bomber defense system known as SAGE.

Operationally obsolete before it was deployed, SAGE (Semi-Automated Ground Environment) nonetheless was an important link between the advanced work on computing and systems management at MIT and the wider world. (SAGE was first operated a year after the Soviets launched the Sputnik satellite in 1957, by which time missiles flying much too fast to intercept had displaced slow-flying bombers as the principal air defense concern.) SAGE aimed to gather and assess a continuous flow of real-time data from radar sites arrayed across North America, with the goal of alerting and directing the response of the Air Force in the case of a Soviet bomber attack. Such a huge effort was beyond even Forrester's grasp, and, just as Bush had spun off the Manhattan Project from his Office of Scientific Research and Development, the MIT administration created the Lincoln Laboratory as a separate branch of the university to manage the massive SAGE project. Forrester's Whirlwind project thus became Lincoln Laboratory's Division 6. Most of the country's leading computer and electronics manufacturers were SAGE contractors, including RCA, Burroughs, Bendix, IBM, General Electric, Bell Telephone Laboratories, and Western Electric. The newly founded Systems Development Corporation wrote the several hundred thousand lines of computer code that the project employed. Estimates for the total cost of SAGE range surprisingly widely, from $4 billion to $12 billion (1950s dollars).[26]

At the center of this fantastic scheme was Forrester's Whirlwind, or more precisely fifty-six of his machines. To produce the Whirlwind duplicates for SAGE, Lincoln Laboratory in 1952 selected a medium-size business machines manufacturer previously distinguished for supplying tabulating machines to the new Social Security Administration and punch cards to the Army. The firm was named, rather grandly, International Business Machines. At the time it began its work on SAGE, IBM ranked fourth among computer manufacturers, behind RCA, Sperry Rand, and Burroughs. IBM would make the most of this military-spawned opportunity, launching itself into the front ranks of computing. IBM was already working on an experimental research computer for the Naval Ordnance Laboratories and had begun work on its first electronic computer, the IBM 701 Defense Calculator. With its participation in SAGE, IBM gained a healthy stream of revenues totaling $500 million across the project's duration. Fully half of IBM's domestic electronic data-processing revenues in the 1950s came from just two military projects: SAGE and the "Bomb-Nav" analog computer for the B-52 bomber. These handsome revenues in turn convinced IBM's president, Thomas J. Watson, Jr., to push the company into the computer age and bankrolled the company's major developmental effort for the System 360 computer (1962–64).

As important as this revenue stream was the unparalleled exposure to state-of-the-art computing concepts and the unconstrained military budgets that permitted the realization of those concepts. With its participation in SAGE, IBM gained first-hand knowledge of MIT's magnetic-core memories and invaluable experience with mass-producing printed-circuit boards. It incorporated the computer-networking concepts of SAGE into its SABRE airline reservation system (the name is revealing: *Semiautomatic* Business-Research *Environment*) which became operational in 1964. Another military undertaking whose influence can be seen in IBM's landmark System 360 was Project Stretch. Beginning in 1955 Stretch involved IBM with the National Security Agency and the Atomic Energy Commission's Livermore nuclear weapons laboratory. The two agencies had radically different computing needs. The AEC lab needed to do numerical computations with high precision, while the NSA needed to manipulate very large databases. By devising a single computer design that satisfied both of these demanding users, IBM was well on the way to a machine, realized with System 360, that satisfied both scientific and business demands. The NSA's Stretch computer, delivered in 1962 as the core of a data-processing complex, served the NSA until 1976, when it was replaced by the renowned Cray I, the first so-called supercomputer.[27]

Even though the commercial success of IBM's System 360 made computing a much more mainstream activity, the military retained its pronounced presence in computer science throughout the 1960s and beyond. This was the result once again of open-ended research budgets and a cadre of technically savvy officers. The center of the military's computing research efforts was a small office known by its acronym IPTO, Information Processing Techniques Office, located within the Pentagon's Advanced Research Project Agency. Founded in 1959, in the wake of the Sputnik crisis, ARPA itself funded research on a wide range of military-related topics, ranging from ballistic missile defense, nuclear test detection, and counter-guerrilla warfare to fields such as behavioral sciences, advanced materials, and computing. The IPTO was far and away the nation's largest funder of advanced computer science from its founding in 1962 through the early 1980s. Its projects strengthened established computing centers at MIT and Carnegie-Mellon and built new computing "centers of excellence" at Stanford, UCLA, Illinois, Utah, Caltech, USC, Berkeley, and Rutgers. Among the fundamental advances in and applications of computer science funded by the IPTO were time-sharing, interactive computer graphics, and artificial intelligence. J. C. R. Licklider, head of the IPTO program in the early 1960s, also initiated work on computer networking that led, after many twists and turns, to the Internet.

In the mid-1960s IPTO saw a looming budget crunch. Each of its leading sponsored research centers wanted a suite of expensive computers, and not even the Pentagon could keep up with their ever-expanding budget requests. The solution, pretty clearly, was to enable researchers at one site to remotely use the specialized computing resources at distant sites. Ever canny, the IPTO's program managers, from the 1960s through the 1980s, sold a succession of Internet-spawning techniques to the military brass as the necessary means to a reliable military communications network. One result of the 1962 Cuban missile crisis was the military's realization that it needed the ability to communicate quickly and reliably across its worldwide field of operations. A break in the fragile communications network might isolate a battlefield commander from headquarters and (as the film *Dr. Strangelove* satirized) lead to all sorts of dangerous mishaps. Alternatively, a single Soviet bomb strike on a centralized switching facility might knock out the entire communication system. A 1964 RAND Corporation report, "On Distributed Communications," proposed the theoretical grounds for a rugged, bombproof network using "message blocks"—later known as "packet switching"—to build a distributed communications system. RAND endorsed this novel technique because it seemed to ensure that military communication networks would survive a nuclear attack. Even if one or more nodes in the distributed network were destroyed, the network itself could identify alternative paths and route a message through to its destination.[28] These concepts became the conceptual core of the Internet.

THROUGHOUT THE DECADES of world war and cold war, the various arenas of military technology (East and West) appeared to many outsiders to be a world unto themselves, cut off from the broader society. This image was reinforced by the esoteric nature of the science involved in military technologies—dealing with the smallest wavelengths and the largest numbers. The pervasive security regimes swallowed up entire fields of science and technology (including some not-so-obviously military ones, such as deep-water oceanography). Western critics faulted the "military-industrial complex" or "national-security state" for being antithetical and even dangerous to a democratic society, with its ideals of openness and transparency.

Through the military-dominated era there was an unsettling tension between the West's individual-centered ideology and its state-centered technologies. Visions of free markets and open democracy were casualties of the secrecy regime of the Manhattan Project, the Oppenheimer-Teller conflict about the hydrogen bomb, President Eisenhower's "military-industrial complex," the strident protests of the Vietnam War (which were perpetuated by the studied deceptions of the Kennedy, Johnson, and Nixon administrations as much as by the strategy and tactics of the war), and the pitched techno-political battles in the Reagan years over the Strategic Defense Initiative. Throughout the Cold War "national security" was a trump card to silence troublesome critics.

On balance, in my view, the West was lucky to be rid of the Cold War, because (as in many conflicts) we took on some of the worst characteristics of our enemies. Faced with the image of a maniacal enemy driven by cold logic unencumbered by the niceties of democratic process, Western governments were all too often tempted to "suspend" open processes and crack down on dissidents. By comparison, dissidents in the Soviet Union found that criticizing the government, especially on military and nuclear matters, was hazardous in the extreme, although, interestingly, criticism of governance and technology pertaining to the environment was permitted. The Soviet Union's state-centered ideology and state-centered technologies "fit" the imperatives of the Cold War better than the West's did. What the Soviet system failed to do was generate economic surpluses large enough to pay for the heavy military expenditures, which dragged down their citizens' economic prospects, bankrupted their governments, and led to their political demise.

What is striking to me is the wide reach and pervasiveness within society of military technologies developed during these decades. Niels Bohr captured something fundamental with his remark about the atom bomb project "turning the whole country into a factory." Developing the bomb involved, in addition to the innumerable military officers and government officials, the managers and workers of the nation's leading corporations; a similar list of well-known enterprises can be compiled for each of the numerous Cold War projects. Together, these military endeavors were not so much an "outside influence" on technology as an all-pervading environment that defined what the technical problems were, how they were to be addressed, and who would pay the bills. While closed-world, command-and-control technologies typified the military era, the post–Cold War era of globalization has generated more open-ended, consumer-oriented, and networked technologies.[29]

8

TOWARD GLOBAL CULTURE

1970–2001

"GLOBALIZATION . . . IS THE MOST important phenomenon of our time. It is a contentious, complicated subject, and each particular element of it—from the alleged Americanization of world culture to the supposed end of the nation-state to the triumph of global companies—is itself at the heart of a huge debate."[1] Examining globalization is a difficult task for a historian at this time, because the dust from its opening flurry has hardly settled. A helpful way of thinking about globalization, it seems to me, is as our generation's industrial revolution, a set of unsettling shifts in society, economics, and culture in which technology appears to be in the driver's seat. We'll see on closer inspection that technology in the era of global culture, no less than in earlier eras, has been both a powerful shaper of political, social, cultural, and economic developments as well as being shaped by these developments.

The arguments about globalization currently simmering, like many arguments involving technology, are largely about contending visions for the future. This is why debates about technology are rarely just about technical details, but rather are about the alternative forms of society and culture that these technical details might promote or deny. As we will see, a vision of globalization is inscribed in the technical details of fax machines and the Internet, even as the commercialization of these technologies, among others, promotes globalization in economics and culture. Globalization, of course, is a strongly contested concept and movement.

Advocates of globalization point out that the rise in world trade, especially dramatic since the 1970s, has increased average incomes, across the world and nation-by-nation. Countries that are open to globalization have done better economically than those that are not. In this view, globalization has rescued millions of people in Asia and Latin America from poverty (while there remain many shortfalls in Africa). Further, advocates maintain that the glare of global publicity has improved working conditions and environmental regulation in countries lacking proper labor and environmental laws. When all else fails, advocates sometimes claim that—in a contemporary version of technological fundamentalism—the global economy is the only game in town. "The global economy in which both managers and policy-makers must now operate . . . has its own reality, its own rules, and its own logic. The leaders of most nation states have a difficult time accepting that reality, following those rules, or accommodating that logic," writes business guru Kenichi Ohmae.[2]

Critics counter that globalization has utterly failed to ease the crushing burdens of the world's poorest citizens and that globalization threatens to increase pollution. For instance, an industrial revolution seems imminent in coal-rich China and threatens to exceed the worst air pollution of industrial-era Britain, while villagers in Brazil have few options but to burn the rain forest and become herders or farmers. In both examples, globalization is implicated in global warming. The high profile of multinational corporations has revived a century's worth of arguments against the excesses of free-market capitalism. Critics also blame globalization for the growing inequality between global "haves" and "have-nots," for the cultural imperialism manifest in American media and fast food, and the antidemocratic power of international entities such as the World Trade Organization. "Satellites, cables, walkmans, videocassette recorders, CDs, and other marvels of entertainment technology have created the arteries through which modern entertainment conglomerates are homogenizing global culture," write Richard Barnet and John Cavanagh, in their essay "Homogenization of Global Culture": "Literally the entire planet is being wired into music, movies, news, television programs, and other cultural products that originate primarily in the film and recording studios of the United States."[3]

The noisy debate on globalization seems to me a healthy one. After all, at issue are our visions for a good life. The debate itself has also modestly altered the "nature" of globalization as it unfolded in the 1990s. The gap between globalization and development was spotlighted in the dramatic street protests at the World Trade Organization's Seattle summit meeting in 1999. One might say that the clouds of tear gas aimed at antiglobalization protesters—in Seattle, Washington, Prague, and Genoa—ended the pie-in-the-sky optimism in which globalizers had been indulging until then. Moreover, there are signs of a change in institutional attitude. Not so long ago the World Bank enforced a narrow measure of economic development. Its sole concern was increasing national GDP levels (an aggregate measure for economic development), and it typically required a recipient country to undergo painful "structural adjustment" of its macroeconomic policies before granting it multibillion-dollar loans. Because the World Bank's structural adjustments typically required cutting government expenditures and/or raising taxes, it gained a frightful reputation among the affected citizens of recipient countries. In the wake of globalization debates, however, the World Bank seems to be embracing broader indexes of social development, giving attention to education, public health, local infrastructure, and the environment.[4]

Whatever the economic and political consequences of globalization, the threat of cultural homogenization concerns many observers. Everyone, it seems, watches CNN and Disney, drinks Coca-Cola, and eats at McDonald's. Are we really becoming more and more alike? There is some support for such a "convergence hypothesis." Look at the consumption of alcohol and the habits of industry. Alcohol consumption during the first industrial revolution was a much-discussed "problem"; factory masters wanted their workers sober and efficient, while many workers preferred to spend their income on beer (see chapter 3). As recently as the mid-1960s the citizens of France, Italy, and Portugal each drank an astounding 120 liters of wine each year, and the Spaniards were not far behind; but the citizens of thoroughly industrial Germany drank one-*sixth* that amount, while producing wine for export. (Making allowances for differential consumption by men, women, and children, one estimates that each adult man in France, Italy, and Portugal drank a liter of wine each day.) By the late 1990s, however, wine consumption had plummeted—by half or more—in each of the four southern European countries, while Germans' wine drinking remained steady. One might say that northern habits of industry and efficiency eroded the southern traditions of heavy wine drinking. If southern Europe gained industrial efficiency, Europe as a whole lost cultural diversity. Today, the worldwide spread of MTV, American television, and American fast food is said to drive out local cultural productions and destroy cultural diversity.

While mindful of the possibilities of convergence, I believe there is greater evidence for a contrary hypothesis. I doubt that globalization—

FIG. 8.1. Coca-Colonization of Europe?

A typical Dutch mobile snack bar, serving French fries with mayonnaise, at Zaanse Schans, 2001. Photograph by Marco Okhuizen, used by permission of Bureau Hollandse Hoogte.

as people around the world experience these changes—is really such a universalizing force as its critics believe it to be. Consider this evidence. In most countries, American television programming fills in the gaps but the top-rated shows are *local* productions. In South Africa the most-watched show is "Generations," a locally produced multiracial soap opera; in France, "Julie Lescaut," a French police series; and in Brazil, "O Clone," a drama produced by TV Globo. For its part MTV, which once boasted "one world, one image, one channel," now creates no fewer than thirty-five different channels (fifteen in Europe alone) in response to local tastes and languages. Four-fifths of the programming on MTV Italy is in fact made in Italy. Even McDonald's, a symbol of globalization, is more complex than it might appear. I suggest below that McDonald's is best understood as a case of active appropriation by local cultures of a "global" phenomenon. The Internet is another construct that appears to be the same across the world only so long as you don't look closely at Trinidadian teenagers' websites or European electronic commerce ventures.[5]

Surprising diversity persists in statistical measures that "ought" to be converging. Taxation rates, unionization rates, and even merchandise exports remain stubbornly different among countries, even in the industrial West. Or consider coffee drinking: Using caffeine-laden beverages to boost working efficiency and make quick meals more satisfying is a product of the industrial era, so one might expect coffee consumption in industrial countries to converge.[6] But in fact, the Swiss and Swedes drink roughly twice as much coffee per capita as Americans do, four times the British and South Koreans, and eight times the Japanese. Whatever it is that keeps hard-working Swiss and hard-working Japanese at their work stations is not equal doses of caffeine from coffee.

The "divergence hypothesis" is also consistent with what we have learned from earlier eras. As we saw in chapters 3 and 5, the paths taken by countries during the first and second industrial revolutions varied according to the countries' political and institutional structures, social and cultural preferences, and the availability of raw materials. It is worth adding here that Japan did industrialize and modernize during the twentieth century but it did not become "Western," as is attested by the many distinctive political, industrial, and financial structures and practices in Japan that to this day baffle visiting Westerners. This chapter first surveys the origins of the global economy, then turns to case studies of the facsimile machine, McDonald's, and the Internet.

The Third Global Economy

Our present-day global economy is not the first or second global economy we have examined in this book, but the third. The first was in the era of commerce (see chapter 2). The Dutch in the seventeenth century created the first substantially "global" economy through multicentered trading, transport, and "traffic" industries that stitched together cottons from India, spices from Indonesia, slaves from Africa, sugar canes in the Caribbean with tulip speculators, stock traders, boat builders, sugar refineries, and skilled artisans in the home country. This early global economy lasted until nationalist economic pressures from France and England overwhelmed it in the early eighteenth century. Early industrialization in England—even though it imported raw materials, like Swedish iron and American cotton, and exported cotton textiles, machine tools, and locomotives—did not create a Dutch-style multicentered trading system.

A second global economy developed in the 1860s and lasted until around the First World War, overlapping with the era of imperialism (see chapter 4). Like present-day globalization, this second global economy involved huge international flows of money, people, and goods. Britain, with its large overseas empire, was the chief player. A key enabling technology was the ocean-going steamship, which by the 1870s was fearsomely efficient and cheaply carried people and goods. In compound steam engines, "waste" steam exiting one bank of cylinders was fed into a second bank of lower-pressure cylinders to extract additional mechanical power. By the 1890s ocean-going steamships employed state-of-the-art triple-stage compound steam engines; some builders even tried out four stages. In these super-efficient engines, the heat from burning two sheets of paper moved a ton of cargo a full mile. Linking efficient steamships with nationwide railroad networks dramatically cut transport costs for all sorts of goods. Even decidedly middling-value products found a ready global market. Butter from New Zealand, sheep from Australia, jute from the Far East, and wheat from North America all flowed in huge volumes to European ports.[7]

The complex tasks of trading, warehousing, shipping, insuring, and marketing these global products led to a sizable and specialized financial superstructure. In these years Chicago became the world's grain-trading center and New York specialized in cotton and railroad securities, while London became the world's investment banker. London, as we noted in chapters 3 and 4, had the world's most sophisticated financial markets and substantial exportable wealth from the first industrial revolution. So active were its international investors that Britain's net outflows of capital averaged 5 percent of GDP from 1880 to 1913, peaking at an astounding 10 percent (while by comparison Japan's net outflows of capital have recently been considered "excessive" at 2–3 percent). In the decade prior to the First World War, Britain's wealthy made foreign direct investments (in factories and other tangible assets) that nearly equaled their domestic direct investments, and the wealthier residents of Europe displayed similar behavior. In fact, foreign direct investments made by investors in Britain, France, Germany, and the Netherlands were substantially *larger* in GDP terms in 1914 than in the mid-1990s. The United States, Canada, Argentina, and Australia were the chief recipients of European investment.

Widespread use of the gold standard assisted the international flow of money, since traders always knew that one British pound sterling equaled exactly five U.S. dollars, while a world without passports facilitated the international flow of people. The industrialization of North American cities and the extension of agriculture across the Great Plains tugged on peasants, small holders, and artisans in Europe facing economic difficulties, religious persecution, or family difficulties. Some 60 million people left Europe for North and South America in the second half of the nineteenth century. This huge transatlantic traffic came to a screeching halt with the First World War, as the Atlantic filled up with hostile ships and submarines. In the early 1920s, a set of blatantly discriminatory laws in the United States finished off immigration from southern and eastern Europe, where the most recent generation of immigrants had come from. Immigration to the United States did not reach its pre-1914 levels until the 1990s.

While wartime pressures and American nativism in the 1920s weakened the second global economy, it was brought low by economic nationalism during the 1930s. Hitler's Germany offered an extreme case of going it alone. Across the industrialized countries, increased tariffs, decreased credits, and various trading restrictions effectively ended the wide-ranging international circulation of money and goods. Import restrictions in the United States culminated in the notorious 1931 Smoot-Hawley tariff, which hiked average U.S. import tariffs to just under 60 percent. This was a deathblow to world trade because the United States was, at the time, the world's leading creditor nation. High U.S. tariffs made it impossible for European debtor nations to trade their goods for U.S. dollars and thus to repay their loans and credits (a hangover from the First World War). Trade among the seventy-five leading countries, already weak in January 1929, during the next four years spiraled down to less than one-third that volume. Country after country defaulted on its loans and abandoned the gold standard. Global movements of people, capital, and goods had looked pretty strong in 1910, but this trend was a dead letter within two decades. Even before the Nazis' brutally nationalistic Four Year Plan of 1936, the second global economy was dead.

Since around 1970 there has been a resurgence of global forces in the economy and in society, but who can say how long it will last. The institutions charged with "managing" the global economy today, including the World Bank, International Monetary Fund, and World Trade Organization, are human creations; they can change and have changed the pace and character of globalization. And the supposed flood tide of globalization now occurring has odd countercurrents. Even in the 1990s, when such countries as Singapore and the Philippines became the globalizers' darlings for their openness to trade and investment, sound credit, active tourism, and transfers from overseas workers, a number of countries actually became *less* globalized. According to an index composed of those measures drawn up by A. T. Kearney, globalization actually *decreased* in Malaysia, Mexico, Indonesia, Turkey, and Egypt.

Fax Machines and Global Governance

The very first electromechanical facsimile machines were something of a global technology in the mid-nineteenth century, and the rapid commercialization of their electronic descendants in the 1970s and 1980s was due in large measure to the emergence of international standards for technology. Like the geographical shifts during the era of science and systems, the global era has seen the geographical relocation of consumer electronics. Fax machines were one of numerous consumer electronics products— transistor radios, televisions, videocassette recorders, and computer-game consoles—that Japanese manufacturers soundly dominated.[8] While U.S. firms accepted the imperatives of military electronics, Japanese firms embraced the possibilities of consumer electronics. One might say that in the United States the military market displaced the consumer market, while in postwar Japan it was the other way around. The structure of the global economy can in part be traced to the different paths taken by each nation's electronics industry.

In one way, Japanese manufacturers had a cultural incentive for facsimile communication. The image-friendly character of fax machines is especially compatible with pictographic languages like Japanese. Indeed it was just after a Chinese telegraph commission visited Boston in 1868 that a young Thomas Edison, impressed by the difficulties of fitting the Chinese language into the digital structure of Morse code telegraphy, began work on an early electromechanical fax machine. While English has only twenty-six characters and was easily translated into a series of telegraphic dots-and-dashes (or digital-era ones and zeros), the standard Japanese system of *kanji* has more than two thousand characters plus forty-six *katakana* phonetic characters used for foreign words. Chinese has even more characters. Edison saw that transmitting pictures by telegraph would be ideal for ideographic languages.

While other would-be inventors of facsimile-transmitting machines typically used raised-type (including the first fax machine patented by Alexander Bain, in 1843), and often used multiple wires, Edison's fax machine used a single telegraph wire. A message written on paper with a common graphite pencil was "read" by his ingenious machine. The electricity-conducting graphite completed a circuit between two closely spaced wires on a pendulum, tracing an arc over the message, and sent a burst of electricity from the transmitting machine to the receiving machine. In the 1930s William G. Finch's radio facsimile deployed an identical conceptual scheme (including synchronized pendulums and chemically treated paper) to successfully transmit text and illustrations. Besides using radio waves as the carrier, Finch's machine also deployed photocells—available since around 1900—to read the image for transmission.

From the 1920s through the 1960s facsimile remained a niche product for specialized users. Facsimile schemes of RCA, Western Union, and AT&T targeted mass-market newspapers seeking circulation-boosting news photographs. The best-known venture in this area was the Associated Press's Wirephoto service, established in 1934. Another specialized user unconstrained by facsimile's high costs was the military. During the Second World War, both the German and the Allied forces transmitted maps, photographs, and weather data by facsimile. Even after the war the Times Facsimile Corporation, a subsidiary of the New York Times Company, found its largest market in the military services. By the late 1950s Western Union, desperate to preserve its aging telegraph network against the onslaught of telephones, had sold 50,000 Desk-Fax units. These machines directly connected medium-sized companies with the nearest Western Union central office to send and receive telegraph messages.

In the 1960s Magnavox and Xerox began selling general-purpose, analog facsimile machines. Their modest sales and poor profits can be traced to their fax machines' slow transmission and corresponding high telephone costs; furthermore, in the absence of commonly accepted standards, most fax machines operated only with machines of their own brand. A further barrier was that Bell Telephone, a regulated monopoly, forbad users to connect non-Bell equipment to its phone lines. (Bell maintained that such a restriction was needed to insure the "integrity" of the telephone system, even as it locked in a lucrative monopoly for Western Electric, its manufacturing arm.) To sell one of its Telecopier machines for use over the public phone system, Xerox needed to add a special box, sold by Western Electric, that connected the Telecopier safely to the phone system. A regulatory change in 1968—the Carterphone decision—eased this restriction, allowing fax machines (in 1978) and eventually also computer modems to be directly connected to the telephone system. Parallel changes occurred in Japan.

The explosive growth of facsimile was made possible by the creation of worldwide technology standards. Only after global standards were in place could fax machines from different manufacturers "interoperate." At the time, in the late 1960s, the leading fax manufacturers—Magnavox, Xerox, and Graphic Services (owned by Burroughs)—each had proprietary standards that only vaguely conformed to the prevailing U.S. standards. Accordingly, sending a message between machines of two different brands was "fraught with error." (Indeed, most facsimile messages were sent on private or leased phone lines between machines of the same brand.) While it was clear that incompatible proprietary standards constituted a barrier to the wider spread of fax technology, companies clung to their own standards for fear of helping a rival or hurting themselves. Enter the principal standards-setting body for facsimile, the CCITT, a branch of the United Nations' International Telecommunication Union, whose headquarters are in Geneva, Switzerland.

The CCITT, or Comité Consultatif International Télégraphique et Téléphonique, was the leading international standards-setting body for all of telecommunications beginning in the 1950s. Its special strength was and remains standards setting by committee. In some instances, when there is a single producer, a single market, or a single country large enough to pull others along, markets or governments set standards. For example, with the stunning success of IBM's System 360 line of computers in the 1960s, that company's proprietary computer standards became de facto international ones. Governments, too, often legislate a certain technological standard, such as North America's 525-line televisions contrasted with Europe's 625-line ones. But for many communications technologies, including facsimile, no single firm or single country could set durable international standards. This fragmentation became increasingly problematic; the collective benefit of common standards in a "network technology" (increased certainty, economies of scale, lowered transaction and legal costs) were reasonably clear and certainly attractive to manufacturers and users alike. It was here that the CCITT—with its quadrennial plenary meetings, on-going formal study groups, scores of working committees, legions of subcommittees, not to mention informal coffee breaks—shaped the standards-setting process for facsimile and the use of the resulting fax technology.

The working groups of CCITT found facsimile standards comparatively easy to agree on since facsimile was a "terminal" technology that relied on the telephone system, rather than an all-embracing system. Decision making among the representatives of the member governments, leading manufacturers, and telephone network operators dealt separately with each of facsimile technology's critical characteristics across successive generations. For fax machines, higher scanning resolutions are compatible with slower transmission rates, at the cost of lengthier phone calls.

The development of "handshaking"—those starting seconds where fax machines scream at one another to find common transmission parameters—liberated fax machines from the constant watch of human operators. With handshaking, a feature that CCITT standardized in 1980, not only could fax machines receive messages by themselves, manufacturers could also add new features to their own line of machines and yet keep them compatible with the installed universe of fax machines. Here was a standard that encouraged innovation while preserving the network advantages of a large installed customer base.

It was CCITT's success with the 1980 standards that made facsimile into a global technology—and relocated the industry to Japan. Handshaking was only part of the achievement. The pre-1980 standards had specified a worldwide standard using amplitude-modulated (AM) signal processing, and transmission times at three minutes per page were close to the theoretical maximum attainable with analog processing. The 1980 standard effected a shift to digital processing that offered a way to "compress" the scanned image; for instance, instead of coding individually thirty-six times for a blank space, a compression routine would code a short message to repeat a blank space thirty-six times. The compressed digital information was then transformed by the fax machine's internal modem into an analog signal and sent over the phone system. Machines conforming to the digital standard could send an entire page in one minute. The achievement of worldwide standards, digital compression, and flexible handshaking, in combination with open access to public telephone systems, created a huge potential market for facsimile. By the late 1990s facsimile accounted for fully half the telephone traffic between Japan and the United States.

With Japanese electronics manufacturers poised to manufacture the digital processing machines, American fax manufacturers in effect declined to make the switch. Already by the late 1960s Matsushita had pioneered flat-head scanning technology, using glass fibers, as well as electronic-tube scanning; in 1971, when the firm started work on solid-state scanning and high-speed (digital) scanning, it had made 296 fax patent applications and had 110 engineers working on various aspects of fax development. An early signal of Matsushita's prominence was that the U.S. firm Litton, once a leading exporter of fax machines, stopped developing its own machines in 1968 and started importing Matsushita's. Ricoh also began its high-speed fax development in 1971, with approximately fifty engineers organized into three teams. In 1973 Ricoh introduced onto the market a digital fax machine capable of 60-second-per-page transmission, while Matsushita answered two years later with its 60-second machine (and by 1986 had achieved a 12-second machine).[9] The distinctive Japanese mixture of technological cooperation with market competition meant that Matsushita and Ricoh, followed closely by Canon, NEC, and Toshiba, emerged as world leaders. By the late 1980s Japanese firms were producing over 90 percent of the world's fax machines.

While fax machines in the late 1980s were mostly used for traditional business activities, such as sending out price lists and product information (not to mention placing lunch orders), a group of inspired secondary-school teachers and students in Europe spotted an exciting possibility in using fax machines for broader cultural purposes. This network of students and teachers, along with some journalists and government officials, is notable not only for creatively using fax technology but also for explicitly theorizing about their culture-making use of technology. Secondary-school students working in the "Fax! Programme" collaborated with student-colleagues across Europe to produce a series of newspapers that were each assembled and distributed, in a single day, by fax.[10]

The Fax! Programme embodied and expressed a "shared desire to transform a European ideal into reality" (39). The idea of using fax machines for building European identity and youth culture originated with the Education and Media Liaison Center of France's Ministry of Education, which was in the middle of a four-year project to boost public awareness of telematics and videotext. (France's famous Minitel system came out of this same context of state support for information technology.) The French media center was particularly impressed that fax technology was then "accessible to a very wide public" at modest cost, that the fax format made it "easy for those participating to do their [own] text and image layout," and that links between schools in different countries were possible because current fax machines could be "found anywhere" and had "maximum compatibility" (18). The vision was to have groups of students, across Europe, compose individual newspaper pages, and fax them in to a central editorial team. "The fax—the 'magical little machine'—is at the symbolic centre of the project" (45). The same day, the editorial team would assemble the contributions into a multilingual student newspaper, compose an editorial on the chosen theme, and arrange for the paper's distribution by fax and in traditional printed form. The first issue, with the theme "a day in the life of a secondary school student," was assembled and published on 3 November 1989 by thirty teams of students from twelve countries, coordinated by an editorial team in Renne, France.

The Fax! Programme quickly developed a life of its own. The fifth issue (with the theme "aged 20 and living in Europe") was the first created entirely by students and on a European theme, while the sixth issue, assembled by students in Bilbao, in the Basque country of Spain, was an international milestone. The Bilbao issue, part of a major educational exhibition, was published in an exceptionally large print run of 80,000 as a supplement to a local daily newspaper. In its first three years (1989–92) the Fax! Programme produced thirty issues and involved 5,000 students from forty countries. An issue was typically composed by students from six countries, often drawing on resources from their schools, local governments, and professional journalists. While self-consciously centered on Europe—defined as the continent from Portugal to Russia and from the United Kingdom to Uzbekistan (and including the French West Indies)—the effort has included "guests" from sixteen countries in Asia, Africa, and the Americas.

Producing a multilingual, intentionally multicultural newspaper has been the students' chief achievement. Each of the participating teams writes its material in their home language and provides a brief summary in the organizing country's language. Students then lay out their text and images, using a suggested template, and compose their page of the collected newspaper—before faxing it to the central editorial team. Besides the main European languages, the newspaper has featured articles in Arabic, Chinese, Japanese, Korean, Nigerian dialects, Occitan, and Vietnamese. Interest on the part of primary school teachers led to four issues of a "junior" fax newspaper, focused on themes of interest to primary school students (e.g., the sports and games we play). While many of the chosen themes have a feel-good quality about them, the student organizers have at times chosen difficult themes, including environmental challenges to the coasts (issue #10), Antarctica after the extension of the Antarctic Treaty (#24), and human rights (#26). Issue 28, organized in Vienna, invited its sixty student contributors to ask themselves who and what was "foreign" to them. The Fax! Programme "is already an important and original vector for the European identity" (34); rather than parroting some bureaucrat's view, students instead "express their vision of Europe, their recognition of difference, their reciprocal curiosity" (50).

The idea of publishing an international student newspaper was certainly more groundbreaking in the 1980s than it may appear in our present Internet-saturated world. What is still notable about the Fax! Programme is the students' ability to identify and develop new cultural formations in a technology then dominated by business activities. The global character of facsimile technology, and especially the possibility of linking students from several different countries, resonated with students intrigued by the evolving process of European integration. Like the Internet visionaries examined below, they successfully translated their idealism about society and technology into tangible results. Fittingly, the European students drew on a technology that had been explicitly designed—through the CCITT's standards—as a global technology.

McWorld or McCurry?

The era of global culture has many faces, but none is better known than Ronald McDonald's, whose fame exceeds that of Santa Claus. McDonald's is without peer as a supercharged symbol of globalization. Born in the 1950s Californian car culture and growing up alongside the baby boomers, McDonald's has emerged as the 800-pound gorilla of the international fast-food jungle. McDonald's is the largest single consumer in the world of beef and potatoes and chicken, and its massive presence has shaped the culture, eating habits, and food industries in no fewer than 120 countries. Sales outside the United States reached 50 percent of the corporation's revenues in 1995. Its self-conscious corporate "system" of special fast-food technology, rigorous training, entrepreneurial franchising, advertising, and active public relations has inspired an entire cottage industry established to scrutinize the company's activities. That it spawned a jargon of "Mc" words testifies to our uneasiness about its impact: "McJobs" are the low-wage, nonunion, service-sector jobs that await teenagers, retirees looking for a little extra spending money, and hapless laid-off factory workers. "McWorld" epitomizes the cultural homogenization and rampant Americanization denounced by many critics of globalization. "McDonaldization" refers to a broader process of the spread of predictability, calculability, and control—with the fast-food restaurant as the present-day paradigm of Max Weber's famous theory of rationalization. And everywhere the discussion turns to globalization—among critics, advocates, or the genuinely puzzled—sooner or later McDonald's is sure to pop up.

Critics of globalization have seized on McDonald's as a potent symbol. José Bové, an otherwise retiring forty-six-year-old French sheep farmer, became an international celebrity in 1999 when he targeted a nearby McDonald's, pulling down half its roof with a tractor and scrawling on what remained, "McDo Dehors, Gardons le Roquefort" (McDonald's Get Out, Let's Keep the Roquefort). While focused on a single McDonald's, his protest had far wider aims. It embodied and expressed the resentment felt by many French people about American economic influence, at that moment focused on the punitive 100 percent American import duties levied against Bové's beloved Roquefort cheese and other fancy French foods, as well as the way American popular culture has challenged a more leisurely and traditional way of life. Big Macs and fancy cheeses were just the tip of a transatlantic trade tangle. Those ruinously high U.S. import duties were a U.S. reaction to the Europeans' banning imports of hormone-treated American beef, a long-running trade dispute only distantly related to McDonald's. As he went off to jail, fist defiantly in the air, Bové promised to persist in "the battle against globalization and for the right of peoples to feed themselves as they choose." French President Jacques Chirac, asserting his patriotic loyalties, is on record that he "detests McDonald's food."[11]

When facing such complaints and protests McDonald's inevitably emphasizes that its restaurants are owned and run by local people, employ local people, and sell locally grown food, but its well-honed communication strategy backfired in the "McLibel" case in Britain. In 1990 McDonald's charged two members of a London protest group named Greenpeace, an offshoot of the international environmental group, with making false and misleading statements in a strongly worded anti-McDonald's pamphlet. As Naomi Klein writes in *No Logo,* "it was an early case study in using a single name brand to connect all the dots among every topic on the social agenda." The activists condemned McDonald's for its role in rain forest destruction and Third World poverty (by cutting down trees and displacing peasants for land in order to raise cattle), animal cruelty, waste and litter, health problems, poor labor conditions, and exploitative advertising aimed at children. While the company eventually won the legal battle, in 1997, the anti-McDonald's activists gained immense television and newspaper publicity. In Britain alone, they distributed some 3 million printed copies of the contested pamphlet, "What's Wrong with McDonald's." And a support group created McSpotlight, a website featuring the original pamphlet, now available in a dozen languages; some 2,500 files on the company, including videos, interviews, articles, and trial transcripts; chat rooms for discontented McDonald's employees; and the latest antiglobalization news. McSpotlight keeps alive such courtroom howlers as the company's claim that Coca-Cola is a "nutritious" food and the assertion that "the dumping of waste [is] a benefit. . . . otherwise you will end up with lots of vast empty gravel pits all over the country."[12]

Advocates of globalization, for their part, point out that the crowds lining up to get into McDonald's restaurants are voting with their feet in favor of American fast food. McDonald's own website extols the 15,000 customers that lined up at the opening of its Kuwait City restaurant in 1994, following the Gulf War. The presence of McDonald's in the conflict-torn Middle East is good news to Tom Friedman, the author of *The Lexus and the Olive Tree* (1999). In this spirited brief on behalf of globalization, Friedman frames the "golden arches theory of conflict prevention." As he puts his theory, gained in a lifetime of munching Quarter Pounders and fries in his wide-ranging international travels for the *New York Times:* "No two countries that both had McDonald's had fought a war against each other since each got its McDonald's." His happy thought is that a country's leaders must think twice about invading a neighboring country when they know that war-making would endanger foreign investment and erode domestic support from the middle class. "People in McDonald's countries don't like to fight wars any more, they prefer to wait in line for burgers," he writes. At the least, the theory is supported by an uncanny pattern. Argentina got its McDonald's in 1986, four years after the Falklands war with Great Britain, but the troublesome hotspots of North Korea, Pakistan, Iran, Iraq, and Syria were still (in 1999) dangerously Mac-free zones. Friedman's unbounded enthusiasm about America's leading role in a globalized world has led him to daffy and wholly unpersuasive sports-team metaphors. He is clearly puzzled by, and at times openly dismissive of, critics of globalization. "In most societies people cannot distinguish anymore between American power, American exports, American cultural assaults, American cultural exports and plain vanilla globalization."[13]

For many reasons, then, McDonald's affords a compelling case study in the economic and cultural dimensions of globalization. The history of McDonald's is drummed into all its trainees and neatly packaged on the company's website. The legend began in 1954 when Ray Kroc discovered the McDonald brothers' hamburger stand in San Bernardino, California, where he saw eight of his Multimixers whipping up milk shakes for the waiting crowds. Quickly, he signed up the brothers and within a year opened the "first" restaurant in Des Plaines, Illinois, "where it all

FIG. 8.2. Capitalist McDonald's in Eastern Europe.

McDonald's in the historic city center of Prague, Czech Republic, May
1999. Photograph by Marleen Daniels, used by permission of Bureau
Hollandse Hoogte.

began." The website emphasizes the human face of the company's growth—
with blurbs on the origins of Ronald McDonald, Big Macs, Happy Meals,
and the company's charity work—saying little about the efforts to find
financing, franchisees, and locations, and entirely passing over the real
reason for its success: "the system" for turning out exactly standardized
hamburgers and French fries. The company's overseas activities began in
1970 with a trial venture in the Netherlands, and within five years it had
full initiatives in Japan, Australia, Germany, France, Sweden, England, and
Hong Kong. To help focus our inquiry we can examine McDonald's in
Russia and in the Far East and ask the following questions: What are the
central elements of the McDonald's system, and how has it exported its

system to other countries? How has McDonald's *changed* the eating habits, food industry, and culture of the foreign countries? And in turn how has the experience of worldwide operations *changed* McDonald's?

The multinational capitalist era in Russia began 31 January 1990 with the opening of the world's largest McDonald's in Moscow. From the start the corporation's efforts to expand into Russia were bound up with Cold War rivalries and superpower symbolism, and complicated by a hundred practical questions about delivering a standard hamburger in a country lacking anything like a Western-style food industry. At the center of McDonald's effort stood the colorful George Cohon, whom Pravda honored in a rare moment of humor as a "hero of capitalist labor." Cohon, as chief of McDonald's Canada, had had his entrepreneurial eye on Russia ever since hosting a delegation of Russian officials at the 1976 Montreal Olympics. In 1979, during his failed effort to secure for McDonald's the food-service contract for the 1980 Moscow Olympics, Cohon told a Canadian audience, "If companies like McDonald's . . . deal with the Russians, this, in a certain way, breaks down the ideological barriers that exist between communism and capitalism."[14] As it happened, Western countries boycotted the Moscow Olympic games in protest of the Soviet invasion of Afghanistan; during the next decade Cohon's extensive personal contacts and some high-level diplomacy eventually landed McDonald's the go-ahead from Moscow's city government to build a string of restaurants.

Cohon's quest had the good luck of accompanying Mikhail Gorbachev's famed *glasnost* and *perestroika* initiatives in the mid-1980s. Having a McDonald's in Moscow would signal that the Soviets were serious about "openness" and "restructuring." Cohon's team focused first on finding high-profile sites ("the Soviets' instincts seemed to be to put us behind the elevator shafts in hotels," he writes) and settling on real-estate prices in a society without proper market mechanisms. They then turned to the far more serious problem of securing food sources that would meet McDonald's stringent standards, especially difficult since the late-Soviet agriculture system was in an advanced state of decay. Meat was a frightful problem. "The existing Soviet abattoirs and processing facilities were . . . like something out of Upton Sinclair."[15] For a time they considered importing food from the huge McDonald's distribution site in Duisburg, West Germany, but decided instead to build from scratch a local complex in a Moscow suburb.

This 100,000-square-foot "McComplex," built at a cost of $21 million, replicates an entire Western-style food-industry infrastructure. It features special lines for processing meat, dairy products, potatoes, and bakery goods, as well as dry storage, freezer storage, and a plant for making ketchup, mustard, and Big Mac sauce. Employing 400 workers, and processing 57 tons of food each day, McComplex accords with the company's policy of using local food supplies wherever possible. When it had trouble making proper French fries from the traditional small, round Russian potato, for instance, the company imported large elongated Russet potatoes from the Netherlands and encouraged Russian farmers to grow them. McComplex also accords with the company's antiunion practices, further riling antiglobalization labor activists.

Soviet citizens experienced the arrival of capitalistic McDonald's in Moscow during the very years when the Soviet economy was crumbling around them. It appeared that Muscovites in droves were voting against communism with their hands and feet. When McDonald's advertised openings for 630 workers, it received 27,000 written applications. And on opening day in Pushkin Square in 1990 McDonald's served a record 30,567 customers who lined up until midnight. The Russian manager of the restaurant, Khamzat Kasbulatov, seized the moment by regaling a television crew with this declaration: "Many people talk about *perestroika,* but for them *perestroika* is an abstraction. Now, me—I can touch my *perestroika.* I can taste my *perestroika.* Big Mac is *perestroika.*" And while cultural conservatives in Russia bemoan McDonald's sway over young people, on balance McDonald's is not widely perceived as a corrosive foreign influence there. In the unsettled days of August 1991, following the attempted coup by old-guard Soviet officers, when the streets of Moscow were filled with "anxious milling crowds," McDonald's was safe. "McDonald's?" the comment went; "They're a Russian company. They're okay."[16]

Yet even as McDonald's deployed its "system"—in building and operating Western-style restaurants, while managing and training workers and fashioning an entire Western-style food supply—the challenges of operating in Russia perceptibly changed the company. The company had fronted an enormous investment, perhaps $50 million, to get the high-profile Russian venture off the ground; and the decade-long groundwork had involved McDonald's executives directly in a tangle of high-level diplomacy, including the contentious issue of Soviet Jews' emigration to Israel. This depth of involvement changed the structure of the company. In most other regions of the world, McDonald's relies on a decentralized network of independent contractors who supply the restaurants with everything from meat to milk shake mix. The fully integrated McComplex in Moscow stands in sharp distinction to that pattern. "This was new for us; we had never been so vertically integrated," writes Cohon.[17] And McDonald's Canada, anticipating licensing fees of 5 percent of total sales, was forced to reinvest the entirety of its accumulated fees in Russia until the ruble eventually became convertible.

McDonald's operations in East Asia, beginning with Japan in 1971, did not have the drama of Pushkin Square and *perestroika*, but the company's presence there has involved it in the region's complex politics and dynamic cultures. Anthropologist James Watson, in *Golden Arches East*, argues that addressing the large issues of cultural transformation in the world requires seriously considering such pervasive everyday phenomena as McDonald's.[18] Based on his colleagues' fieldwork in China, Taiwan, Korea, Japan, and his own in Hong Kong, Watson reports complex ethnographic findings that do not fit the notion that McDonald's is corrupting local cultures or turning them into pale copies of America, as the critics of cultural imperialism often charge. Watson and his colleagues find instead a complex process of appropriation, in which McDonald's is one agent among many and in which McDonald's itself is transformed.

McDonald's has certainly altered Asian eating habits. It arrived in Tokyo in 1971, Hong Kong in 1975, Taipei in 1984, Seoul in 1988, and Beijing in 1992. Except in Korea, where beef was already eaten, a hamburger was literally an unknown food before McDonald's appeared on the scene. In Taiwan, owing almost entirely to McDonald's, young people have enthusiastically taken up French fries. In Japan, it is now permitted to break the taboo of *tachigui*, "eating while standing." (Most Japanese avoid violating another prohibition by only partially unwrapping hamburgers and holding the wrapped portion, so that their fingers do not directly touch the food.) Through its promotions McDonald's has transplanted to Asia the (Western) concept of an individual's annual birthday celebration. And McDonald's strives to teach its Asian customers, and others around the world, about standing in line, carrying your own food, seating yourself, and taking away your own garbage.

Yet, in the strong currents of globalization McDonald's has not been an unmoved mover. Asian consumers have subverted the very concept of "fast food," which in the United States is understood to be food quickly delivered and quickly devoured. In many parts of Asia, consumers have turned their local McDonald's into a leisure center or after-school club. Even in super-crowded Hong Kong, McDonald's customers on average take a leisurely twenty to twenty-five minutes to eat their meals, while during off-peak times in Beijing the average mealtime stretches out to fifty-one minutes, whereas in the United States the fast-food industry standard is a scant eleven minutes. From 3:00 to 6:00 in the afternoon, McDonald's restaurants in Hong Kong—including seven of the ten busiest McDonald's in the world—fill up with students seeking a safe place to relax with their friends, time away from the supervision of teachers and parents, and a bit of space in the world's most densely packed city. "The obvious strategy is to turn a potential liability into an asset: 'Students create a good atmosphere which is good for our business,' said one manager as he watched an army of teenagers—dressed in identical school uniforms—surge into his restaurant."[19] And in a sharp departure from the company's pervasive labor-saving practices (it's hard to believe in the days of frozen French fries, but long ago at each individual restaurant, fresh Russet potatoes were stored, washed, peeled, cut, soaked, dried, and fried), each McDonald's in Beijing hires ten extra women whose job it is to help customers with seating.

McDonald's corporate strategy of localization not only accommodates local initiatives and sensibilities but also, as the company is well aware, blunts the arguments of its critics. McDonald's International takes a 50 percent ownership stake in each overseas restaurant, while its local partner owns 50 percent. Accusations of economic or cultural imperialism are simply less compelling when a local partner pockets half the profits, has substantial operating autonomy (setting wage rates and choosing menu items), and takes substantial pride in navigating the shoals of the local business environment. A nationalistic Korean marketing manager maintains that his local McDonald's is a *Korean* business, not an American multinational, since it is conducted entirely by Koreans, who have the necessary savvy to deal effectively with the country's complex banking system and government bureaucracies. Localization is a matter, then, not of opportunism or chance but of long-standing corporate policy. McDonald's strives to "become as much a part of the local culture as possible," according to the president of McDonald's International. "People call us a multinational. I like to call us *multilocal*." Much the same point was made by a troop of Japanese Boy Scouts, traveling abroad, who "were pleasantly surprised to find a McDonald's in Chicago."[20]

Not even the appearance of a standard "American" menu, down to the last perfectly placed pickle, should be automatically assumed to be an act of cultural hegemony or corrupting Americanization. In Beijing the menu is self-consciously standard because the Chinese franchisee is striving for an "authentic" American experience. (Across Asia, Watson's ethnographers report that what is achieved is Americana *as constructed by* the local culture, a far more complex and open-ended process than the one-way control claimed by many cultural-imperialism theorists, who typically assert that exact copies of America are transferred abroad.) In Japan, the legendary Den Fujita, a supremely flamboyant entrepreneur who established McDonald's Japan, positioned his new venture as a genuine Japanese company while offering a standard American menu. His tactic was to label the hamburger not as an American import, which would have inflamed Japanese national sensibilities, but as a "revolutionary" food. His campaign led to some goofy publicity. "The reason Japanese people are so short and have yellow skins is because they have eaten nothing but fish and rice for two thousand years," he told reporters. "If we eat McDonald's hamburgers and potatoes for a thousand years, we will become taller, our skin will become white, and our hair blond."

In his localizing campaign, Fujita brazenly flouted McDonald's corporate formula for starting restaurants. To begin, instead of opening restaurants in McDonald's tried-and-true suburban locations, his first one was in Tokyo's hyperexpensive Ginza shopping district. His previous career in importing luxury handbags gave Fujita the contacts to secure 500 square feet of precious space from a former client, a long-established department store in a prime location. Fujita squeezed an entire restaurant into the tiny site, one-fifth the size of most McDonald's restaurants, by custom designing a miniature kitchen and offering only stand-up counters. His real challenge was erecting his restaurant, a complex job typically taking three months, in the thirty-nine hours mandated by his agreement with the department store. The contract required Fujita to start and complete his restaurant's construction between the department store's closing on a Sunday evening and its reopening on Tuesday morning. Seventy workmen practiced setting up the restaurant three times at an off-site warehouse before shaving the time down to an acceptable thirty-six hours. ("But where's the store?" asked a mystified McDonald's manager just two days before Fujita's grand opening on 20 July 1971.) In the succeeding eighteen months Fujita opened eighteen additional restaurants, as well as starting the Tokyo branch of Hamburger University. In fact, the Japanese mastery of McDonald's notoriously detailed "system" actually rattled McDonald's president Fred Turner. "In Japan, you tell a grill man only once how to lay the patties, and he puts them there every time. I'd been looking for that one-hundred-percent compliance for thirty years, and now that I finally found it in Japan, it made me very nervous."[21] And all this with the American-standard hamburger menu!

Whether we are more likely, then, to get a homogenized "McWorld" or a hybrid "McCurry" depends somewhat on the trajectory of McDonald's and its resident managers and much on consumers' appropriating or even transforming its offerings. We may well end up with a culturally bland "McWorld" if the company's localized menu and offerings are only a temporary measure to gain entrée into a new country. Australia seems to fit this pattern. When launching McDonald's in the early 1970s, the local franchisee initially tailored the menus to Australian tastes. His offerings included English-style fish-and-chips (rather than the standard fish sandwich), a special fried chicken product that was unknown in the United States yet made up 30 percent of Australian sales, and hamburgers in line with Aussie's decided preference for lettuce, tomato, and mayonnaise, but no perfectly placed pickles. (When customers were sold the standard hamburger, the franchisee recalls, "we had pickles all over the store—sticking on the ceilings and walls.") Yet over time McDonald's was able to woo Aussies to its standard American menu—and these localized offerings were put aside.[22]

On the other hand, we are likely to get a zesty "McCurry" world to the extent that various cultures and cuisines actively engage McDonald's. In Japan, for instance, a locally inspired teriyaki burger is now a permanent menu item. Vegetarian burgers are popular at certain upscale McDonald's in the Netherlands, while the company requires alternatives to beef hamburgers and pork breakfast sausages for McDonald's that operate in countries where religious traditions forbid the eating of cows or pigs or meat altogether. In India, McDonald's serves a mutton-based Maharaja Mac as well as Vegetable McNuggets, while in Malaysia and Singapore the company satisfied Muslim authorities' stiff criteria for a *halal* certificate, requiring the total absence of pork. (Sometimes slippage occurs. Vegetarian patrons of McDonald's in the United States blasted the company for having loudly announced in 1990 that it would use "all vegetable" oil, rather than beef lard, for its French fries while quietly slipping beef extracts into the fries themselves.) My own fieldwork confirms that beer is widely available in German, Danish, Swedish, and Dutch McDonald's, while the delicious icy vegetable soup called gazpacho appears in Spain. (Even American-as-apple-pie Disney, facing huge financial losses, was forced to sell red wine to its patrons at Disneyland-Paris.) In a true "McCurry" world, these multilocal innovations would circulate to the United

States. Is it too much to hope in the United States for a vegetarian burger, with tamari-flavored fries, and a glass of cold beer?

Internet Culture

As McDonald's found to its distress with McLibel, many aspects of our public culture have moved onto the Internet. A long list of multinational companies, including McDonald's, Nike, Shell, and Monsanto, have found that taking out expensive advertisements in newspapers or on television no longer secures the hearts and minds of the public. McSpotlight represents a growing wave of activist websites that challenge once-dominant commercial institutions. "The most comprehensive collection of alternative material ever assembled on a multinational corporation is gaining global publicity itself, and is totally beyond McDonald's control," writes John Vidal.[23] Unraveling the many threads of the Internet and assessing its interactions with global cultures is a difficult task, since "the Internet" consists of many different components—email, chat rooms, World Wide Web sites, intranets, and e-commerce.

Understanding the influence of the Internet requires us to grapple with two rival viewpoints on its origins and nature. Many popular accounts offer the comforting version embodied in the "civilian origins" story. One of the most comforting is Katie Hafner and Matthew Lyon's pioneering *Where Wizards Stay Up Late: The Origins of the Internet* (1996), which locates the Internet's roots in the civilian computer-science subculture of the 1960s and emphatically denies significant military influence. "The project had embodied the most peaceful intentions . . . [and] had nothing to do with supporting or surviving war—never did," they write.[24]

Despite Hafner and Lyon's dismissal, the Internet's military origins are pronounced. Many of the important technical milestones—the RAND concept of packet switching (1964), the Navy-funded Alohanet that led to the Ethernet (1970–72), the Defense Department's ARPANET (1972), the diverse military communication needs that led to the "*inter*network" concept (1973–74), and the rapid adoption of the now-ubiquitous TCP/IP internetworking protocols (1983)—were exclusively funded or heavily promoted or even outright mandated by the military services. "Every time I wrote a proposal I had to show the relevance to the military's applications . . . but it was not at all imposed on us," recalled one top researcher. Vinton Cerf, a leading computer scientist who headed the Defense Department's IPTO in the mid-1970s and who is widely credited as the "father" of the Internet, reflected on the two faces of military funding in a 1990 interview. Janet Abbate in her history of the Internet summarizes his view: "Although Principal Investigators at universities acted as buffers between their graduate students and the Department of Defense, thus allowing the students to focus on the research without necessarily having to confront its military implications, this only disguised and did not negate the fact that military imperatives drove the research." A former high ARPA official told computer historian Paul Edwards, "We knew exactly what we were doing. We were building a survivable command system for nuclear war."[25]

Overall, we can discern three phases in the Internet story: the early origins, from the 1960s to mid-1980s, when the military services were prominent; a transitional decade beginning in the 1980s, when the National Science Foundation became the principal government agency supporting the Internet; and the commercialization of the Internet in the 1990s, when the network itself was privatized and the World Wide Web came into being. During each of these stages, its promoters intended the "Internet" to be a "global" technology, not merely with worldwide sweep but also connecting diverse networks and communities and thereby linking people around the world.

In the phase of military prominence, the heady aim of being global was sold to the military services as a way to build a robust, worldwide communications system. In their "civilian origins" interpretation, Hafner and Lyon skip right over the 1960 technical paper by Paul Baran, the RAND researcher who invented the concept of packet switching, who describes a "survivable" communication system: "The cloud-of-doom attitude that nuclear war spells the end of the earth is slowly lifting from the minds of the many. . . . It follows that we should . . . do all those things necessary to permit the survivors of the holocaust to shuck their ashes and reconstruct the economy swiftly."[26] (Not everyone finds such vintage RAND-speak to be in good taste.) As we noted in chapter 7, the Defense Department's Advanced Research Projects Agency, or ARPA, used Baran's concepts to build a distributed network linking the fifteen ARPA-funded computer science research centers across the United States. This computer network, known as ARPANET, was the conceptual and practical prototype of the Internet. While an emerging networking community saw the ARPANET as an exciting experiment in computer science, ARPA itself carefully justified it to the Pentagon and Congress in strictly economic and military terms. For example in 1969 the ARPA director told Congress that the ARPANET "could make a factor of 10 to 100 difference in effective computer capacity per dollar," while two years later the message to

Congress was that the military would benefit enormously from the ability to transmit over the ARPANET "logistics data bases, force levels, and various sorts of personnel files."[27]

The groundwork for interconnecting different networks also took place under ARPA's guidance and owed much to military goals. By the mid-1970s, ARPA was running, in addition to ARPANET, two other packet-based networks: PRNET and SATNET. PRNET evolved from ARPA's project to link the University of Hawaii's seven campuses using "packet radio," the sending of discrete data packets by radio broadcast; ARPA then built a packet-radio network, PRNET, in the San Francisco region. (PRNET was the conceptual basis of the widespread Ethernet concept, and of the fortune of Robert Metcalfe, the founder of 3Com. Ethernets transmit packets of data that are "broadcast" between computers connected by cables.[28]) The third significant ARPA network was SATNET, which used satellite links to rapidly transmit seismic data collected in Norway to analysis sites in Maryland and Virginia. With SATNET speed was of the essence, since the data might indicate Soviet nuclear explosions. ARPA's dilemma was that ARPANET was a point-to-point network built with near-perfect reliability and guaranteed sequencing, while PRNET was a broadcast network with lower reliability, and SATNET experienced delays as data packets bounced between ground stations and satellites.

The *inter*-networking concept responded to the military's need to integrate its technically diverse, geographically expansive communication networks. The internet conception resulted from an intense collaboration between Vinton Cerf, a Stanford computer scientist who had helped devise the ARPANET protocols, and Robert Kahn, a program manager at ARPA. In 1973 they hit upon the key concepts—common host protocols within a network, special gateways between networks, and a common address space across the whole—and the following year published a now-classic paper, "A Protocol for Packet Network Intercommunication." Although this paper is sometimes held up as embodying a singular Edisonian "eureka moment," Cerf and Kahn worked very closely for years with an international networking group to test and refine their ideas. In 1976 Cerf became ARPA's manager for network activities, and a year later Kahn and Cerf successfully tested an internet between PRNET in California, the Norway-London-Virginia SATNET, and the multinode ARPANET. "All of the demonstrations that we did had military counterparts," Cerf later commented. "They involved the Strategic Air Command at one point, where we put airborne packet radios in the field communicating with each other and to the ground, using the airborne systems to sew together fragments of Internet that had been segregated by a simulated nuclear attack."[29]

Not only did Kahn and Cerf advance the Internet as a solution to a military problem; the military services vigorously promoted the internetworking concepts they devised for operating and linking these networks. At the time, there were at least three rival protocols for linking different computers into a network, but Kahn and Cerf's internet protocols—the now-ubiquitous transmission control protocol (TCP) and Internet Protocol (IP), combined as TCP/IP—won out over these rivals largely because of the strong support it received from the Defense Department. The ARPANET, a single network, was built around a protocol named NWP. Its switchover to the TCP/IP Internet protocols was mandated in 1982 by the Defense Communication Agency (DCA), which had taken over the running of the ARPANET and was concerned with creating a single, integrated military communications system. It was a classic case of a forced choice among technologies. As of 1 January 1983, DCA directed the network operators to reject any transmissions that were not in TCP. ARPA also directly funded its contractors to implement the TCP/IP protocols on Unix-based computers, and it spent an additional $20 million to fund other computer manufacturers' implementation of its standard with their machines. "All the major computer companies took advantage of this opportunity," writes Abbate.[30]

The direct military impetus that had built the ARPANET and fostered the internet concept did not persist long into the 1980s. ARPA for its part, having already by 1975 given up the day-to-day operation of the ARPANET to the DCA, took up the Reagan-era Strategic Defense Initiative. The military services split off their secret and sensitive activities to the newly created MILNET in 1983, which was reconnected to ARPANET, by then dominated by universities. The military's concern with interconnecting diverse computer networks, embodied in TCP/IP, turned out to cope brilliantly with two massive developments in computing during the 1980s. The personal-computer revolution expanded computing from the restricted province of a few hundred researchers and hobbyists into a mainstream activity of the millions, while the deliberate civilianization of the Internet by the National Science Foundation (NSF) created a mass network experience. Both of these developments powerfully shaped the culture of the Internet.

While computing in the 1960s had meant room-sized "mainframe" computers, followed by refrigerator-sized "minicomputers" in the 1970s, computing by the 1980s increasingly meant "microcomputers," or personal computers, small enough and cheap enough for an individual to own. While small computers in the era of Radio Shack's TRS-80s had required luck and patience, IBM's entry into personal computing in 1981, and the many clones permitted by its nonproprietary operating system, turned computing increasingly into a mainstream activity. Many universities and businesses found that their computing needs could be more cheaply met not by purchasing computing power from a networked mainframe, ARPANET's original inspiration, but by purchasing dozens or hundreds or even thousands of personal computers. The ready availability of local-area networks, or LANs, such as Robert Metcalfe's Ethernet and IBM's Token Rings, meant that these numerous computers could be locally networked and thus enabled to share software and files. Ethernets fairly exploded across the 1980s. By the mid-1990s there were five million Ethernet LANs, and each could be connected to the Internet. The parallel expansion of computer-specific proprietary networks—such as DECNET, the Unix-derived USENET, and the IBM-based BITNET—added even more potential Internet users.

Closely related to the massive spread of computing was the invention of electronic mail, which for years has formed the predominant use of the Internet. Email is an excellent example of how computer users transformed ARPANET into something that its designers had never foreseen. ARPANET, one recalls, was conceived to share data files and to efficiently use expensive computers. An early ARPA blueprint explicitly dismissed email: sending messages between users was "not an important motivation for a network of scientific computers." Yet a strong suggestion that ARPA's vision had missed something big came in some anomalous data from MIT. The small switching computer that linked MIT to the ARPANET seemed always busy; yet surprisingly few files actually went out over the ARPANET. It turned out that MIT users, via the ARPANET switching computer, were sending messages to one another. Soon enough ARPA converted to email, a consequence of the enthusiasm of its director, Stephen Lukasik; he made email his preferred means of communicating with subordinates, who quickly got the message that if they wanted to talk to the boss they needed email, too. Email was initially piggybacked onto preexisting protocols for sending data files (e.g., the file transfer protocol, or ftp). Meantime, the sprouting of special-subject email lists, a practice in which a single computer kept a file of users' addresses, created the notion of a "virtual community" of like-minded individuals interacting solely through computers. All of this unforeseen activity was quietly tolerated by ARPA, since ARPANET was in fact underutilized and the heavy network traffic sent out by large membership lists like "sf-lovers" provided useful experiments. Beginning in the early 1980s, the Internet community specified a simple mail transfer protocol, or SMTP, and a parallel protocol for transferring news files, both of which are widely used today.[31]

The second broad development of the 1980s was the "civilianization" of the Internet. While ARPA's contractors at the leading computer research centers were the only authorized users of the ARPANET, the NSF-funded expansion of computing and computer science in the 1970s had created a much larger community of potential network users. In 1980 the NSF approved a $5 million proposal by computer scientists outside ARPA's orbit to create CSNET. CSNET was initially conceived with the rival X.25 protocol then gaining ground in Europe, but Vinton Cerf offered to set up connections between CSNET and ARPANET if CSNET would adopt TCP/IP instead. When it began in 1982, CSNET consisted of 18 full-time hosts plus 128 hosts on PhoneNet (which used telephones to make network connections only when needed), all of which were linked to the ARPANET. CSNET was soon also connected to the computer-science networks that had sprung up in Japan, Korea, Australia, Germany, France, England, Finland, Sweden, and Israel. Commercial use of the network was strictly forbidden, however, which reinforced the idea that cyberspace was a free collection of individuals rather than an advertisement-strewn shopping mall.[32]

CSNET was just the start of the NSF-led phase of the Internet. The earlier ARPA vision of networking central computing facilities guided the NSF's supercomputing program of the mid-1980s. The NSF planned and built six supercomputing centers that were linked, along with nine additional regional centers, by a super-fast "backbone." Altogether the NSF spent $200 million creating this NSFNET, which became the hub of an "information infrastructure." Soon, NSFNET was joined to smaller regional networks. For example, forty universities in New England formed one regional consortium; universities in the southeast, northwest, Rocky Mountain states, and New York State area, among others, formed independent regional consortia, often with NSF assistance; in turn, all of them were linked to the NSFNET. NSFNET initially used a non-ARPA protocol, but it too was converted to TCP/IP during a five-year upgrade that began in 1987. As a creation of the federal government, NSFNET further reinforced the noncommercial appearance of cyberspace.

A good example of how the Internet gained its seemingly effortless "global" character is the so-called domain-name system, or DNS. Most users experience this system, which emerged in the mid-1980s, as the two halves of email addresses ("user" and "host.domain") joined by the ubiquitous "at" sign (@). Early users of email may remember that it often took three or four tries with different addresses, through different network gateways, to reach a correspondent who happened to be on a different network than yours. In this era, a central "host file" was kept of *all* approved host names, correlated with their numerical addresses. As the number of hosts grew, the "host file" became unmanageably large; just sending copies of it out to each of the individual host computers could overwhelm the network. With the spread of the domain-name system, any single user can be addressed with one simple address. More important, the DNS established an address space that is massively expandable and yet can be effectively managed *without* any single center. Initially, ARPA created six defined domains (.edu, .gov, .mil, .net, .org, and .com). A consortium responsible for that area manages each domain, and it can assign names within its domain as it pleases. In turn, for example within the top-level .edu domain, each educational institution is assigned a subdomain and can assign subnames as it pleases. Subdomains, like suburbs, can sprawl across the prairie; the longest official host name in 1997 was challenger.med.synapse.uah.ualberta.ca. With DNS there is no need for a central "host table" with an impossibly large number of all host names and numerical addresses; rather, each domain is responsible for keeping track of the host names within its domain (while each successive subdomain does the same). And, even though it irritates some non-U.S. users, the DNS system was able to effortlessly add "foreign" countries, by establishing a two-letter country domain (like .uk and .ru). Absent some similar decentralized addressing scheme, it is difficult to see how the Internet could have grown so rapidly while maintaining reliable routing across networks and countries.

The entirely civilian, resolutely noncommercial Internet lasted five years and two months. When the twenty-year-old ARPANET was closed down in February 1990, NSFNET took over the Internet as a government-owned and -operated entity. In turn, in April 1995 the NSFNET was replaced by today's for-profit system of "internet service providers" (or

ISPs). All along, NSF was interested in the Internet mostly as an infra-structure for research. Enforcing the "acceptable use" policy required by its congressional paymasters, which strictly forbad commercial activities, became ever more troublesome as more businesses and commercial ventures came on line. The path toward privatization actually began in 1987, when the NSF awarded a five-year contract for upgrades and operations to MERIT, a regional consortium based in Michigan, and its private-sector partners, IBM and MCI. In 1990, MERIT spun off a nonprofit corporation named Advanced Network Services, which in turn spun off a *for*-profit company that began selling commercial network services. Several other regional networks hatched similar privatization plans; three of these networks formed Commercial Internet Exchange in 1991. Finally, the large telecommunications companies, including MCI, Sprint, and AT&T, began offering commercial Internet services.

What evolved through 1995 were two parallel Internets. One was linked by the NSFNET, which strictly prohibited all commercial activities; while the other was formed by the for-profit companies, and onto it rushed all manner of money-making ventures. After 1995, a truly "inter-network" Internet emerged, with private companies (ISPs) not only wiring homes and schools and businesses but also providing the high-speed backbones that carried the bulk of long-distance data traffic. In turn have come gateways connecting the networks of the big telecommunications companies, as well as overseas networks. This array of networks, backbones, and gateways creates the Internet writ large.

The global character of the Internet is perhaps most fully expressed in the World Wide Web, the multimedia extravaganza that meets most computer users today when they search for weather reports, deals on eBay, free music (while it lasts), or the latest pictures of Britney Spears, the planet Jupiter, or their grandchildren. The Web is, at least conceptually, nothing more than a sophisticated way of sending and receiving data files (text, image, sound, or video). It consists of a protocol (hypertext transfer protocol, or http), allowing networked computers to send and receive these files; an address space (uniform resource locator, or URL), which permits remote computers to access all files; the language of the Web (e.g., hypertext markup language, or html), which allows combination of text, images, and other elements onto a single page; along with a Web browser able to interpret and display the files. The Web relies on the existing infrastructure of the Internet, and since its public release in 1991 has become

FIG. 8.3. Internet Café.

Cybercafé in Amsterdam's historic Waag building, constructed in 1488 as a city gate. It was rebuilt in 1617 as the city's customs house and for 200 years also provided rooms for the city's guilds. More recently, it has been a museum and a new-media center. Currently it is one of Europe's leading experimental sites for wireless Internet access. Photograph courtesy of Bureau Hollandse Hoogte.

the Internet's chief source of traffic, displacing telnet (remote login) in 1994 and ftp transfers in 1995. Its global character owes much to the vision and forceful presence of its creator, Tim Berners-Lee.

Tim Berners-Lee has had the paradoxical good fortune to turn a brilliant idea into a full-time career without ever going through the trouble of starting a company and to hold a prestigious named chair at MIT without ever writing a Ph.D. thesis. (He has six honorary doctorates as of 2001 and has won numerous awards and honors.) Born in 1955, Berners-Lee grew up with computers at home. His mother and father were both mathematicians who worked on programming Britain's state-of-the-art Ferranti Mark. From them, he says, he first learned about humans' abilities to create and control computers. After earning an honors degree in physics from Queen's College, Oxford, in 1976 and at the same time building his first computer, he took a series of software and consulting jobs. One of these was a six-month stay in 1980 at CERN, the European high-energy physics laboratory. While there he wrote a program for his personal use, called Enquire, to store information and easily make links. This was the conceptual core of the WWW. He returned to CERN in 1984 and five years later formally proposed a global hypertext project, to be known as the World Wide Web. Within a year CERN had a version of the WWW program running on its network; CERN released it to the world in 1991. Berners-Lee came to MIT in 1994 and has since then directed the World Wide Web Consortium, which sets standards for the Web's user and developer communities.

From the start, Berners-Lee built in to the Web a set of global and universal values. These values were incorporated into the design at a very deep level. "The principles of universality of access irrespective of hardware or software platform, network infrastructure, language, culture, geographical location, or physical or mental impairment are core values in Web design," he wrote in a recent essay on the architecture of the Web. And it is not merely that all computers should be able to access the Web. "The original driving force was collaboration at home and at work," he writes of the philosophy behind the Web. "The idea was, that by building together a hypertext Web, a group of whatever size would force itself to use a common vocabulary, to overcome its misunderstandings, and at any time to have a running model—in the Web—of their plans and reasons." While earlier hypertext systems, such as Apple's HyperCard, had allowed users to create links between different image, text, and sound documents, these schemes were limited to pointing to documents on a local filing system. Often they used a central database to keep track of all links. The result was a reliable but *closed* system to which outsiders could not establish links. As Berners-Lee commented, "there was no way to scale up such a system to allow outsiders to easily contribute information to it. . . . I sacrificed that consistency requirement [of centrally stored and reliable links] to allow the Web to work globally."[33]

While Berners-Lee states that "the original idea of the Web [was] being a creative space for people to work together," he has an even more ambitious goal for the Web. The second goal, dependent on achieving the first goal of human communication through shared knowledge, is that of machine-understandable information. He has sketched the following scenario: A company or group using the Web in its work will create a map, in cyberspace, of all the dependencies and relationships that define how the project is going. A set of links and data might describe how a government agency relates to other agencies, to nongovernmental organizations, or private companies. Berners-Lee sees the possibility of using software "agents" to help analyze and manage what the agency is doing. This might involve the agent's dealing with massive amounts of data, taking up the agency's routine processes, and helping manage the large scale of our human systems. "If we can make something decentralised, out of control, and of great simplicity, we must be prepared to be astonished at whatever might grow out of that new medium."[34]

THESE EXAMPLES—worldwide financial flows, fax machines, McDonald's, and the Internet—taken together indicate that globalization is both a fact of contemporary life and a historical construction that emerged over time. This dual character has obvious implications for us as citizens and scholars. For the moment, globalization is a strong force that orients people's thinking about technology and culture. To the extent that people believe it to be real it *is* real. All the same, globalization was not and is not a permanent fact of nature; indeed, it may not last long beyond 11 September 2001.

Globalization was and is materialized through technologies. One can say without exaggeration that global values were consciously and explicitly "designed" into fax machines, McDonald's, and the Internet. Fax technology's open standards, international standards-setting, and Japanese manufacturers all created in fax machines an impressive ability to operate across different phone systems and between different cultural systems. Fax machines were not always this way: for nearly a half-century (1930s–1970s) electronic facsimile was a closed, proprietary technology. Similarly, McDonald's is explicit about its corporate capitalist strategies of localization and cultural responsiveness. And finally "global values" were built into the Internet from its origins in the 1960s onward through the emergence of the World Wide Web in the 1990s.

This situation, of globalization as both solid fact and emergent creation, presents us as citizens both a paradox and a challenge. The paradox is that at the same moment when consumers' values are actively being solicited by businesses, governments, and nongovernmental organizations, and in theory the "consumer is king" everywhere, it is also a time when many citizens feel powerless to deal with their changing world. Globalization, in the form of multinational corporations and international agencies, often enough takes the rap. It is certainly cold comfort to suggest to a family whose breadwinner has been done out of a job by "foreign competition" that, well, at least, imported toys from China are cheaper or imported grapes from Chile are more affordable. Or to suggest to another family that when their country's currency markets are roiled by jitters in London they should be grateful for the (temporary) inflow of foreign investment capital.

Indeed, the certainty evident during the 1990s that globalization would continue and expand, seemingly without borders, ended with the attacks on 11 September 2001. Whatever one makes of the resulting "war on terrorism," it seems inescapable that the nation-state is, contrary to the globalizers' utopian dreams, alive and thriving as never before. McDonald's will not deliver greater security in airports. More to the point, just as the reduction of military spending after the Cold War forced technology companies to turn toward civilian markets, including global media and entertainment and manufacturing, the sharp military build-up that began in the United States during 2002 has already reversed this trend. A national security–oriented technological era may be in the offing.[35] It would be strange indeed if the September 11th attackers—acting in the name of antimodern ideologies—because of the Western nations' national security–minded and state-centered reactions, brought an end to this phase of global modernity.

9

THE QUESTION OF TECHNOLOGY

POETS AND PUNDITS and prognosticators have approached the question of technology in a variety of settings and from a range of viewpoints, but all of them, I think, share the assumption that if we could just figure out what technology *is*—what it is good for, when and how it changes society—then we would learn something of fundamental importance about our society and our future. I believe that the account in this book supports an engaged middle-ground stance toward technology, resisting the undue pessimism of some writers and rejecting the unwarranted optimism of others. Both of these extreme positions lead us away from a serious engagement with the problems and potentials of technology; the cultural pessimists go overboard with the suggestion, if not the forthright assertion, that the future would be more desirable without new technology, while the technological optimists seem to embrace any future so long as it is packed with new technology. Neither of these extreme positions describes a world that I want to live in.[1]

What is technology? The word *technology* has a specific if largely unwritten history, but in this book I have used the term in a broad way.[2] Indeed, my underlying goal has been to show the *variety* of technologies, to describe how technologies have changed across time, and also to underscore that they have interacted deeply with societies and cultures. I have not found a simple definition of *technology* that conveys the variety of its forms or, least of all, emphasizes the social and cultural interactions that I believe are crucial to understanding what technology is.[3] In the chapters on courts, industry, science and systems, and modernism, I recounted the historical origins of attributes we presume to be the essence of "modern technology."

Science and Economics

In our own time, and largely as a result of the science-and-systems era, most people believe *science* to be an essential part of technology, that scientific discoveries are the most important driver of technological innovations. There is indeed a case for the centrality of science: the link between the basic laws of electricity and Edison's electric lighting system and between the insights of structural organic chemistry and the synthetic dye industry; the employment of thousands of physicists and chemists in industrial research, which has been widely publicized by government and corporate promoters; and modern artists and architects playing on the presumption that science, technology, rationality, and social progress were a tidy package. The notion that social change starts with science and science-based technologies is a key tenet of modernist thinking.

However, the centrality of science to technology is often overstated. Scientific theories had little to do with technological innovation during the eras of industry, commerce, and courts. Scientific discoveries did not drive the technologies of gunpowder weapons or palace building in the Renaissance, wooden shipbuilding or sugar-refining in the era of Dutch commerce, or even the steam-driven cotton textile factories of the industrial revolution. Indeed, acknowledging the fact that steam engines of several types came first while the science of thermodynamics resulted from measuring and analyzing them, scientist-historian L. J. Henderson famously wrote, "Science owes more to the steam engine than the steam engine owes to Science."

More recently, in the military era, we saw how the savvy atomic scientists at Los Alamos drew on the presumed primacy of physics to grab the lion's share of credit for the Manhattan Project even though 95 percent of the atom bomb project was better described as engineering and mass-production industry. And science is only one part, along with engineering, economics, and state subsidies, of computer and microelectronics technologies. Yet even today, the brazen motto of the 1933 Chicago World's Fair—"Science Finds—Industry Applies—Man Conforms"[4]—lives on as an article of faith for many scientists and policy analysts and as a commonplace in popular discourse.

The burden of my evidence suggests that science, while useful in many technical fields, is not an essential aspect of technology. Besides scientists, there are too many other active parties—engineers, financiers, government officials, workers, and consumers—just as intimately involved with making and using technologies. In fact, since scientists today often investigate the properties of human-made substances such as low-temperature superconductors, high-frequency semiconductors, or the tiny spherical carbon shells known as Buckyballs (rather than naturally occurring materials), one could even invert the presumed primacy of science to say that "*technology* finds" entirely new fields for science. Biotechnology firms now employ many thousands of scientists, but on close inspection their techniques for cloning cows look more like informed tinkering than scientific theorizing. Scientists' studies of nature, from tiny subatomic particles through global climate change to the vastness of the universe, are entirely dependent on specialized technologies.

A complementary viewpoint to the linear science-industry-society model defines technology as a desirable instrument of *economic growth.* In the eras of commerce, industry, and science and systems, technological innovations were drivers of economic growth and structural change. In the seventeenth century, generating wealth was a chief preoccupation of Dutch shipbuilders, who developed the distinctive fluyts, whose hulls were shaped to carry maximum cargoes while their topdecks were shaped to minimize transit taxes, and the factorylike herring busses. Generating wealth was preeminently the object of the Dutch import-and-export processing complexes. The British industrial revolution was built around a set of metal, textile, and power technologies that engineers and industrialists consciously tuned to increasing production, cutting costs, and boosting profits; while later, in the science-and-systems era, there emerged the new science-based industries (electricity, chemicals, pharmaceuticals, photography, refrigeration, steel, radio, telephones).

The social, political, and cultural changes that accompany technological change are no less important than the economic changes, even if they are less readily quantified and served up with mathematical models or plugged into computer databases. Renaissance court culture not only created and shaped many technologies, such as massive building projects, incessant war-making, and dynastic displays, but it also was created by and shaped by these technological efforts. "The movement of wealth through the art market . . . transformed . . . the structure of society," writes Richard Goldthwaite in his study of Florence's building industry. "That flow, in short, was the vital force that generated the Renaissance."[5] The quantifiable aspects of Renaissance-era economic growth may have been elsewhere (in the woolen and banking industries), but the era's court-based technologies were crucial to the emergence of Renaissance culture.

Similarly, commercial technologies helped constitute Dutch society and culture, not just impel its economic growth. Commerce and commercial technologies permeated Dutch society and culture, extending from the top strata of merchants and financiers, through the middle strata of those who owned mills and boats, down to agricultural workers milking cows for cheese exports. Members of the elite invested in shares of the great trading companies and bought elegant townhouses, while decidedly average Dutch citizens possessed decent oil paintings and, for a time, speculated in dodgy tulip futures. The "embarrassment of riches," identified by cultural historian Simon Schama as a core theme of Dutch culture, depended on the unusual material plenty generated by and through commercial technologies.

By comparison to Dutch commerce, at least, the early industrial culture in Britain featured a narrower distribution of wealth. Britain was, in aggregate terms, a relatively wealthy country, but the industrial revolution did not generate substantial wealth for British workers until after the 1840s. What is more, while a perspective centered on quantifying economic growth might spotlight the industrial era's economic successes, it invariably ignores the era's polluted rivers, cramped housing, and stinking privies. These too were aspects of industrial society.

The wrenching cultural changes and environmental damage that have accompanied globalization are overlooked by writers who focus only on the aggregate gains in wealth. And this is a real-world problem. Development agencies whose policies focus only on aggregate economic measures (such as per capita GDP figures) persistently disregard the environmental hazards and cultural strains in "undeveloped" regions or "Third World" countries. Much of the frank resentment today aimed at the World Bank, International Monetary Fund, and World Trade Organization stems from their conceptual blindness to the negative aspects of technology in social and cultural change.[6] Technological changes—whether in seventeenth-century Holland, nineteenth-century India, or twenty-first-century Thailand—generate a complex mix of economic, social, and cultural changes. I maintain that this insight is more valuable than having a simple definition of a complex term.

Variety and Culture

What is technology good for? In short, many things—and with a far greater variety of social and cultural consequences than we typically think. It is precisely this unacknowledged variety that requires us to gain insight into technology's potential for changing society and culture. Acquiring this insight is difficult, because we lack a full picture of the technological alternatives that once existed as well as knowledge and understanding of the decision-making processes that winnowed them down. We see only the results and assume, understandably but in error, that there was no other path to the present. Yet it is a truism that the victors write the history, in technology as in war, and the technological "paths not taken" are often suppressed or ignored.[7]

Taking a long-term perspective, we have seen historical actors at various times embracing technologies toward many different ends: building impressive palaces or asserting dynasties in the Renaissance courts; generating wealth or improving industries or maintaining far-flung imperial possessions; embracing modern culture, contesting the Cold War, or even fighting globalization. Some of these actions and ends, such as commissioning Renaissance-era sculptures to legitimate political rule, might startle readers who have previously assumed that technology is only about economic growth and material plenty. Other activities made possible by technology, such as the transatlantic slave trade in the era of commerce, the slaughter of native peoples during the era of imperialism, or the poison-gas chambers built by German industrialists and engineers under Hitler, present some of the principal moral quandaries of the modern age. This vibrant diversity of technological ends and the visions of a good life they embody, as well as the manifestly evil purposes toward which people have developed technology, suggests that the "question of technology" is far deeper, more complex, and more troublesome than we typically recognize.

Indeed, without trying to minimize the economic results of technology (the analysis of which leaves much to be desired),[8] I believe it is a more pressing problem to understand far better than we do the varied social, political, and cultural consequences of technology. As we have seen repeatedly, decisions about technologies are made to promote, enhance, reinforce, or preserve certain social and cultural arrangements. Leonardo's sculptures, automata, and war machines helped create a distinctive Renaissance court culture, just as the modern architects' embrace of science and rationalism, steel and concrete promoted a distinctive modernist culture. These technologies created culture.

It follows that technologies are not neutral tools. The very presence of a certain technique or technology can alter the goals and aims of a society as well as the ways people think in articulating their ideas. An obvious case in point is the representational technologies, ranging from Renaissance-era geometrical perspective and moveable-type printing through to today's Internet. Each of these technologies has made vivid changes in how people think and communicate with others and has brought about major cultural movements, including court culture and patronage, and contemporary cyberculture.

A more subtle and yet more pervasive example of technology's interactions with the goals and aims of society resides in the *process* of technical change. Frequently, technical change is vigorously promoted by enthusiastic and powerful actors who create new demands—and new technologies to meet them—rather than simply responding to "demands" set by society. One can think of the dynamics of technical change in terms similar to the dynamism of capitalist economies; in both, agents of change not only create new products and services but also create new demands for them, typically through marketing or advertising, but also through the regulation, promotion, and procurement activities of government. And of course the state itself—whether the precarious city-states of the Renaissance or the powerful nation-states of the military era—can be an actor in technical change as well as be transformed by technical change.

Technologies in the age of imperialism illustrate how the process of technical change can alter the goals and aims of entire societies. In the mid-nineteenth century the ready availability of distance-spanning technologies such as telegraphs, railroads, and steamships framed the discussions about imperialism in Britain (and soon enough in other imperialist countries) in terms never quite seen before. "Day by day the world perceptibly shrinks before our eyes. Steam and electricity have brought all the world next door. We have yet to re-adjust our political arrangements to the revolution that has been wrought in time and space," wrote one prominent London newspaper in the mid-1880s. "Science has given to the political organism a new circulation, which is steam, and a new nervous system, which is electricity," suggested another writer of the time. "These new conditions make it necessary to reconsider the whole colonial problem."[9]

These authors recognized that the once-formidable barriers to British control of overseas possessions, including great distances and uncertain communication, had lessened. The discovery of quinine to fight malaria reduced another type of barrier to imperialism, that of endemic tropical disease. Many British people, having seen that telegraphs had "saved" their colonial rule in India, especially during the 1857 Indian mutiny, believed that railroads, undersea telegraphs, and other imperial technologies were necessary to the same end.

These distance-spanning technologies appeared to make the imperialist venture too cheap to ignore, displacing a host of political and social questions about whether it was a good idea in the first place. In *The Imperial Idea and Its Enemies,* A. P. Thornton documents how difficult it was for opponents of imperialism to drive their critical point home, despite the manifest uncertainties thrown up by the Indian Mutiny, the costly Boer Wars, the disastrous First World War, and the rise of nationalism abroad and democracy at home. "Statesmen, journalists, and preachers came to every question of policy or morality bound by the silent influence of a half-uttered thought: 'Come what may, the Empire must be saved.'"[10] Europeans for the most part followed blindly where Britain had led, notably in the notorious land grab that carved up the continent of Africa into European colonies. American imperialism gained steam after the 1898 war with Spain.

Yet it is important to remember that each of these technologies—telegraphs, railroads, steamships, quinine—did not exist in a state of perfection prior to the imperial venture. Technologies off the shelf did not "cause" imperialism. Rather, they were themselves selected, developed, and promoted by colonial officials and by engineers or industrialists keen on selling to the colonial state. In this way, then, imperial-era technologies were both *products of* and *forces for* imperialism. In like measure, the imperial state was both created by and a force behind the evolving imperial technologies. Indeed, the era of imperialism illustrates in a particularly clear way what is true of each era examined in this book—technologies have interacted deeply with social, cultural, and economic arrangements.

Imperialism also raises in a particularly blunt form the question of who has ready access to and practical control over technologies—and who does not. In the imperialist era, no balanced parliamentary debate made the decision that Britain would become industrial while India would remain agricultural (and thereby a source of raw materials for British industry as well as a captive market for industrial goods). There was instead a blatant asymmetry of access to decision makers and centers of policy and power, not only in formal terms (lobbying Parliament in London or buying Indian railway shares in London) but also in the access to and content of technical education and the designs for railways. In earlier eras, there had also been substantial formal and informal barriers to accessing technology, notably during the court, industrial, and military eras, in which governments or rulers erected barriers around technologists.[11] By comparison, the eras of commerce and global culture have featured fewer restrictions on and wider access to technology.

Full and informed access to the content of technology may be just as crucial to technological development as influence on high-level policy decisions. Power does flow from the end of a gun; Europeans' deadly machine guns in the colonial wars proved that point. But there is an important dimension of power that resides in things, in the built world, and in the knowledge about that world that people have access to or are excluded from. One can see this in a generation of heart-felt experiments in attempting to "transfer" technologies across socioeconomic and cultural boundaries; the all-too-frequent failures strongly suggest that technologies cannot be crated up and sent off like Christmas presents, ready to be unwrapped and instantly put into service. On close examination, one can see that "transferred" technologies often need subtle refinements if they are to be adapted to a new environment in ways that the original designers could never have foreseen. Social arrangements that do not grant agency to local implementers of technology are consequently predisposed to failure.

The conceptual muddle surrounding these questions of technology transfer can be cleared up with Arnold Pacey's useful notion of "technology dialogue," an interactive process which he finds is frequently present when technologies successfully cross cultural or social barriers. The desirability of this conceptual and cultural openness (a two-way process of curiosity and mutual learning) firmly contradicts the proposition that technology must be "under the control" of power structures. Historically, the most coercive power structures have had little success with wide-ranging technical innovation.[12] Conversely, without anything like a well-organized modern nation-state, technology transfer succeeded in Renaissance Italy, commercial Holland, and industrial England. We tend to think that British technologists in the industrial era invented new and revolutionary cotton-textile machines, but if we think in terms of "technology dialogue," it may be just as accurate to say that they brilliantly adapted earlier machines and techniques (from Italy, France, and India) to the peculiarly British conditions of access to low-wage labor, a ready source of imported cotton, and large domestic and export markets. And, much closer to home, McDonald's has taken this basic insight about the desirability of local adaptation and dialogue and embraced it as explicit corporate policy, in the aim of being a "multilocal" company highly attuned to cultural sensibilities and culinary preferences. Its decentralized and networked organization is by design quite different from the corporate power center imagined by many antiglobalization activists.

Displacement and Change

How does technology change society? With the benefit of history and hindsight, it is easier to identify the implicit choices about society, politics, and culture that are bound up with choices about technologies. We have seen that *displacement* is one important dynamic of the interaction between technology and culture. Displacement occurs when a set of technology decisions has the effect of displacing alternatives or precluding open discussion about alternatives in social development, cultural forms, or political arrangements. Something of this sort occurred with the hemmed-in debate in Britain over imperialism. During the military era, the heavy governmental investment in nuclear power, especially in the United States, France, and the Soviet Union, displaced investigation into—let alone development of—such alternatives as wind and solar energy.[13] Political arrangements were also in play but not openly discussed. Recall the atomic bombing of Japan and how, in effect, the Manhattan Project's military leader commandeered presidential power: "I didn't have to have the President press the button on this affair." Its critics contend that the campaign to embrace a nuclear future displaced a traditional vision of democracy and society. Their argument is that given the danger inherent in making, moving, and storing nuclear fuel and nuclear waste, a nuclear-powered society (whatever its formal politics) requires a strong central state, with highly coercive security powers, acting over an extraordinarily long duration of time. Even if networked computers make possible a radical decentralization of political and social arrangements, nothing of the sort will occur due to the imperatives of nuclear-era state security.

In another instance, Renaissance-era patrons of technology were in effect displacing technologists from improving industry or generating wealth when they kept them engaged in court activities. Similarly, by concentrating on technologies of commerce, the Dutch technologists displaced a court-centered society while enhancing a capitalistic, but nonindustrial pattern of economic growth and cultural development. And British imperial officials and industrialists, by their explicit designs for railways and public works and technical education, displaced an industrial future for India, directing the colony to be agricultural instead of developing its manufacturing and mechanical capabilities.[14]

Displacement can be seen in technology-related aesthetics, too. The fascination of modernist architects and their patrons with the "modern" building materials of glass and steel and concrete created innumerable city streets, office blocks, and public buildings with a high degree of similarity, or, if you prefer, insufferable monotony. In the name of technology, modernists, as I pointed out, waged a determined campaign and carefully positioned their institutional efforts to create a modern culture. For roughly fifty years, a certain technical perspective on modern architecture displaced alternative, more eclectic approaches. Of course, with the rise of postmodern architecture, which embraces varied colors, materials, and historical styles, there is a counter-displacement of the once-dominant modernism.

During each of the eras examined in this book, the asserting of a distinct purpose for technology, and a desirable direction for social and cultural development, displaced alternative purposes and directions. Displacement, then, is how societies, through their decisions about technologies, orient themselves toward the future and, in a general way, direct themselves down certain social and cultural paths rather than other paths. Of course, the paths and the technologies are frequently contested, sometimes in open political debate or social protest (such as the upheavals surrounding Manchester during the industrial revolution). And often enough, societies experience unintended consequences of technology decisions (such as the open-border global culture built from the military-spawned Internet). Yet not all paths are continually contested and not all consequences are unintended. I would simply observe that, over long stretches of time, there are durable patterns in technologies, societies, and cultures. These long-duration patterns, or eras, of technology can be seen as one foundation for our cultural norms.

Displacement does not occur on a level playing field, and so we must again consider the relation of technology to power. In *Capital*, Karl Marx asserted, "It would be possible to write quite a history of the inventions, made since 1830, for the sole purpose of supplying capital with weapons against the revolts of the working-class."[15] Chapters in that history would certainly recount the efforts of leading British technologists such as James Nasmyth and Richard Roberts to combat the "refractory class" of workers by inventing "self-acting" machines that "did not strike for wages."

Power comes in many forms, and capitalists have not had a monopoly on power or on technology. During the Renaissance it was the quasi-feudal courts—not the capitalistic wool merchants—that most effectively deployed technology, shaping it in the image of court culture. During the industrial revolution capitalistic brewers and cotton spinners and machine builders, rather than women spinning thread at home, engaged the era's technologists. During the "command economy" of the military era, powerful nation-states set the agenda for atomic weapons and command-and-control systems, while industrial managers, under both capitalism and socialism, largely followed the nation-state's direction. Under globalization, capitalists gained power while the nation-state seemed to wither away. An indicator that we may be facing an imminent national-security era is the return to prominence of the nation-state.

Despite the many powerful actors that have set the agenda for technologies at a high level, it is a serious mistake to dismiss technology as a tool that can only be wielded by the powerful. Indeed, we need to ask the corresponding question: Can technologies be used by nondominant actors to advance their alternative agendas? To begin, we have seen numerous instances in which historical actors whose power was frankly precarious used technology to gain, enhance, modify, or sustain their cultural or political power. In the Renaissance, the patronage of technological projects to build churches, palaces, statues, or other high-profile public works often enhanced the political legitimacy of an unsteady and insecure ruler. Cosimo de' Medici's many buildings were "worthy of a crowned king," declared a fellow Florentine palace-builder.[16] But Cosimo was *not* a crowned king, and in fact he ruled uneasily, and his family's dynasty was exiled for nearly two decades. An identical insecurity pervaded the Sforza family's rule in Milan and a succession of scandal-ridden popes. These rulers were not all-powerful beforehand; rather, one might say, they consolidated their power and legitimated their rule through their active patronage of technology and other public works.

Closer to home, let's reconsider the story of fax machines in the late 1980s and the European students' discovery of their international community-building potential, an instance I believe of genuine and spontaneous cultural innovation. It is no surprise that the Council of Europe

gave wide publicity and official sanction to this grass-roots effort at creating a positive notion of a unified Europe. This occurred just when the
momentum of the project of "building Europe" got an institutional boost
from the Maastricht Treaty of 1992. The European Union continues its
quest for political legitimacy today through its active patronage of public works projects, promotional films, educational initiatives, and publicity campaigns—not to mention the launch of the Euro as a single
transnational currency.[17] Each of these actors, then, used the commissioning and display of technologies to create, enhance, or sustain power.
One might say they *became* dominant through their success in mobilizing technology.

A second reason for looking closely at the technology-power nexus
is the possibility that *non*-dominant groups in society will effectively mobilize technology.[18] Workers, with few exceptions since Marx's time, have
rarely considered the selective uptake or cautious embrace of technology
to be a means for furthering their own aims; they have, in a roundabout
way, too often proved Marx right in claiming that technology is the
weapon solely of capital. The most famous example of workers' embrace
of technology comes from the Scandinavian auto workers who reconfig-
ured their factory environment to break the monotonous production-line
routines of Fordism.

Further examples of nondominant groups' using technology to advance a nonelite agenda come readily to mind once you start looking.
From the 1960s onward, citizens' movements in many countries have
pushed governments and industries to provide safer cars, cleaner water,
broader access to medical treatments, and greater transparency in government decision making. Women's groups have for decades formed a
strong and effective lobby for contraceptive technologies. Today's antiglobalization movement depends on the Internet to circumvent established and expensive media outlets, while antiglobalization protests are
conducted with the latest in communications technology. Each of these
are instances where nonelite groups have exerted meaningful influence on
technological changes—and hence on social and cultural developments.

Especially interesting shifts in the environmental movement's atti-
tudes toward modern technology are presently in play. For many decades,
from at least the 1960s onward, environmentalists pinned the blame for
pollution squarely on capitalism and technology; and they typically rejected science, technology, and capitalism as a matter of principle. Many
environmentalists still retain faith in some version of a small-is-beautiful

stance that is first and foremost distant from modern technology. But recently, under the banner of "ecological modernization," some true-blue environmentalists, especially in Europe, have shifted to a cautious and selective embrace of technology and market mechanisms. Advocates of ecological modernization view the institutions of modernity, including science, technology, industry, and the nation-state, "not only as the main *causes* of environmental problems, but also as the principal *instruments* of ecological reform."[19] The new diagnosis coming from ecological modernization is that dealing effectively with the environmental crisis will require serious engagement with technology.

In this book we have seen similar instances of social movements' embracing technologies with the aim of bringing about desirable social, political, or cultural changes. From a Dutch perspective, one could say that it was the effective deployment of technologies—printing presses, ship designs, and muskets—that won the Netherlands' battle to break free of imperial Spain, easily the most powerful state in Europe at the time. Or consider, as a question of power, the "railroad republicanism" of Paul Kruger, president of the Transvaal region in southern Africa, who for a time used his independently funded railroads to stand up against the entire might of the British Empire. In India, Britain's imperialist railways were built only after protracted negotiations with local native rulers—not to say that these talks took place on a level playing field. "I consented with great reluctance,"[20] was one native Indian ruler's description of permitting the railroad to cross his domain. Powerful entities do not automatically get precisely what they want.

We can see the similar challenges to elite agendas around us today. Ask the World Trade Organization, supposed by some to be a powerful global cabal, about the email-organized street protests that disrupted its summit meetings in Seattle and Genoa and derailed its globalizing agenda. And remind powerful McDonald's about the grass-roots campaign against its beef extract–laced "vegetarian" French fries or for that matter about its McLibel suit, archived on the Internet, where the company's preposterous courtroom arguments are published by antiglobalization activists for all the world to see. Each of these is an example of nondominant elements' effectively using technologies to contest the power of dominant groups. Technology, to repeat, is not a neutral tool. Social movements that, as a point of principle, distance themselves from technology (as the environmental movement did for decades) are depriving themselves of a crucial strategic resource.

Disjunctions and Divisions

Tensions about technology surround us. A leading source of our current anxieties about technology is the troubling gap that often exists between what we think and what we see, or, in slightly more precise terms, the *disjunction* between our normative expectations for technology and what we observe and experience in the world around us. We perceive technology to be changing quickly, I believe, because we are continually confronted with social and cultural and economic changes in which technology is clearly involved. Often, these changes are unsettling ones. And, frequently enough, "technology" serves as a shorthand term to refer to a host of complex social and political processes underlying the change.[21] Nevertheless, it is a mistake to follow the commonplace conviction that technology by itself "causes" change, because technology is not only a *force for* but also a *product of* social and cultural change.

In the twentieth century, the once-distinctive eras of technology have piled up on one another. One might say that during the eras of courts, commerce, industry, and imperialism, it was reasonably easy to see the dominant uses for technology and the dominant values built into it. But, in the twentieth century, it became increasingly difficult to determine whether technology was about science, modernism, war, or global culture. And then, just when it appeared an era of globalization was taking durable form, came the attacks of 11 September 2001.

The wide-ranging responses by Western nations, above all by the United States, to the September 11th attacks simply do not fit into a pattern of globalization. It may be too soon to say that the era of global culture has ended, but consider what we have seen since those attacks: the dramatic resurgence of the nation-state, assuming vastly expanded powers at home and abroad; the abrupt end of the conception of global culture as open and peaceful that prevailed in the 1990s; the boiling up of nationalistic trade policies; and a painfully heightened awareness of the profound differences that divide the peoples of the world, of which more in a moment. Global trade continues, but the rosy vision of a peaceful world, economically integrated and culturally harmonious, knitted together by information technology, is dead. We still have good reason to believe that the shape of technology will to a large degree mold our future. But, now, what will be the forces that shape technology?

There is a wider area of disjunction involving technology, of course, and that is the role of technology in the world's geographical, economic, and cultural *divisions*. Many thoughtful people worry that the most pressing problem is the "digital divide" between the rich societies that have ready access to computers, telecommunications, and the Internet, and the poor societies that do not. In rich countries of the industrial West itself there are many such "divides," yet the closer one looks the less pressing seems to be the question of access per se and the more pertinent the matter of how access is employed.

Consider several recent examples of how access to information technologies is being accomplished in India. This country, despite its persistent poverty, is one where small-scale fishermen are using cell phones to drive hard bargains with harbors to get the best prices for their boatloads of fish, where villagers too poor to buy a computer for each and every family might share an Internet kiosk (or cell phone) among an entire village, and where a world-class software industry is growing up around the southern city of Bangalore. Indeed, the continual circulation of software engineers between Bangalore, Silicon Valley, and other world centers in itself constitutes an impressive high-technology dialogue. While certainly there are many Indian citizens lacking ready access to technology and reasonable control over its use, India does not fall hopelessly off the digital divide. On the contrary, because of its strong investments in education and its vibrant tradition of democracy, India seems well positioned to surmount whatever digital divides it may face.

A region with a much more severe disjunction—between its dominant cultural norms and the varied potentials of twenty-first-century technology—resides in the long arc that stretches from Palestine across the Middle East through the disputed Pakistan-India border and eastward to the huge populations of China and Indonesia. One hesitates to offer any sweeping generalizations, especially since the region is rife with political battles and conceptual disputes.[22] Yet, clearly, many of the region's countries are experiencing an excruciating disjunction between their traditional cultural and political forms and the wider world of modern technologies.

This internal disjunction is compounded by the external *division* between the Moslem-Arab worldview and the Western worldview, made evident by the September 11th attacks. Most people in the West find it intolerable that Moslem fundamentalists, acting in the name of antimodern religious beliefs, would mount deadly attacks on such symbols of economic and political power as the World Trade Center and the Pentagon. In turn, Moslem fundamentalists, and the governments that have funded, tolerated, and in effect protected them, find the presence in their region of determined modernizers and aggressive Westernizers to be equally intolerable and culturally arrogant.

Across much of this large and varied region, in my view, the case is quite the opposite of India, in that the social, cultural, and political systems of the region seem to be out of step with the practical needs for technology dialogue that exist in these countries. Many have instituted wide-ranging censorship on their citizens' access to the Internet.[23] Making any long-term and durable headway on the region's political difficulties will require addressing a number of problems involving technologies and cultures. These little-reported situations include water shortages in the arid lands of the Near East, the dilemma of "development" and modernization in the tradition-bound petroleum-rich countries, the frank danger of an Indian-Pakistani nuclear exchange over their contested border, and unstable nation-states (their possible creation in Palestine, reconstruction in China, and probable dissolution in Indonesia). How will the choices made in addressing these problems affect a global society with reasonable prospects for peace?

It is an especially pressing concern that scholars and citizens in the West know all too little about the details and dynamics of how modern technologies are interacting with traditional social forms. This is true not only for the Middle East, Asia, and Africa but also for native peoples in North and South America. We can see some of the questions we need to ask by recalling the insights of Michael Adas's *Machines as the Measure of Men*.[24] Adas explains how Europeans came to believe in their cultural superiority over the peoples of Africa, India, and China. Whereas Europeans had been curious and largely respectful of these peoples and cultures early in the eighteenth century, they became increasingly chauvinistic during the course of industrialization in the nineteenth century—and as a result of the material and technological divisions that the industrialization process opened up. Eventually Europeans, given their palpable technological leadership, believed themselves possessed of a more fundamental (and highly problematic) cultural and social superiority. Moreover, the imperial era of rabid colonization inevitably generated resentments among the peoples of Africa, Asia, and the Americas, who were often forcibly incorporated within Western technological modernity. These resentments persist, and they have festered. At the very least, the optimistic view that reigned during the height of globalization—of an entire world peacefully, and profitably, interconnected—has come full stop to an end.

The boundless technological optimism of the pie-in-the-sky global-izers was always a bit overwrought, but I hope that we do not slide into a cynical disengagement with the problems of technology. The dynamics of *displacement, disjunction,* and *division* suggest the magnitude of the challenges before us. There are also models of new thinking about tech-nology to be explored, promising instances of new approaches to tech-nology in developmental thinking, philosophy of technology, and envi-ronmental thinking. These are bellwethers of a cautious embrace of technology as a means to effect desirable cultural and social changes. En-gineering as a professional institution seems aware of the pressing need for fundamental changes in education and practice.[25]

A powerful case can always be made against the naïve and uncritical embrace of technology, reminding ourselves of how much mischief such approaches have caused in the past. Yet, confronting such pressing prob-lems as we face with pollution and privacy and poverty in the absence of a determined engagement with technology seems senseless. The moment is ripe for an embrace of technology by would-be reformers, social move-ments, and citizens' groups seeking to activate this potent means of shap-ing the future.

NOTES

PREFACE

1. Quoted on liftoff.msfc.nasa.gov/academy/history/VonBraun/Germany. html (7 February 2002). (All dates that follow Web addresses are accession dates.)

2. In the United States, military R&D funding peaked in 1987, fell about one-third (in constant, inflation-adjusted dollars) with the end of the Cold War, and remained at a lower plateau through the 1990s. The pronounced military build-up since 11 September 2001 is altering this picture. Specifically, the Bush White House's record-sized military budgets with a heavy emphasis on so-called high-tech weapons could set off a new, national security–dominated era for technology—and might bring down the curtain on the globalization era. Specifically, the American Association for the Advancement of Science is forecasting a military R&D budget for 2004 to substantially *exceed* the Cold War peak of 1987 (www.aaas.org/spp/rd/trdef04p.gif [13 May 2003]).

3. See setiathome.ssl.berkeley.edu.

CHAPTER 1. TECHNOLOGIES OF THE COURT, 1450–1600

1. The Medici dynasty was founded by Giovanni di Bicci de' Medici (1360–1429), who became banker to the papacy and amassed a large fortune. Medici rule in nominally republican Florence began in 1434 and extended until the family's political "exile" during 1494–1512. In 1513 the first of two Medici popes was elected, and Medici rule returned to Florence. A successor to the second Medici pope named the head of the Medici dynasty as first duke of Florence in 1537 and as first grand duke of Tuscany in 1569. In 1495 Machiavelli joined his father's confraternity, which helped launch his public career three years later with his election as second chancellor to the city; his public career ended in 1512 with the return of the Medici.

2. For court-based technical activities including "royal factories" and "imperial workshops" in China, Ottoman Turkey, Mughal India, and the Persian Empire, see Arnold Pacey, *Technology in World Civilization* (Cambridge: MIT Press, 1990), 82–88. In the cases he discusses, Pacey notes that government control over manufacturing was meant not so much to increase trade but "to supervise pro-

duction of quality textiles and furniture in support of a magnificent style of court life" (quote on 82).

3. Paolo Galluzzi, "The Career of a Technologist," in Paolo Galluzzi, ed., *Leonardo da Vinci: Engineer and Architect* (Montreal Museum of Fine Arts, 1987), 41–109, quote on 41.

4. Ross King, *Brunelleschi's Dome* (New York: Penguin, 2001); Anthony Grafton, *Leon Battista Alberti* (New York: Hill & Wang, 2000), quote on 72. For recent photos from inside the sphere atop the cathedral, see www.vps.it/propart/segreti_web/duom4e.htm (27 September 2002).

5. Galluzzi, "Career of a Technologist," 50.

6. Galluzzi, "Career of a Technologist," quote on 62 (letter to Ludovico); Pamela Taylor, ed., *The Notebooks of Leonardo da Vinci* (New York: NAL, 1971), quote on 207 ("time of peace").

7. Martin Clayton, *Leonardo da Vinci* (London: Merrell Holberton, 1996), quote on 50.

8. Galluzzi, "Career of a Technologist," quote on 68 ("injury to friends"); Taylor, *Notebooks of Leonardo*, quote on 107 ("evil nature").

9. Galluzzi, "Career of a Technologist," quote on 80.

10. Bert S. Hall, *Weapons and Warfare in Renaissance Europe* (Baltimore: Johns Hopkins University Press, 1997).

11. A story of Matteo Bandello quoted in Martin Kemp, *Leonardo da Vinci: The Marvellous Works of Nature and Man* (Cambridge: Harvard University Press, 1981), 180. The Sforza horse was finally cast from Leonardo's model in 1999.

12. David Mateer, ed., *Courts, Patrons, and Poets* (New Haven: Yale University Press, 2000), quotes on 136 ("propaganda") and 138 ("seven boys"); Mark Elling Rosheim, "Leonardo's Lost Robot," *Achademia Leonardi Vinci* 9 (1996): 99–108; Kemp, *Leonardo da Vinci*, 152–53, 167–70.

13. Galluzzi, "Career of a Technologist," quote on 83 ("duke lost"); Irma A. Richter, ed., *The Notebooks of Leonardo da Vinci* (Oxford: Oxford University Press, 1952), quote on 326 ("flooding the castle").

14. The Leonardo-Machiavelli collaboration is reconstructed in Roger D. Masters, *Fortune Is a River* (New York: Free Press, 1998). Jerry Brotton, *The Renaissance Bazaar: From the Silk Road to Michelangelo* (Oxford: Oxford University Press, 2002), quote on 68.

15. Richter, *Notebooks of Leonardo da Vinci*, quote on 382; Clayton, *Leonardo da Vinci*, 129.

16. Galluzzi, "Career of a Technologist," 91.

17. Kemp, *Leonardo da Vinci*, 347–49.

18. The artist Raphael, quoted in Peter Burke, *The European Renaissance: Centres and Peripheries* (Oxford: Blackwell, 1998), 69.

19. Grafton, *Leon Battista Alberti*, 3–29, 71–109, 189–224, quote on 28.

20. Grafton, *Leon Battista Alberti*, quote on 84. Grafton points out that the Alberti family's exile from Florence was lifted only in 1428, some time after which

Leon Battista first entered the city. Thus, he was unlikely to have seen Brunelleschi's show box spectacle in person; he did dedicate to Brunelleschi the Italian version of his painting treatise.

21. Joan Gadol, *Leon Battista Alberti* (Chicago: University of Chicago, 1969), 26 n9, 205 n82.

22. Stephen F. Mason, *A History of the Sciences* (New York: Collier, 1962), 111.

23. Lynn White, "The Flavor of Early Renaissance Technology," in B. S. Levy, ed., *Developments in the Early Renaissance* (Albany: SUNY Press, 1972), quote on 38.

24. Printing in the Far East follows Maurice Daumas, ed., *A History of Technology and Invention* (New York: Crown, 1962) 1:285–89. Evidence on whether printing in the Far East *might* have influenced Gutenberg in Europe is evaluated in Albert Kapr, *Johann Gutenberg: The Man and His Invention* (Aldershot, England: Scolar Press, 1996), 109–22.

25. Colin Clair, *A History of European Printing* (London: Academic Press, 1976), 6–14; Daumas, *History of Technology and Invention*, 2:639 (on typography). On dates, I have followed Kapr, *Johann Gutenberg*, 29–99, quote on 84 ("use of a press").

26. Kapr, *Johann Gutenberg*, 142–79, 197–210, 259–66, quote on 259.

27. Clair, *History of European Printing*, quotes on 13 and 14.

28. Elizabeth L. Eisenstein, *The Printing Revolution in Early Modern Europe* (Cambridge: Cambridge University Press, 1983), 145–48.

29. Eisenstein, *Printing Revolution*, 170–75.

30. Elizabeth L. Eisenstein, *The Printing Press as an Agent of Change*, 2 vols. (Cambridge: Cambridge University Press, 1979), 1:408.

31. On Plantin, see Clair, *History of European Printing*, 195–206; Francine de Nave and Leon Voet, "The Plantin Moretus Printing Dynasty and Its History," in Christopher and Sally Brown, eds., *Plantin-Moretus Museum* (Antwerp: Plantin-Moretus Museum, 1989), 11–53; Eisenstein, *Printing Press*, quote on 1:444.

32. Eisenstein, *Printing Revolution*, 43.

33. Pacey, *Technology in World Civilization*, quote on 7.

34. See Dieter Kuhn, *Science and Civilisation in China*, vol. 5 part 9: *Textile Technology: Spinning and Reeling* (Cambridge: Cambridge University Press, 1988), 161, 184, 210, 348, 352, 356, 363, 366–68. By comparison, a drawing by Leonardo (on 164) fairly leaps out of the page with its three-dimensional realism. Francesca Bray (*Technology and Gender: Fabrics of Power in Late Imperial China* [Berkeley: University of California Press, 1997], 45, 214, 227) and Arnold Pacey (*Technology in World Civilization*, 27) also provide instances of incomplete illustrations of key spinning technologies.

35. Eugene Ferguson, *Engineering and the Mind's Eye* (Cambridge: MIT Press, 1992), 107–13.

36. Pamela O. Long, "The Openness of Knowledge: An Ideal and Its Context in Sixteenth-Century Writings on Mining and Metallurgy," *Technology and Culture* 32 (April 1991): 318–55, quote on 326. A version of this article appears in Long's *Openness, Secrecy, Authorship: Technical Arts and the Culture of Knowledge from*

Antiquity to the Renaissance (Baltimore: Johns Hopkins University Press, 2001), chap. 6.

37. Long, "Openness of Knowledge," 330–50.

38. Agricola quoted in Eisenstein, *Printing Revolution*, 193.

39. Ferguson, *Engineering and the Mind's Eye*, 115.

40. Galileo quoted in Eisenstein, *Printing Revolution*, 251. Galileo's consulting work for Florence in 1630 is discussed in Richard S. Westfall, "Floods along the Bisenzio: Science and Technology in the Age of Galileo," *Technology and Culture* 30 (1989): 879–907. The Medici figure prominently in Mario Biagioli's *Galileo, Courtier* (Chicago: University of Chicago Press, 1993).

41. Luca Molà, *The Silk Industry of Renaissance Venice* (Baltimore: Johns Hopkins University Press, 2000), quote on 37, 121. In early modern Europe demand for mechanical clocks was generated by "the numerous courts—royal, princely, ducal, and Episcopal" as well as by urban merchants, according to David Landes, *Revolution in Time: Clocks and the Making of the Modern World* (Cambridge, Mass.: Belknap Press, 1983), 70. See the Note on Sources for additional sources on glassmaking, shipbuilding, and court technologies.

42. Pacey, *Technology in World Civilization*, 82–88; Noel Perrin, *Giving up the Gun: Japan's Reversion to the Sword, 1543–1879* (Boston: Godine, 1979).

CHAPTER 2. TECHNIQUES OF COMMERCE, 1588–1740

1. The finances of the immensely rich Strozzi and three other prominent families are anatomized in Richard A. Goldthwaite's *Private Wealth in Renaissance Florence* (Princeton: Princeton University Press, 1968), 40–61. During the 1420s and 1430s, the Strozzi family had its assets mostly in real estate holdings: the father Simone's estate consisted of real estate holdings (64%), business investments (23%), and state funds (7%); his son Matteo's estate was real estate holdings (70%), with additional investments in state funds (13%) and in wool manufacture (17%). Matteo's son Filippo built a large fortune in banking and commerce, both in Italy and across Europe; in 1491–92 his estate consisted of real estate, including a palace (14%), personal property, including household furnishings and precious metals (11%), cash (45%), and business investments (30%).

2. In 1575 and 1596, and at least four times in the following century, the Spanish crown declared itself bankrupt; between 1572 and 1607, Spanish troops mutinied forty-six separate times. A relative decline affected the fortunes of Italian city-states and courts. In the 1620s the Swedish crown (as noted below) lost control of its finances to Dutch copper merchants.

3. Karel Davids and Jan Lucassen, eds., *A Miracle Mirrored: The Dutch Republic in European Perspective* (Cambridge: Cambridge University Press, 1995), quote on 1. For the long-term "secular trends," see Fernand Braudel, *The Perspective of the World* (New York: Harper & Row, 1984), 71–88.

4. Jonathan I. Israel, *Dutch Primacy in World Trade, 1585–1740* (Oxford: Clarendon Press, 1989), quote on 13. This chapter draws heavily on this work.

5. William H. McNeill, *The Pursuit of Power* (Chicago: University of Chicago Press, 1982), 126–41, quotes on 137–39.

6. Joel Mokyr, *Industrialization in the Low Countries, 1795–1850* (New Haven: Yale University Press, 1976), quote on 9.

7. On de Geer and the Swedish iron and copper industries, see Jonathan I. Israel, *The Dutch Republic* (Oxford: Clarendon Press, 1995), 271–75; E. E. Rich and C. H. Wilson, eds., *Cambridge Economic History of Modern Europe* (Cambridge: Cambridge University Press, 1977), vol. 5: 245–49, 484–87, 495. On Eskilstuna, see Maths Isacson and Lars Magnusson, *Proto-industrialisation in Scandinavia: Craft Skills in the Industrial Revolution* (Leamington Spa, U.K.: Berg, 1987), 89–108.

8. Rich and Wilson, *Cambridge Economic History,* vol. 5: 148–53; Israel, *The Dutch Republic,* 16–18, 116–19; Jan Bieleman, "Dutch Agriculture in the Golden Age, 1570–1660," in Karel Davids and Leo Noordegraaf, eds., *The Dutch Economy in the Golden Age* (Amsterdam: Netherlands Economic History Archives, 1993), 159–83.

9. Israel, *Dutch Primacy in World Trade,* 21.

10. Richard W. Unger, *Dutch Shipbuilding before 1800* (Assen, Netherlands: Van Gorcum, 1978), 29–47, quote on 25.

11. Karel Davids, "Technological Change and the Economic Expansion of the Dutch Republic, 1580–1680," 79–104 in Davids and Noordegraaf, *Dutch Economy,* 81–82; Arne Kaijser, "System Building from Below: Institutional Change in Dutch Water Control Systems," *Technology and Culture* 43 (2002): 521–48.

12. Israel, *Dutch Primacy in World Trade,* 21–22; Leo Noordegraaf, "Dutch Industry in the Golden Age," in Davids and Noordegraaf, *Dutch Economy,* 142–45; Carlo Cipolla, *Before the Industrial Revolution,* 3rd edition (New York: W. W. Norton, 1994), 254.

13. John E. Wills, Jr., *Pepper, Guns and Parleys: The Dutch East India Company and China, 1622–1681* (Cambridge: Harvard University Press, 1974), 10, 19–21, quote on 20.

14. Israel, *Dutch Primacy in World Trade,* 75–76, quote on 75.

15. Israel, *Dutch Primacy in World Trade,* 73–79, quote on 73 n8.

16. Peter M. Garber, *Famous First Bubbles: The Fundamentals of Early Manias* (Cambridge: MIT Press, 2000), 82.

17. Paul Zumthor, *Daily Life in Rembrandt's Holland* (Stanford: Stanford University Press, 1994), quote on 173 ("so many rules"); Mike Dash, *Tulipomania* (New York: Crown, 1999), quote on 141 ("intoxicated head"). In 1590 Haarlem residents consumed an estimated 300 liters of beer per person per year; see Richard W. Unger, "The Scale of Dutch Brewing, 1350–1600," *Research in Economic History* 15 (1995): 261–92. For children at an Amsterdam orphanage during 1639–59, the daily ration of beer amounted to 270 liters per person per year; see Anne McCants, "Monotonous but not Meager: The Diet of Burgher Orphans in Early Modern Amsterdam," *Research in Economic History* 14 (1994): 69–119. A beer keg is prominent in a ca. 1660 oil painting depicting the feeding of orphans; see Richard W. Unger, *A History of Brewing in Holland, 900–1900* (Leiden: Brill, 2001), 248.

18. Israel, *The Dutch Republic,* 318–27.

19. P'eng Sun-I, "Record of Pacification of the Sea" (1662–64) quoted in Wills, *Pepper, Guns and Parleys,* quote on 29; Israel, *Dutch Primacy in World Trade,* 71–73, 76, 86.

20. Israel, *Dutch Primacy in World Trade,* 101–6.

21. Israel, *Dutch Primacy in World Trade,* 171–87, quote on 177.

The Tokugawa era in Japanese history (1603–1867) suggests the extreme measures needed to preserve a court-based culture in an age of commerce. The arrival of firearms, carried by shipwrecked Portuguese sailors in 1543, into a society dominated by competing warlords had made for a volatile situation. Japanese artisans quickly learned to make immense numbers of firearms, and battles between rival warlords soon involved tens of thousands of firearms (far in excess of the number in Western Europe). As Noel Perrin makes clear in his masterful *Giving up the Gun: Japan's Reversion to the Sword, 1543–1879* (Boston: Godine, 1979), the proliferation of firearms directly threatened the status of the *samurai,* or warrior class. Low-class soldiers could simply shoot dead a high-class *samurai* before he could do any damage with his formidable sword. In 1603, the warlord Ieyasu Tokugawa asserted central control on the country and actually achieved it over the next two decades. The warring rivals who had torn the country apart with firearms were turned into subservient *daimyo* and given feudal lands to administer. On religious pretext, the troublesome firearms were confiscated and melted down. Japanese merchants were forbidden to conduct trade with or travel to foreign countries. Europeans, having brought firearms, as well as Christianity, which further disrupted Japanese society and culture, also fell under a cloud of suspicion. The Spanish and Portuguese missionaries were especially offensive. By contrast, the Dutch kept their religion to themselves and were permitted to stay (and trade) even after the forcible expulsion of the other Europeans in the 1620s and 1630s. These drastic measures to preserve its court-centered culture made Japan a "closed society" through the mid-nineteenth century.

22. Israel, *Dutch Primacy in World Trade,* 156–60; Pieter Emmer, "The West India Company, 1621–1791: Dutch or Atlantic?" in his *The Dutch in the Atlantic Economy* (Aldershot, England: Ashgate, 1998), chap. 3.

23. Israel, *Dutch Primacy in World Trade,* 160–70, quote on 163, 186, 255, 319–27; Pieter Emmer and Ernst van den Boogart, "The Dutch Participation in the Atlantic Slave Trade, 1596–1650," in Emmer, *The Dutch in the Atlantic Economy,* chap. 2; Carla Rahn Phillips, *Six Galleons for the King of Spain: Imperial Defense in the Early Seventeenth Century* (Baltimore: Johns Hopkins University Press, 1986), 3–7.

24. Mokyr, *Industrialization in the Low Countries,* quote on 2; Israel, *Dutch Primacy in World Trade,* 114–16; Cipolla, *Before the Industrial Revolution* (3rd ed.), 249–59.

25. Israel, *Dutch Primacy in World Trade,* 187–91, quote on 188.

26. Israel, *Dutch Primacy in World Trade,* 35–36, quote on 36 (on linen bleaching), 116–17, 190–96, quote on 194 ("technical innovations"); Carla Rahn Phillips

and William D. Phillips, Jr., *Spain's Golden Fleece* (Baltimore: Johns Hopkins University Press, 1997), 260, 303.

27. Israel, *Dutch Primacy in World Trade,* 260–64, quote on 262.

28. Phillips and Phillips, *Spain's Golden Fleece,* 169–78, 193–209, quote on 204.

29. Israel, *Dutch Primacy in World Trade,* 116, 264–66.

30. Israel, *Dutch Primacy in World Trade,* 266–69; Karel Davids, "Windmills and the Openness of Knowledge: Technological Innovation in a Dutch Industrial District, the Zaanstreek, c. 1600–1800," paper to Society for the History of Technology (October 2001).

31. Israel, *Dutch Primacy in World Trade,* quote on 269. The varied activities of Dutch guilds, learned societies, and towns—sometimes promoting technical innovation, sometimes resisting it—are evaluated in Karel Davids, "Shifts of Technological Leadership in Early Modern Europe," in Davids and Lucassen, *A Miracle Mirrored,* 338–66.

32. The Dutch economic decline is detailed in Israel, *The Dutch Republic,* 998–1018.

33. Davids, "Technological Change," 81–83.

34. Israel, *Dutch Primacy in World Trade,* 102, 292–358.

35. Zumthor, *Daily Life in Rembrandt's Holland,* 194–99, quote on 195; Cipolla, *Before the Industrial Revolution* (3rd ed.), 259.

CHAPTER 3. GEOGRAPHIES OF INDUSTRY, 1740–1851

1. E. J. Hobsbawm, *Industry and Empire* (London: Weidenfeld & Nicolson, 1968; reprint, London: Penguin, 1990), quote on 34; David S. Landes, *The Unbound Prometheus* (Cambridge: Cambridge University Press, 1969), quote on 1–2. For sectoral growth rates of real output (percent per annum), see Maxine Berg, *The Age of Manufactures, 1700–1820,* 1st edition (Oxford: Oxford University Press, 1986), 28.

2. Berg, *Age of Manufactures* (1986), quote in (unpaged) preface. For data on value added by sector, see Maxine Berg, *The Age of Manufactures,* 2nd edition (London: Routledge, 1994), 38. On handloom weavers, see Geoffrey Timmons, *The Last Shift* (Manchester: Manchester University Press, 1993), 25–28, 110, 220.

3. Raymond Williams, *Culture and Society, 1780–1950* (New York: Harper & Row, 1966), xi–xvi and passim.

4. L. D. Schwarz, *London in the Age of Industrialisation: Entrepreneurs, Labour Force and Living Conditions, 1700–1850* (Cambridge: Cambridge University Press, 1992), quotes on 1 (Braudel), 231 ("a storm"). The "gentlemanly capitalism" thesis is scrutinized in M. J. Daunton, "'Gentlemanly Capitalism' and British Industry, 1820–1914," *Past and Present* 122 (1989): 119–58; the thesis is defended in P. J. Cain and A. G. Hopkins, *British Imperialism, 1688–2000,* 2nd edition (Harlow: Longman, 2002), 114–50, passim.

5. Around 1800, London's total of 290 steam engines outnumbered those in Manchester, Leeds, or Glasgow (respectively home to 240, 130, and 85); see

Richard L. Hills, *Power from Steam: A History of the Stationary Steam Engine* (Cambridge: Cambridge University Press, 1989), 299 n78.

6. Data from Eric J. Evans, *The Forging of the Modern State: Early Industrial Britain, 1783–1870* (London: Longman, 1983), 407–9; E. A. Wrigley, "A Simple Model of London's Importance in Changing English Society and Economy 1650–1750," *Past and Present* 37 (1967): 44–70; David R. Green, *From Artisans to Paupers: Economic Change and Poverty in London, 1790–1870* (London: Scolar Press, 1995), Defoe quote on 15. Populations in 1851, in thousands: London 2,362, Liverpool 376, Glasgow 345, Manchester 303, Birmingham 233, Edinburgh 194, Leeds 172, Sheffield 135, Bristol 137, Bradford 104, Newcastle 88, Hull 85, Preston 70.

7. Roy Porter, *London: A Social History* (Cambridge: Harvard University Press, 1994), 134.

8. M. Dorothy George, *London Life in the Eighteenth Century* (New York: Capricorn Books, 1965, orig. 1925), quote on 323 n2; data from 1851 census, here and elsewhere in this chapter, from Schwarz, *London in the Age of Industrialisation,* 255–58.

9. Michael W. Flinn, *The History of the British Coal Industry* (Oxford: Clarendon Press, 1984), 2:217, table 7.2; Defoe quoted in Porter, *London,* 138.

10. George, *London Life,* quote on 167.

11. This section relies on Peter Mathias, *The Brewing Industry in England, 1700–1830* (Cambridge: Cambridge University Press, 1959). The aggregate output statistics—which show beer brewing growing at an anemic 0.21 percent per year in 1700–1760 and shrinking at 0.10 percent per year in 1760–70—utterly fail to locate the industrialization of brewing.

12. Mathias, *The Brewing Industry in England,* 53–62, quotes on 61, 62.

13. T. R. Gourvish and R. G. Wilson, *The British Brewing Industry, 1830–1980* (Cambridge: Cambridge University Press, 1994), 30–35. Gourvish and Wilson assume that children under 15 (35% of the population) and tea-totallers (3 million in 1899) drank no alcohol, and estimate that women drank half as much as men, calculating that in the peak year of 1876 the average adult male consumed 103 gallons of beer a year (16 pints a week).

14. Gourvish and Wilson, *British Brewing Industry,* 87, 226–66; Mathias, *Brewing Industry in England,* 554–58.

15. Mathias, *Brewing Industry in England,* 36, 551–58.

16. Eric Robinson and A. E. Musson, *James Watt and the Steam Revolution* (New York: Kelley, 1969), quote on 88 (Watt); Mathias, *Brewing Industry in England,* 78–98.

17. Mathias, *Brewing Industry in England,* 41–42, 48–53, 106–9. For ancillary industries, see Philip Scranton, *Endless Novelty: Specialty Production and American Industrialization, 1865–1925* (Princeton: Princeton University Press, 1997).

18. Mathias, *Brewing Industry in England,* 102–6, 117–33.

19. Gourvish and Wilson, *British Brewing Industry,* 30–35, quote on 12. Per

capita consumption of beer in England and Wales was 28.4 gallons in 1820–24, 35.4 gallons in 1835–39, and 30.5 in 1840–44, a figure maintained until 1860.

20. Schwarz, *London in the Age of Industrialisation*, 255–58 (1851 census data); *Report on Handloom Weavers* (1840) quoted in Ivy Pinchbeck, *Women Workers and the Industrial Revolution, 1750–1850* (London: Frank Cass, 1930), quote on 179.

21. George, *London Life*, quotes on 74 and 345 n32; Schwarz, *London in the Age of Industrialisation*, 255 (census data); Linda Clarke, *Building Capitalism: Historical Change and the Labour Process in the Production of the Built Environment* (London: Routledge, 1992), 82–84, 174–77, 207–17, 243–48; Hermione Hobhouse, *Thomas Cubitt: Master Builder* (London: Macmillan, 1971), 23–25, 96–102.

22. Keith Burgess, "Technological Change and the 1852 Lock-Out in the British Engineering Industry," *International Review of Social History* 14 (1969): 215–36, on 233; Green, *From Artisans to Paupers*, 27–32, 59 n30.

23. Peter Linebaugh, *The London Hanged* (Cambridge: Cambridge University Press, 1992), 371–401, quote on 380; Carolyn Cooper, "The Portsmouth System of Manufacture," *Technology and Culture* 25 (1984): 182–225.

24. David Jeremy, *Transatlantic Industrial Revolution* (Cambridge: MIT Press, 1981), 67.

25. V. A. C. Gatrell, "Labour, Power, and the Size of Firms in Lancashire Cotton in the Second Quarter of the Nineteenth Century," *Economic History Review* 30 (1977): 95–139, statistics on 98; Steven Marcus, *Engels, Manchester, and the Working Class* (New York: Random House, 1974), quote on 46 n30 ("only one chimney"); W. Cooke Taylor, *Notes of a Tour in the Manufacturing Districts of Lancashire* (London: Frank Cass, 1842; reprint, New York: A. M. Kelley, 1968), quote on 6 ("mighty energies").

26. Sidney J. Chapman, *The Lancashire Cotton Industry* (Manchester: Manchester University Press, 1904; reprint, Clifton, N.J.: A. M. Kelley, 1973), quotes on 2 (Defoe), 13 (George Crompton).

27. See the classic account of home and factory spinning in Pinchbeck, *Women Workers*, 129–56, 183–201. Pinchbeck has been an invaluable source for many more recent treatments, including Sally Alexander, *Women's Work in Nineteenth-Century London* (London: Journeyman Press, 1983); Deborah Valenze, *The First Industrial Woman* (New York: Oxford University Press, 1995); and Katrina Honeyman, *Women, Gender and Industrialisation in England, 1700–1870* (London: Macmillan, 2000).

28. Pinchbeck, *Women Workers*, 148.

29. Pinchbeck, *Women Workers*, 157–82; for data on Manchester mills, see Roger Lloyd-Jones and M. J. Lewis, *Manchester and the Age of the Factory* (London: Croom Helm, 1988).

30. Lloyd-Jones and Lewis, *Manchester and the Age of the Factory*, 67.

31. Pinchbeck, *Women Workers*, 186 (data from 151 Lancashire cotton mills in 1834); Mary Freifeld, "Technological Change and the 'Self-Acting' Mule: A Study

of Skill and the Sexual Division of Labour," *Social History* 11 (October 1986): 319–43; G. N. von Tunzelmann, *Steam Power and British Industrialization to 1860* (Oxford: Clarendon Press, 1978), 185 (employees and horsepower); Lloyd-Jones and Lewis, *Manchester and the Age of the Factory*, *Gazette* quoted on 201. In 1833 the three largest mills in Manchester each employed 1,400 workers, eight mills had 500–900, an additional eight had 300–500, and seventeen mills had 100–300; see Chapman, *Lancashire Cotton Industry*, 58.

32. Martin Hewitt, *The Emergence of Stability in the Industrial City: Manchester, 1832–67* (Aldershot, England: Scolar Press, 1996), 31 (employment data); Timmons, *The Last Shift*, 20, table 1.1 (on power looms).

33. Samuel Smiles, *Industrial Biography* (1863), available at sailor.gutenberg. org/index/by-author/smo.html (January 2002); see the analysis in Freifeld, "Technological Change."

34. Nasmyth quoted in L. T. C. Rolt, *A Short History of Machine Tools* (Cambridge: MIT Press, 1965), 113.

35. A. E. Musson and Eric Robinson, *Science and Technology in the Industrial Revolution* (Toronto: University of Toronto Press, 1969), quote on 507 (from Nasmyth); Lloyd-Jones and Lewis, quote on 134 (on unrest over price of labour); Chapman, *Lancashire Cotton Industry*, quote on p. 198 ("social war").

36. Marcus, *Engels, Manchester, and the Working Class*, 30–44, quote on 46 ("chimney of the world"); Taylor, *Notes of a Tour in Lancashire* [1842], quote on 2 ("forest of chimneys"); John Kasson, *Civilizing the Machine* (New York: Grossman, 1976), 59–60.

37. Marcus, *Engels, Manchester, and the Working Class*, 30–44; Disraeli quoted in Robert J. Werlin, *The English Novel and the Industrial Revolution* (New York: Garland, 1990), 74, 84 n1.

38. Page numbers cited in the text below for quotations from Friedrich Engels, *The Condition of the Working Class in England* (1845) refer to the edition translated and edited by W. O. Henderson and W. H. Chaloner (Stanford: Stanford University Press, 1958).

39. W. O. Henderson and W. H. Chaloner, "Friedrich Engels in Manchester," *Memoirs and Proceedings of the Manchester Literary and Philosophical Society* 98 (1956–57): 13–29; Karl Marx and Friedrich Engels, "Manifesto of the Communist Party" (1848) in Robert C. Tucker, ed., *The Marx-Engels Reader*, 2nd edition (New York: W. W. Norton, 1978), quote on 476.

40. Nathaniel Hawthorne, *Our Old Home* (1863) in Sylvia Pybus, ed., '*Damned Bad Place, Sheffield*' (Sheffield: Sheffield Academic Press, 1994), quote on 126.

41. Pinchbeck, *Women Workers*, 275–76; Evans, *Forging of the Modern State*, 165.

42. Geoffrey Tweedale, *Steel City: Entrepreneurship, Strategy and Technology in Sheffield, 1743–1993* (Oxford: Clarendon Press, 1995), 49–50.

43. Tweedale, *Steel City*, 53.

44. Geoffrey Tweedale, in *Sheffield Steel and America* (Cambridge: Cambridge University PRess, 1987), expresses some skepticism about an apparent exception

to the network or cluster model, Greaves's Sheaf Works, built in 1823 for £50,000, where supposedly "one grand end was kept in view, namely that of centralizing on the spot all the various processes through which iron must pass . . . until fashioned into razor, penknife or other article of use" (51).

45. Tweedale, *Sheffield Steel*, 63–67, quote (from Bessemer, *Autobiography*) on 64.

46. Sidney Pollard, *A History of Labour in Sheffield* (Liverpool: Liverpool University Press, 1959), 50–54, quote on 51.

47. Pollard, *History of Labour in Sheffield*, 53–54, 83.

48. Pollard, *History of Labour in Sheffield*, 62–65, 328. Longer hours of work by Paris sewermen around 1900 resulted in "deleterious effects on their health" according to Donald Reid, *Paris Sewers and Sewermen* (Cambridge: Harvard University Press, 1991), 220 n24.

49. Pollard, *History of Labour in Sheffield*, 14–15.

50. Pollard, *History of Labour in Sheffield*, 5–23, 93–100.

51. See, for example, Jeremy, *Transatlantic Industrial Revolution*; Ken Alder, *Engineering the Revolution: Arms and Enlightenment in France, 1763–1815* (Princeton: Princeton University Press, 1997); Thomas J. Misa, *A Nation of Steel: The Making of Modern America, 1865–1925* (Baltimore: Johns Hopkins University Press, 1995).

52. Maths Isacson and Lars Magnusson, *Proto-industrialisation in Scandinavia: Craft Skills in the Industrial Revolution* (Leamington Spa, U.K.: Berg, 1987); Svante Lindqvist, *Technology on Trial: The Introduction of Steam Power Technology into Sweden, 1715–1736* (Stockholm: Almqvist & Wiksell, 1984), 23–33. A "miscellaneous" category brings the total models to 212.

CHAPTER 4. INSTRUMENTS OF EMPIRE, 1840–1914

1. Charles Bright, *Telegraphy, Aeronautics and War* (London: Constable, 1918), 56.

2. This is the argument of Daniel Headrick, *The Tools of Empire: Technology and European Imperialism in the Nineteenth Century* (Oxford: Oxford University Press, 1981).

3. Lance E. Davis and Robert A. Huttenback, *Mammon and the Pursuit of Empire: The Economics of British Imperialism* (Cambridge: Cambridge University Press, 1988).

4. On steam in India see Headrick, *Tools of Empire*, 17–57, 129–56.

5. Zaheer Baber, *The Science of Empire: Scientific Knowledge, Civilization, and Colonial Rule in India* (Albany: SUNY Press, 1996), 138–53.

6. Satpal Sangwan, "Technology and Imperialism in the Indian Context: The Case of Steamboats, 1819–1839," in Teresa Meade and Mark Walker, eds., *Science, Medicine, and Cultural Imperialism* (New York: St. Martin's Press, 1991), 60–74, quote on 70.

7. Captain William Hall to Thomas Peacock, May 1841, quoted in Headrick, *Tools of Empire*, 50.

8. Martin Booth, *Opium* (New York: St. Martin's Press, 1996), 103–73, quote on 146.

9. *Encyclopaedia Britannica*, 11th edition (1910), 14:389; Shrutidev Goswami, "The Opium Evil in Nineteenth Century Assam," *Indian Economic and Social History Review* 19 (1982): 365–76.

10. Bright, *Telegraphy, Aeronautics and War*, 56.

11. This section follows Mel Gorman, "Sir William O'Shaughnessy, Lord Dalhousie, and the Establishment of the Telegraph System in India," *Technology and Culture* 12 (1971): 581–601; Saroj Ghose, "Commercial Needs and Military Necessities: The Telegraph in India," in Roy MacLeod and Deepak Kumar, eds., *Technology and the Raj: Western Technology and Technical Transfers to India, 1700–1947* (New Delhi: Sage, 1995), 153–76, quote on 156.

12. Edwin Arnold, *The Marquis of Dalhousie's Administration of British India*, (London, 1862) 2:241–42, as quoted in Michael Adas, *Machines as the Measure of Men* (Ithaca: Cornell University Press, 1989), 226.

13. Quotes in Vary T. Coates and Bernard Finn, *A Retrospective Technology Assessment: Submarine Telegraphy* (San Francisco: San Francisco Press, 1979), 101.

14. Ghose, "Commercial Needs and Military Necessities," quote on 168. For different reasons, battlefield commanders might resent the control that telegraphs permitted. The Prussian field marshall and chief-of-staff H. K. Moltke wrote of the Austro-Prussian War (1866), "No commander is less fortunate than he who operates with a telegraph wire stuck into his back" (as quoted in Martin van Creveld, *Command in War* [Cambridge: Harvard University Press, 1985], 146). I am indebted to Ed Todd for this citation.

15. Daniel Headrick, *The Invisible Weapon: Telecommunications and International Politics, 1851–1945* (Oxford: Oxford University Press, 1991), 17–115; Headrick, *Tools of Empire*, 130.

16. Peter Harnetty, *Imperialism and Free Trade: Lancashire and India in the Mid-Nineteenth Century* (Vancouver: University of British Columbia Press, 1972), quote on 6.

17. K. N. Chaudhuri, "The Structure of the Indian Textile Industry in the Seventeenth and Eighteenth Centuries," in Michael Adas, ed., *Technology and European Overseas Enterprise* (Aldershot, England: Variorum, 1996), 343–98; Harnetty, *Imperialism and Free Trade*, 7–35.

18. Harnetty, *Imperialism and Free Trade*, quotes 66–67.

19. Arun Kumar, "Colonial Requirements and Engineering Education: The Public Works Department, 1847–1947," in MacLeod and Kumar, *Technology and the Raj*, 216–32; Daniel Headrick, *The Tentacles of Progress: Technology Transfer in the Age of Imperialism, 1850–1940* (Oxford: Oxford University Press, 1988), 315–45.

20. Daniel Thorner, *Investment in Empire: British Railway and Steam Shipping Enterprise in India, 1825–1849* (Philadelphia: University of Pennsylvania Press, 1950; reprint, New York: Arno, 1977), quotes on 11.

21. Thorner, *Investment in Empire*, quotes on 49 ("first consideration"), 96

("extension"). Military and administrative priorities also informed the routing of later lines; for cases studies see Bharati Ray, "The Genesis of Railway Development in Hyderabad State: A Case Study in Nineteenth Century British Imperialism," *Indian Economic and Social History Review* 21 (1984): 45–69, on 54–55; Mukul Mukherjee, "Railways and Their Impact on Bengal's Economy, 1870–1920," *Indian Economic and Social History Review* 17 (1980): 191–209, on 193–94.

22. Headrick, *Tentacles of Progress*, quote on 71.

23. Another reason "the 'import component' in India's permanent way [railroad] remained unusually high" was that "a very small proportion of the subcontinent's iron requirements was locally manufactured." See Ian Derbyshire, "The Building of India's Railways: The Application of Western Technology in the Colonial Periphery, 1850–1920," in MacLeod and Kumar, *Technology and the Raj*, 177–215, quote on 189.

24. Daniel Thorner, "Great Britain and the Development of India's Railways," *Journal of Economic History* 11 (1951): 389–402, quote on 392.

25. Creed Haymond, *The Central Pacific Railroad Co.* (San Francisco: H. S. Crocker, 1888), quote on 43 (Indian problem). A pronouncement made by General William T. Sherman in 1867 left little room for compromise: "The more we can kill this year, the less will have to be killed in the next war, for the more I see of these Indians the more convinced I am that they all have to be killed or maintained as a species of paupers." Quoted in John Hoyt Williams, *A Great and Shining Road* (New York: Times Books, 1988), 152.

26. Donald W. Roman, "Railway Imperialism in Canada, 1847–1868," and Ronald E. Robinson, "Railways and Informal Empire," in Clarence B. Davis and Kenneth E. Wilburn, Jr., eds., *Railway Imperialism* (Boulder: Greenwood, 1991), 7–24, 175–96; A. A. Den Otter, *The Philosophy of Railways: The Transcontinental Railway Idea in British North America* (Toronto: University of Toronto Press, 1997), quote on 204.

27. William E. French, "In the Path of Progress: Railroads and Moral Reform in Porfirian Mexico," in Davis and Wilburn, *Railway Imperialism*, 85–102; Robinson, "Railways and Informal Empire," quote on 186 ("tariff regulations"); David M. Pletcher, *Rails, Mines, and Progress: Seven American Promoters in Mexico, 1867–1911* (Port Washington: Kennikat Press, 1972), quote on 1 ("our India").

28. Pletcher, *Rails, Mines, and Progress*, 313.

29. M. Tamarkin, *Cecil Rhodes and the Cape Afrikaners* (London: Frank Cass, 1996), 8.

30. A similar tactic—contesting the imperialist's railway plans by building alternative lines with financing directly from private London money—was used with some success by India's princely state of Hyderabad; see Ray, "Genesis of Railway Development," 56–60.

31. Kenneth E. Wilburn, Jr., "Engines of Empire and Independence: Railways in South Africa, 1863–1916," in Davis and Wilburn, *Railway Imperialism*, 25–40.

32. W. Travis Hanes III, "Railway Politics and Imperialism in Central Africa,

1889–1953," in Davis and Wilburn, *Railway Imperialism,* 41–69; John Edward Glab, "Transportation's Role in Development of Southern Africa" (Ph.D. diss., American University, 1970), 62–69.

33. As one contemporary critic of imperialism, Hilaire Belloc, parodied the situation: "I shall never forget the way / That Blood stood upon this awful day / Preserved us all from death. / He stood upon a little mound, / Cast his lethargic eyes around, / And said beneath his breath: / Whatever happens, we have got / The Maxim Gun, and they have not." Quoted in Howard Bailes, "Technology and Imperialism: A Case Study of the Victorian Army in Africa," *Victorian Studies* 24 (Autumn 1980): 83–104, quote on 88.

34. E. J. Hobsbawm, *Industry and Empire* (London: Weidenfeld & Nicolson, 1968; reprint, London: Penguin, 1990), 178–93; Patrick K. O'Brien, "The Costs and Benefits of British Imperialism, 1846–1914," *Past and Present* 120 (1988): 163–200.

CHAPTER 5. SCIENCE AND SYSTEMS, 1870–1930

1. See the discussion of the word *technology* in my essay "The Compelling Tangle of Modernity and Technology" in Thomas Misa et al., eds., *Modernity and Technology* (Cambridge: MIT Press, 2003), 1–30.

2. Peter Mathias, *The Brewing Industry in England, 1700–1830* (Cambridge: Cambridge University Press, 1959), quote on 65.

3. Henk van den Belt, "Why Monopoly Failed: The Rise and Fall of Société La Fuchsine," *British Journal for the History of Science* 25 (1992): 45–63.

4. Jeffrey Allan Johnson, *The Kaiser's Chemists: Science and Modernization in Imperial Germany* (Chapel Hill: University of North Carolina Press, 1990), 216 n7.

5. Henk van den Belt and Arie Rip, "The Nelson-Winter-Dosi Model and Synthetic Dye Chemistry," in Wiebe Bijker, Trevor Pinch, and Thomas Hughes, eds., *The Social Construction of Technological Systems* (Cambridge: MIT Press, 1987), 135–58, quotes on 143–44.

6. Van den Belt and Rip, "Nelson-Winter-Dosi Model," quotes on 151.

7. Johnson, *Kaiser's Chemists,* 33.

8. Carl Duisberg, *Meine Lebenserinnerungen* (Leipzig, 1933), 44; also in van den Belt and Rip, "Nelson-Winter-Dosi Model," quote on 154.

9. Johnson, *Kaiser's Chemists,* quote on 34.

10. Van den Belt and Rip, "Nelson-Winter-Dosi Model," quote on 155.

11. John Joseph Beer, *The Emergence of the German Dye Industry* (Urbana: University of Illinois Press, 1959), 134.

12. Wilfred Owen, "Dulce et Decorum Est" (1917). Various manuscript versions are on-line at www.hcu.ox.ad.uk/jtap/warpoems.htm#12 (16 May 2003).

13. L. F. Haber, *The Poisonous Cloud: Chemical Warfare in the First World War* (Oxford: Clarendon Press, 1986), 22–40, 217, 228, 243.

14. Peter Hayes, *Industry and Ideology: IG Farben in the Nazi Era* (Cambridge: Cambridge University Press, 1987), 17 (1930 stats), 359, 361, 370.

15. Paul Israel, *From Machine Shop to Industrial Laboratory: Telegraphy and*

the Changing Context of American Invention, 1830–1920 (Baltimore: Johns Hopkins University Press, 1992), quote on 138.

16. Thomas P. Hughes, *American Genesis* (New York: Viking, 1989), 29.

17. Thomas P. Hughes, *Networks of Power: Electrification in Western Society, 1880–1930* (Baltimore: Johns Hopkins University Press, 1983), 24.

18. Edison to Puskas, 13 November 1878, Edison Archives, as quoted in Hughes, *Networks of Power,* 33.

19. Robert Friedel and Paul Israel, *Edison's Electric Light: Biography of an Invention* (New Brunswick: Rutgers University Press, 1986), 137.

20. Edison patent 223,898 (granted 27 January 1880), from Friedel and Israel, *Edison's Electric Light,* quote on 106.

21. Hughes, *Networks of Power,* 33–38, 193.

22. Friedel and Israel, *Edison's Electric Light,* quote on 182.

23. Hughes, *Networks of Power,* quote on 45.

24. Louis Galambos and Joseph Pratt, *The Rise of the Corporate Commonwealth* (New York: Basic Books, 1988), 10–11.

25. Hughes, *Networks of Power,* 46.

26. W. Bernard Carlson, *Innovation as a Social Process: Elihu Thomson and the Rise of General Electric, 1870–1900* (Cambridge: Cambridge University Press, 1991), quote on 284 n20. The next five paragraphs draw on Carlson's book.

27. Carlson, *Innovation as a Social Process,* 253–57, 290.

28. Hughes, *Networks of Power,* 107–9; Carlson, *Innovation as a Social Process,* 261–63, 285.

29. Carlson, *Innovation as a Social Process,* 290–301, quotes on 292 (Edison) and 297–98 (Fairfield).

30. A remarkably high number of the largest U.S. companies were founded during the 1880–1930 period; see the data in Harris Corporation, "Founding Dates of the 1994 *Fortune* 500 U.S. Companies," *Business History Review* 70 (1996): 69–90.

31. Cited in George Basalla, *The Evolution of Technology* (Cambridge: Cambridge University Press, 1988), 128. A comprehensive study is David A. Hounshell and John Kenly Smith, Jr., *Science and Corporate Strategy: Du Pont R&D, 1902–1980* (Cambridge: Cambridge University Press, 1980).

32. George Wise, "A New Role for Professional Scientists in Industry: Industrial Research at General Electric, 1900–1916," *Technology and Culture* 21 (1980): 408–29, quote on 422.

33. Karl L. Wildes and Nilo A. Lindgren, *A Century of Electrical Engineering and Computer Science at MIT, 1882–1982* (Cambridge: MIT Press, 1985), 32–66, quote on 43. See also W. Bernard Carlson, "Academic Entrepreneurship and Engineering Education: Dugald C. Jackson and the MIT-GE Cooperative Engineering Course, 1907–1932," *Technology and Culture* 29 (1988): 536–67.

34. Vannevar Bush, *Pieces of the Action* (New York: William Morrow, 1970), quote on 254; Wildes and Lindgren, *A Century of Electrical Engineering at MIT,* 48, 70.

35. Wildes and Lindgren, *A Century of Electrical Engineering at MIT,* quote on

57. In 1933 the electrical engineering curriculum was reorganized to focus on electronics, electromagnetic theory, and energy conversion as basic common concepts (see 106).

36. Wildes and Lindgren, *A Century of Electrical Engineering at MIT,* 49–54, 62–66, quote on 63.

37. Wildes and Lindgren, *A Century of Electrical Engineering at MIT,* 75–77, 96–105, quote on 100 (on GE's Doherty), 103 (on terminating the network analyzer).

CHAPTER 6. MATERIALS OF MODERNISM, 1900–1950

1. Umberto Boccioni, Carlo Carrà, Luigi Russolo, Giacomo Balla, and Gino Severini, "Manifesto of the Futurist Painters" (1910) www.futurism.org.uk/manifestos/manifesto02.htm (3 July 2002). The "great divide" between the ancient and modern worlds is analyzed in Thomas Misa et al., *Modernity and Technology* (Cambridge: MIT Press, 2003); Bruno Latour, *We Have Never Been Modern* (Cambridge: Harvard University Press, 1993), chap. 2; and Jeffrey L. Meikle, "Domesticating Modernity: Ambivalence and Appropriation, 1920–40," in Wendy Kaplan, ed., *Designing Modernity: The Arts of Reform and Persuasion, 1885–1945* (New York: Thames & Hudson, 1995), 143–67.

2. Bruno Taut, *Modern Architecture* (London: The Studio, 1929), quote on 4; Reyner Banham, *Theory and Design in the First Machine Age* (Cambridge: MIT Press, 1980), 94 (Loos citations).

3. For a parallel story about concrete, see Amy Slaton, *Reinforced Concrete and the Modernization of American Building, 1900–1930* (Baltimore: Johns Hopkins University Press, 2001).

4. On the change from hand- to machine-blown glassmaking, see Pearce Davis, *The Development of the American Glass Industry* (Cambridge: Harvard University Press, 1949; reprint, New York: Russell & Russell, 1970; Harvard Economic Studies, vol. 86); Ken Fones-Wolf, "From Craft to Industrial Unionism in the Window-Glass Industry: Clarksburg, West Virginia, 1900–1937," *Labor History* 37 (Winter 1995–96): 28–49; and Richard John O'Connor, "Cinderheads and Iron Lungs: Window-glass Craftsmen and the Transformation of Workers' Control, 1880–1905," (Ph.D. diss., University of Pittsburgh, 1991).

5. Taut, *Modern Architecture* (1929), 6.

6. R. W. Flint, ed., "The Birth of a Futurist Aesthetic" in *Marinetti: Selected Writings* (New York: Farrar, Straus & Giroux, 1973), 81.

7. Filippo Marinetti, "Founding Manifesto of Futurism" (1909), in Pontus Hulten, ed., *Futurismo e Futurismi* (Milan: Bompiani, 1986), quotes 514–16; Boccioni quoted in Banham, *Theory and Design,* 102. For a photograph of Marinetti's car in the ditch, see F. T. Marinetti, *The Futurist Cookbook* (San Francisco: Bedford Arts, 1989), 10.

8. Boccioni et al., "Manifesto of the Futurist Painters."

9. Umberto Boccioni, "Technical Manifesto of Futurist Sculpture" (1912), in Hulten, *Futurismo e Futurismi,* 132–33, quotes on 433.

10. Antonio Sant'Elia, "Manifesto of Futurist Architecture" (1914), in Hulten, *Futurismo e Futurismi*, 418–20.

11. Sant'Elia, "Manifesto of Futurist Architecture," 418–20; Banham, *Theory and Design*, 127–37, quotes on 129. Here I have quoted from Sant'Elia's original text (from Banham); Marinetti later changed *new* and *modern* to *Futurist*, upgraded *iron* to *steel*, and added several paragraphs at the beginning and, most famously, at the end. Città Nuova is featured in Vittorio Magnago Lampugnani, ed., *Antonio Sant'Elia: Gezeichnete Architektur* (Munich: Prestel, 1992), 7–55, 166–97.

12. R. J. B. Bosworth, *Mussolini* (New York: Oxford University Press, 2002), 123–69.

13. Marinetti, *Futurist Cookbook*, 119 (car crash menu). Such dishes as "Italian Breasts in the Sunshine" and "Nocturnal Love Feast" express a hedonistic and flagrantly antitraditional attitude to bodies and sex, which was (grotesquely) theorized in Valentine de Saint-Point's "Futurist Manifesto of Lust" (1913), in Hulten, *Futurismo e Futurismi*, 503–5.

14. Banham, *Theory and Design*, quote on 155.

15. Carsten-Peter Warncke, *The Ideal as Art: De Stijl 1917–31* (Cologne: Taschen, 1998), Mondrian quotes on 66; Banham, *Theory and Design*, 148–53, quotes 150, 151; James Scott, *Seeing Like a State* (New Haven: Yale University Press, 1998), quote on 392 n55 ("After electricity").

16. Theo van Doesburg, "The Will to Style: The New Form Expression of Life, Art and Technology," *De Stijl*, vol. 5, no. 2 (February 1922): 23–32; no. 3 (March 1922): 33–41; reprinted in Joost Baljeu, *Theo van Doesburg* (London: Studio Vista, 1974), 115–26, quotes on 119 and 122.

17. Banham, *Theory and Design*, 139–47, quotes on 141 ("decoration and ornament"), 142 ("modern villa"), and 147 ("normal tool" and "old structural forms").

18. Banham, *Theory and Design*, quotes on 158 ("I bow the knee") and 160–62.

19. John Willett, *The New Sobriety, 1917–1933: Art and Politics in the Weimar Period* (London: Thames & Hudson, 1978), quote on 120.

20. Banham, *Theory and Design*, quote on 283.

21. Warncke, *De Stijl*, 120–45; Theo van Doesburg, "Towards Plastic Architecture," *De Stijl*, ser. 12, nos. 6–7 (1924): 78–83; reprinted in Baljeu, *Theo van Doesburg*, 142–47, quote on 142.

22. Gropius quoted in Banham, *Theory and Design*, 281.

23. Gropius quoted in Banham, *Theory and Design*, 282.

24. Willett, *New Sobriety*, quote on 123.

25. Thomas P. Hughes, *American Genesis* (New York: Viking, 1989), 312–19; Willett, *New Sobriety*, 120–22.

26. The Berlin building campaign is illustrated in Taut, *Modern Architecture*, 100–132.

27. Willett, *New Sobriety*, 124–26.

28. Willett, *New Sobriety*, 125–27.

29. Willett, *New Sobriety*, 127–29, quote on 128.

30. On the Museum of Modern Art's campaign to create and shape the International Style and its expression at the 1939 New York World's Fair, see Terry Smith, *Making the Modern: Industry, Art, and Design in America* (Chicago: University of Chicago Press, 1993), 385–421.

31. Hitler's preference for "authentic" German architecture was not a repudiation of technological modernism. As Michael Allen puts it, "The Four Year Plan, the Hermann Göring–Werke, Volkswagen, SS concentration-camp industry, Albert Speer's war production ministry, and the rantings of Adolf Hitler displayed a consistent fascination with techniques of modern organization and standardized mass production—in short, for the capital-intensive, concentrated industry that was new to the twentieth century." See Michael Thad Allen, "Modernity, the Holocaust, and Machines without History," in Michael Thad Allen and Gabrielle Hecht, eds., *Technologies of Power* (Cambridge: MIT Press, 2001), 175–214, quote on 186.

32. For biographical details, see Lihotzky's summary of her own career and Lore Kramer's appreciation, "The Frankfurt Kitchen: Contemporary Criticism and Perspectives," in Angela Oedekoven-Gerischer et al., eds., *Frauen im Design: Berufsbilder und Lebenswege seit 1900*, vol. 1 (Stuttgart: Design Center Stuttgart, 1989), 148–73.

33. Mary Nolan, *Visions of Modernity* (New York: Oxford University Press, 1994), quote on 208 ("the household").

34. For this and the following paragraphs I owe much to Susan R. Henderson, "A Revolution in the Women's Sphere: Grete Lihotzky and the Frankfurt Kitchen," in Debra Coleman, Elizabeth Danze, and Carol Henderson, eds., *Architecture and Feminism* (New York: Princeton Architectural Press, 1996), 221–53. On the German government's Home Economics section of the National Productivity Board see Nolan, *Visions of Modernity*, 206–26.

35. Oedekoven-Gerischer et al., *Frauen im Design*, 159 ff., quote on 165; Peter Noever, ed., *Die Frankfurter Küche* (Berlin: Ernst & Sohn, n.d.), 44–46. A 1929 governmental report on housewives' views of the Frankfurt kitchen, while praising a number of the kitchen's design elements, noted a "certain degree of overorganization," especially in the rigidly designed storage units, and also that the kitchen was "too long and narrow" for more than one person to work in effectively (*Frauen im Design*, 167).

CHAPTER 7. THE MEANS OF DESTRUCTION, 1936–1990

1. Alan Milward, *War, Economy, and Society, 1939–1945* (Berkeley: University of California Press, 1977), 180. Stuart W. Leslie, *The Cold War and American Science* (New York: Columbia University Press, 1993), 8. See Henry Etzkowitz, "Solar versus Nuclear Energy: Autonomous or Dependent Technology?" *Social Problems* 31 (April 1984): 417–34. On rival digital and analog computers in Project SAGE, see Thomas P. Hughes, *Rescuing Prometheus* (New York: Pantheon, 1998), 40–47.

On the military's role in shaping computer-controlled machine tools, see David F. Noble, *Forces of Production* (New York: Knopf, 1984).

2. Paul Josephson, *Red Atom* (New York: Freeman, 2000).

3. Robert Pool, *Beyond Engineering* (Oxford: Oxford University Press, 1997), 72–81; Daniel Greenberg, *The Politics of Pure Science* (New York: NAL, 1971), 170–206; William J. Broad, *Teller's War: The Top-Secret Story behind the Star Wars Deception* (New York: Simon & Schuster, 1992). James Bamford, *The Puzzle Palace* (New York: Penguin, 1983), 109, estimated that the National Security Agency's consolidated cryptography budget was as large as $10 billion. Estimates of the NSA's current budget range between $4 and $10 billion (www.fas.org/irp/nsa/nsabudget.html [1 February 2002]).

4. William Carr, *Arms, Autarky and Aggression: A Study in German Foreign Policy, 1933–1939* (New York: W. W. Norton, 1972); Milward, *War, Economy, and Society,* 23–30, 184–206.

5. Martin van Creveld, *Technology and War* (New York: Free Press, 1989), 217–32.

6. Mark Walker, *German National Socialism and the Quest for Nuclear Power, 1939–1949* (Cambridge: Cambridge University Press, 1989), 129–52. Documents recently published at the Niels Bohr Archive (www.nba.nbi.dk [7 February 2002]) make clear that Heisenberg did *not* sabotage the German atomic bomb effort from inside, as portrayed in journalist Thomas Powers' *Heisenberg's War* (1993) or playwright Michael Frayn's celebrated *Copenhagen.*

7. Walker, *German National Socialism,* 153–78.

8. Leslie R. Groves, *Now It Can Be Told* (New York: Harper & Row, 1962), 20.

9. Edward Teller, *The Legacy of Hiroshima* (Garden City: Doubleday, 1962), quote on 211.

10. On Hanford see Richard G. Hewlett and Oscar E. Anderson, Jr., *The New World, 1939–1946,* vol. 1 of *A History of the United States Atomic Energy Commission* (University Park: Pennsylvania State University Press, 1962), 180–226; and David A. Hounshell and John Kenly Smith, Jr., *Science and Corporate Strategy: DuPont R&D, 1902–1980* (Cambridge: Cambridge University Press, 1988), 338–46.

11. Gar Alperovitz, *Atomic Diplomacy: Hiroshima and Potsdam* (New York: Simon & Schuster, 1965), quote on 14.

12. Winston S. Churchill, *Triumph and Tragedy* (Boston: Houghton Mifflin, 1953), quote on 639; Michael Sherry, *The Rise of American Air Power* (New Haven: Yale University Press, 1987), 330.

13. Sherry, *Rise of American Air Power,* quote on 341 (Groves).

14. "The bomb was also to be used to pay for itself, to justify to Congress the investment of $2 billion, to keep Groves and Stimson out of Leavenworth prison," writes Richard Rhodes, *The Making of the Atomic Bomb* (New York: Simon & Schuster, 1986), 697. Spencer R. Weart and Gertrud Weiss Szilard, eds., *Leo Szilard* (Cambridge: MIT Press, 1978), quote on 184 (Szilard); Groves, *Now It Can*

Be Told, quote on 70; Hewlett and Anderson, *New World*, 339–40; www.gmu.edu/academic/pcs/polking.html (14 May 2003).

15. Bernard J. O'Keefe, *Nuclear Hostages* (Boston: Houghton Mifflin, 1983), quote on 97.

16. Richard G. Hewlett and Jack M. Holl, *Atoms for Peace and War, 1953–1961* (Berkeley: University of California Press, 1989), 196–98, 419–23, 493–95; Thomas P. Hughes, *American Genesis* (New York: Viking, 1989), 428–42.

17. The figures for Safeguard and the U.S. nuclear program are expressed in constant 1996 dollars, correcting for inflation. By this measure, the nuclear plane project cost $7 billion. See the Brookings Institution's "Atomic Audit" at www.brook.edu/fp/projects/nucwcost/weapons.htm (27 July 1999).

18. Charles Weiner, "How the Transistor Emerged," *IEEE Spectrum* 10 (January 1973): 24–35, quote on 28; Daniel Kevles, *The Physicists* (New York: Knopf, 1977), 302–20. Kevles writes (308), "The production costs for the Rad Lab radar systems alone amounted to some $1.5 billion; the combined bill for radar, proximity fuzes, and rockets far exceeded the $2 billion spent on the atom bomb." Proximity fuzes were small, on-board radar sets that triggered artillery shells to explode at a set distance from their targets.

19. Lillian Hoddeson, "The Discovery of the Point-Contact Transistor," *Historical Studies in the Physical Sciences* 12, no. 1 (1981): 41–76, quote on 53.

20. Thomas J. Misa, "Military Needs, Commercial Realities, and the Development of the Transistor, 1948–1958," in Merritt Roe Smith, ed., *Military Enterprise and Technological Change* (Cambridge: MIT Press, 1985), 253–87, quote on 268.

21. Misa, "Military Needs," quote on 282.

22. M. D. Fagen, ed., *A History of Engineering and Science in the Bell System: National Service in War and Peace, 1925–1975* (Murray Hill, N.J.: Bell Telephone Laboratories, 1978), quote on 394.

23. Kenneth Flamm, *Creating the Computer* (Washington, D.C.: Brookings Institution, 1988), 259–69.

24. Nancy Stern, *From ENIAC to UNIVAC* (Bedford, Mass.: Digital Press, 1981); William Aspray, *John von Neumann and the Origins of Modern Computing* (Cambridge: MIT Press, 1990); I. Bernard Cohen, "The Computer: A Case Study of Support by Government," in E. Mendelsohn, M. R. Smith, and P. Weingart, eds., *Science, Technology and the Military*, vol. 12 of *Sociology of the Sciences* (Dordrecht: Kluwer, 1988), 119–54.

25. Kent C. Raymond and Thomas M. Smith, *Project Whirlwind: The History of a Pioneer Computer* (Bedford, Mass.: Digital Press, 1980), Forrester quote on 42.

26. Hughes, *Rescuing Prometheus*, 15–67.

27. Charles J. Bashe et al., *IBM's Early Computers* (Cambridge: MIT Press, 1986); Emerson W. Pugh, *Memories that Shaped an Industry: Decisions Leading to IBM System/360* (Cambridge: MIT Press, 1984).

28. Arthur L. Norberg and Judy E. O'Neill, *Transforming Computer Tech-*

nology: Information Processing for the Pentagon, 1962–1986 (Baltimore: Johns Hopkins University Press, 1996), 160–61.

29. For the influence of closed-world thinking on national security strategies, computer development, and the metaphors and models of several scientific fields, including cognitive psychology and artificial intelligence, see Paul N. Edwards, *The Closed World: Computers and the Politics of Discourse in Cold War America* (Cambridge: MIT Press, 1996). One can see the "end" of the military era in the U.S. national accounting for research: military R&D spending peaked in 1987, fell around a third during the early 1990s, and remained at a lower plateau until 2000, while nonfederal R&D spending (mostly industrial) rose during these years. For data see the National Science Foundation's "National Patterns of R&D Resources," www.nsf.gov/sbe/srs/nsf99335/htmstart.htm (15 June 1999). As discussed in the introduction, since 11 September 2001, U.S. military R&D budgets have increased sharply, to all-time highs.

CHAPTER 8. TOWARD GLOBAL CULTURE, 1970–2001

1. John Micklethwait and Adrian Wooldridge, *A Future Perfect: The Essentials of Globalization* (New York: Crown Business, 2000), viii; Roland Robertson, "Mapping the Global Condition," *Theory, Culture, and Society* 7 (1990): 15–30.

2. Kenichi Ohmae, *The Evolving Global Economy* (Boston: Harvard Business School, 1995), quote on xiv. Ohmae was the cofounder of the Tokyo offices of the consultants McKinsey and Company.

3. Richard Barnet and John Cavanagh, "Homogenization of Global Culture," in Jerry Mander and Edward Goldsmith, eds., *The Case against the Global Economy* (San Francisco: Sierra Club Books, 1996), 71–77, quote on 71.

4. See the World Bank's briefing paper "Assessing Globalization" at www.worldbank.org/economicpolicy/globalization/key_readings.html#ag (15 May 2003). In recent years the bank seems to be making a greater effort at transparency as well as modestly altering its policies to embrace wider measures of development. This change, however modest, may not survive the increasing national security emphasis of the Bush administration in the wake of the attacks on 11 September 2001.

5. See "Think Local: Cultural Imperialism Doesn't Sell," *Economist* (13 April 2002): 12 ("Survey of Television" section); "Biker, Baker, Television Maker: The European Way of the Web," *New York Times* (18 April 2001): D1, D14–15; Don Slater, "Modernity under Construction: Building the Internet in Trinidad," in Thomas Misa et al., *Modernity and Technology* (Cambridge: MIT Press, 2003), 139–60. More generally, on technology and appropriation see Mikael Hård and Andrew Jamison, eds., *The Intellectual Appropriation of Technology: Discourses on Modernity, 1900–1939* (Cambridge: MIT Press, 1998).

6. Sidney Mintz, *Sweetness and Power: The Place of Sugar in Modern History* (New York: Viking Penguin, 1985).

7. Daniel Headrick, *The Tentacles of Progress: Technology Transfer in the Age of Imperialism, 1850–1940* (New York: Oxford University Press, 1988), 29.

8. Simon Partner, *Assembled in Japan: Electrical Goods and the Making of the Japanese Consumer* (Berkeley: University of California Press, 1999).

9. Information on Japanese firms' development of digital facsimile is from Ken Kusunoki, "Incapability of Technological Capability: A Case Study on Product Innovation in the Japanese Facsimile Machine Industry," *Journal of Product Innovation Management* 14 (1997): 368–82.

10. This section follows the evaluation report: J. Agnes, *The Fax! Programme; Three Years of Experimentation (June 1989–October 1992): A Teaching Aid for Opening up to Europe* (Strasbourg: Council of Europe, 1994). Page numbers for quotes are given in the text.

11. Roger Cohen, "Fearful over the Future, Europe Seizes on Food," *New York Times* (29 August 1999), www.nytimes.com/library/review/082999europe-food-review.html (18 September 2000); Suzanne Daley, "French See a Hero in War on 'McDomination,'" *New York Times* (12 October 1999), www.nytimes.com/library/world/europe/101299france-protest.html (18 September 2000).

12. Naomi Klein, *No Logo* (New York: Picador, 1999), quote on 388. A comprehensive account of the lawsuit is John Vidal, *McLibel: Burger Culture on Trial* (New York: New Press, 1997), quotes on 115 ("nutritious") and 198 ("gravel pits"). The "gravel pit" quote is on-line at www.mcspotlight.org/case/trial/transcripts/941205/29.htm (19 July 2001) and Coca-Cola as "nutritious" at www.mcspotlight.org/case/trial/transcripts/941208/54.htm (19 July 2001).

13. Thomas L. Friedman, *The Lexus and the Olive Tree* (New York: Farrar, Straus and Giroux, 1999), quotes on 195, 196, and 309. Friedman blasted anti-globalization critics in his unusually shrill column, "Senseless in Seattle," *New York Times* (1 December 1999), www.nytimes.com/library/opinion/friedman/120199frie.html (2 October 2000).

14. George Cohon, *To Russia with Fries* (Toronto: McClelland & Stewart, 1999), quote on 132.

15. Cohon, *To Russia with Fries,* quotes on 133 and 179–80.

16. Cohon, *To Russia with Fries,* quotes on 195 ("Big Mac is *perestroika*") and 281 ("Russian company"). Stiff criticism of McDonald's in Russia does exist; see the *Green Left Weekly*'s interview with Vadim Damier, a member of the Initiative of Revolutionary Anarchists and the League of Green Parties, "Moscow Left Demonstrates against McDonald's," jinx.sistm.unsw.edu.au/greenlft/1991/33/33p10.html (3 June 2001).

17. Cohon, *To Russia with Fries,* quotes on 176. The vertical integration of McDonald's U.K. was the result of the established food industry's unwillingness to accommodate McDonald's special standards, rather than an inability to do so (as was the case in Russia); see John F. Love, *McDonald's: Behind the Arches* (New York: Bantam, 1995; original edition 1986), 440–41.

18. James L. Watson, ed., *Golden Arches East: McDonald's in East Asia* (Stanford: Stanford University Press, 1997), viii.

19. Watson, *Golden Arches East,* 56, 93, quote on 106. Such leisurely socializa-

tion is also visible each afternoon at my nearby McDonald's in northwest Chicago, where large gatherings of older men who have recently arrived from southeastern Europe chat pleasantly. They simply ignored a sign (in English) stating, "Guests may take no longer than 30 minutes to finish their meals"; but that sign is gone.

20. Watson, *Golden Arches East,* 141, quotes on 12 and 181.

21. Love, *McDonald's: Behind the Arches,* 423, 424, 426.

22. Love, *McDonald's: Behind the Arches,* quote on 432.

23. Vidal, *McLibel,* quote 178.

24. Katie Hafner and Matthew Lyon, *Where Wizards Stay Up Late: The Origins of the Internet* (New York: Simon & Schuster, 1996), 10.

25. Janet Abbate, *Inventing the Internet* (Cambridge: MIT Press, 1999), quotes on 77; Paul Edwards, "Infrastructure and Modernity: Force, Time, and Social Organization in the History of Sociotechnical Systems," in Misa et al., *Modernity and Technology* (on nuclear war). Edwards in *The Closed World: Computers and the Politics of Discourse in Cold War America* (Cambridge: MIT Press, 1996) emphasizes the military origins of digital computing.

26. Hafner and Lyon, *Where Wizards Stay Up Late,* 56; Paul Baran, "Reliable Digital Communications Systems Using Unreliable Network Repeater Nodes" (RAND Corporation report P-1995), in Abbate, *Inventing the Internet,* 7–21, quote on 222 n5. Baran's conceptual work was published in a twelve-volume study, *On Distributed Communications* (Rand Report Series, 1964).

27. Abbate, *Inventing the Internet,* 43–81, quotes on 76 (congressional testimony).

28. Urs von Burg, *The Triumph of the Ethernet: Technological Communities and the Battle for the LAN Standard* (Stanford: Stanford University Press, 2001).

29. Abbate, *Inventing the Internet,* 113–32, quotes on 131–32.

30. Abbate, *Inventing the Internet,* 133–45, 147–79, quote on 143. The rivals to TCP/IP were X.25 (promoted by Europe's telecommunication industry) and the Open Systems Interconnection (supported by the ISO standards-setting body).

31. Abbate, *Inventing the Internet,* 83–111, quote 108 ("not an important motivation").

32. This section follows Abbate, *Inventing the Internet,* 181–218.

33. For this section on the World Wide Web I have drawn on Tim Berners-Lee, "Web Architecture from 50,000 Feet," quote on 1 (universality of access), www.w3.org/DesignIssues/Architecture.html (23 July 2001); Tim Berners-Lee, "Realising the Full Potential of the Web," www.w3.org/1998/02/Potential.html (23 July 2001); Herb Brody, "The Web Maestro: An Interview with Tim Berners-Lee," *Technology Review* (July 1996): 34–40.

34. Berners-Lee, "Realising the Full Potential of the Web," quote on 5. For the wider goal of machine intelligence, see Tim Berners-Lee, *Weaving the Web* (New York: HarperBusiness, 1999), 157–209; and Tim Berners-Lee, James Hendler, and Ora Lassila, "The Semantic Web," *Scientific American* (May 2001), www.sciam.com/2001/0501issue/0501berners-lee.html (6 August 2001).

35. The American Association for the Advancement of Science is *forecasting* a record high military R&D budget for 2004, surpassing even the Cold War peak (www.aaas.org/spp/rd/04pch6.htm [14 May 2003]).

CHAPTER 9. THE QUESTION OF TECHNOLOGY

1. Among commentators who are unduly pessimistic about technology I would include Jacques Ellul, Bill Joy, and some antiglobalization activists who strain to see cultural destruction wrought by technology wherever there is American activity or influence. Among the rival faction of those who are unwarrantedly optimistic about technology I would number George Gilder, most cyber-utopians, and many popular writers who publish books with titles of the form "how the invention of 'X' changed the world."

2. See, for example, Ruth Oldenziel, *Making Technology Masculine: Men, Women and Modern Machines in America, 1870–1945* (Amsterdam: Amsterdam University Press, 1999), chap. 1; Eric Schatzberg, "From Knowledge to Object: The Contested Meanings of Technology" (presentation to the EHESS Seminar, 12 December 2001).

3. I discussed four definitions of *technology* in my essay "Theories of Technological Change: Parameters and Purposes," *Science, Technology and Human Values* 17 (1992): 3–12. I discussed how the meaning of the word has shifted over time as well as the liabilities of essentialist definitions of the term in "The Compelling Tangle of Modernity and Technology," in Thomas Misa et al., eds., *Modernity and Technology* (Cambridge: MIT Press, 2003), 1–30.

4. Cheryl R. Ganz, "Science Advancing Mankind," *Technology and Culture* 41 (2000): 783–87.

5. Richard A. Goldthwaite, *The Building of Renaissance Florence* (Baltimore: Johns Hopkins University Press, 1980), quote on 425.

6. See Amartya Sen, *Development as Freedom* (New York: Knopf, 1999); and Haider Khan, "Technology, Modernity, and Development," in Misa et al., *Modernity and Technology*, 327–57.

7. Historical studies that cover a long span of time often are the best means of seeing clearly how technical choices entail cultural forms and social developments; see, for example, Otto Mayr, *Authority, Liberty, and Automatic Machinery in Early Modern Europe* (Baltimore: Johns Hopkins University Press, 1986); Cecil O. Smith, Jr., "The Longest Run: Public Engineers and Planning in France," *American Historical Review* 95 (1990): 657–92; Francesca Bray, *Technology and Gender: Fabrics of Power in Late Imperial China* (Berkeley: University of California Press, 1997); and Noel Perrin, *Giving up the Gun: Japan's Reversion to the Sword, 1543–1879* (Boston: D. R. Godine, 1979). Notable successes in dealing with the "paths not taken" include David Noble, *Forces of Production: A Social History of Industrial Automation* (New York: Knopf, 1984); Ken Alder, *Engineering the Revolution: Arms and Enlightenment in France, 1763–1815* (Princeton: Princeton

University Press, 1997); and Eric Schatzberg, *Wings of Wood, Wings of Metal: Culture and Technical Choice in American Airplane Materials, 1914–1945* (Princeton: Princeton University Press, 1999).

8. See the discussion of economics and technical change in Thomas J. Misa, *A Nation of Steel* (Baltimore: Johns Hopkins University Press, 1995), 262–82.

9. Winfried Baumgart, *Imperialism: The Idea and Reality of British and French Colonial Expansion, 1880–1914* (New York: Oxford University Press, 1982), quotes on 21.

10. A. P. Thornton, *The Imperial Idea and Its Enemies* (New York: St. Martin's Press, 1966), 54.

11. During the Renaissance, Alberti wrote extensively about how difficult and time consuming it was to become and remain a player in court culture, with one's mind always on how to please one's courtly patron; he did not advocate mixing in with the commoners of the time. During the industrial era, there were economic and legal barriers to technology. Arkwright restricted access to his patented spinning technology through licensing agreements stipulating access only by large, 1,000-spindle mills. Secrecy concerns pervaded the military era. A U.S. Navy security officer once told me, with a twinkle in his eye, "We have five levels of security: classified, secret, top secret, and two others that are so secret that I can't even tell you their names."

12. On "technology dialogue" see Arnold Pacey, *Technology in World Civilization* (Cambridge: MIT Press, 1990). Coercive state structures have frequently directed technical innovation down certain paths, but they have rarely been flexible in changing technical paths. The Soviet Union's success with early (Sputnik-era) rocket technology, contrasted with its complete failure with microelectronic technologies, provides a case in point.

13. Henry Etzkowitz, "Solar versus Nuclear Energy: Autonomous or Dependent Technology?" *Social Problems* 31 (April 1984): 417–34.

14. Peter Harnetty, *Imperialism and Free Trade: Lancashire and India in the Mid-Nineteenth Century* (Vancouver: University of British Columbia Press, 1972), 6.

15. Karl Marx, *Capital* (vol. 1, chap. 15), quote at www.marxists.org/archive/marx/works/1867-c1/ch15.htm#a128 (10 March 2002).

16. Goldthwaite, *The Building of Renaissance Florence*, quote on 84.

17. Cris Shore, *Building Europe: The Cultural Politics of European Integration* (London: Routledge, 2000).

18. The phenomenon of so-called oppositional agency is discussed in Andrew Feenberg, *Questioning Technology* (London: Routledge, 1999), chap. 5. An excellent example is discussed in Susan J. Douglas, "Oppositional Uses of Technology and Corporate Competition," in William Aspray, ed., *Technological Competitiveness* (New York: IEEE Press, 1993), 208–24.

19. Arthur P. J. Mol, "The Environmental Transformation of the Modern Order," in Misa et al., *Modernity and Technology*, 303–26.

20. Bharati Ray, "The Genesis of Railway Development in Hyderabad State: A Case Study in Nineteenth-Century British Imperialism," *Indian Economic and Social History Review* 21 (1984): 45–69, quote on 54.

21. The point is discussed in my essay "Retrieving Sociotechnical Change from Technological Determinism," in Merritt Roe Smith and Leo Marx, eds., *Does Technology Drive History? The Dilemma of Technological Determinism* (Cambridge: MIT Press, 1994), 115–41.

22. As an example of the difficulties in generalizing about the cultures of the Middle East, test the generalizations about Islam made by Bernard Lewis in *What Went Wrong?: Western Impact and Middle Eastern Response* (Oxford: Oxford University Press, 2002) against the strident criticisms of Lewis made by Edward Said in *Orientalism* (New York: Vintage, 1979), 314–21, and in *Culture and Imperialism* (New York: Vintage, 1994), 37–38.

23. See the Human Rights Watch report from 1999, "The Internet in the Mideast and North Africa: Free Expression and Censorship," www.hrw.org/advocacy/internet/mena/, and the international map of press restrictions at www.freedomhouse.org/pfs2000/pfsmap2k.pdf.

24. Michael Adas, *Machines as the Measure of Men* (Ithaca: Cornell University Press, 1989).

25. For instances of this new approach to technology, advocating a cautious embrace of it, see the essays by Andrew Feenberg (on philosophy of technology), Haider Khan (on development policies), and Arthur Mol (on environmental thinking) in Misa et al., eds., *Modernity and Technology* (Cambridge: MIT Press, 2003). Engineering education is undergoing a sea change in accreditation standards; see criterion 3 in www.abet.org/images/Criteria/eac_criteria_b.pdf and NSF's engineering reform efforts at www.eng.nsf.gov/eec/. See also Robert Pool's concluding chapter in his *Beyond Engineering* (Oxford: Oxford University Press, 1997).

Notes on Sources

LISTED HERE are those works that I have drawn on repeatedly, on matters of fact and of interpretation. To avoid cluttering the text, I have generally created source notes only for direct quotes, numerical data, or particular points. The authors below are the crucial bedrock sources.

Scholarship on Renaissance technology once began with Bertrand Gille's classic *Engineers of the Renaissance* (Cambridge: MIT Press, 1966), but today the first stop, for Leonardo at least, is Paolo Galluzzi's "The Career of a Technologist," in *Leonardo da Vinci: Engineer and Architect* (Montreal Museum of Fine Arts, 1987), published in an indispensable volume edited by Galluzzi that brings together recent scholarship and magnificent illustrations. The consequences for engineering of Renaissance-era perspective informs Eugene Ferguson's *Engineering and the Mind's Eye* (Cambridge: MIT Press, 1992). An extensive bibliography is at galileo .imss.fi.it/news/mostra/ebibli.html (13 June 2002). A. Richard Turner's *Inventing Leonardo* (New York: A. A. Knopf, 1993) looks critically at Leonardo myth making.

A neat synthesis of recent fine-arts scholarship is Peter Burke's *The European Renaissance: Centres and Peripheries* (Oxford: Blackwell, 1998), while Jerry Brotton's *The Renaissance Bazaar: From the Silk Road to Michelangelo* (Oxford: Oxford University Press, 2002) takes a global view of European developments. More pertinent to my concerns were Mary Hollingsworth's *Patronage in Renaissance Italy* (Baltimore: Johns Hopkins University Press, 1994); David Mateer's *Courts, Patrons, and Poets* (New Haven: Yale University Press, 2000); and Stella Fletcher's *The Longman Companion to Renaissance Europe, 1390–1530* (Harlow, England: Longman, 2000). On Alberti as a court-based scholar, architect, and engineer see Anthony Grafton's *Leon Battista Alberti* (New York: Hill & Wang, 2000). For detail on Alberti's technical activities, see the classic study by Joan Gadol, *Leon Battista Alberti* (Chicago: University of Chicago Press, 1969).

Pamela Long's *Openness, Secrecy, Authorship* (Baltimore: Johns Hopkins University Press, 2001) surveys technological knowledge since antiquity, with several chapters dealing with authorship and patronage during the Renaissance. Bert Hall's *Weapons and Warfare in Renaissance Europe* (Baltimore: Johns Hopkins University Press, 1997) corrects many mistaken impressions about the "gunpowder revolution." For specific topics, see Richard A. Goldthwaite's *The Building of Renaissance Florence* (Baltimore: Johns Hopkins University Press, 1980) and also his *Private Wealth in Renaissance Florence* (Princeton: Princeton University Press, 1968), Luca Molà's *The Silk Industry of Renaissance Venice* (Baltimore: Johns Hopkins University Press, 2000), and W. Patrick McCray's *Glassmaking in Renaissance Venice* (Aldershot, England: Ashgate, 1999). For shipbuilding see Frederic Chapin Lane's classic *Venetian Ships and Shipbuilders of the Renaissance* (Baltimore: Johns Hopkins University Press, 1992; originally published 1934); and Robert C. Davis's social history *Shipbuilders of the Venetian Arsenal* (Baltimore: Johns Hopkins University Press, 1991).

For Gutenberg see Albert Kapr's definitive *Johann Gutenberg: The Man and His Invention* (Aldershot, England: Scolar Press, 1996), supplemented with Elizabeth L. Eisenstein's two-volume study *The Printing Press as an Agent of Change* (Cambridge: Cambridge University Press, 1979) or her one-volume illustrated *The Printing Revolution in Early Modern Europe* (Cambridge: Cambridge University Press, 1983).

THE KEY SOURCE for Dutch commerce in the seventeenth century is Jonathan Israel's *Dutch Primacy in World Trade, 1585–1740* (Oxford: Clarendon Press, 1989), at once a masterful synthesis of the literature and a goldmine of data. Israel provides a detailed view of Dutch culture and politics in the 1,200 pages of *The Dutch Republic: Its Rise, Greatness, and Fall, 1477–1806* (Oxford: Clarendon Press, 1995). A portrait of Dutch society from rich to poor is Paul Zumthor, *Daily Life in Rembrandt's Holland* (Stanford: Stanford University Press, 1994).

Other crucial sources include Karel Davids and Jan Lucassen, eds., *A Miracle Mirrored: The Dutch Republic in European Perspective* (Cambridge: Cambridge University Press, 1995), Karel Davids and Leo Noordegraaf, eds., *The Dutch Economy in the Golden Age* (Amsterdam: Netherlands Economic History Archives, 1993); and Joel Mokyr, *Industrialization in the Low Countries, 1795–1850* (New Haven: Yale University Press, 1976). Compare Richard W. Unger, *Dutch Shipbuilding before 1800* (Assen: Van

Gorcum, 1978) with Carla Rahn Phillips, *Six Galleons for the King of Spain: Imperial Defense in the Early Seventeenth Century* (Baltimore: Johns Hopkins University Press, 1986). For the Dutch tulip mania, enliven the somber evaluation of Peter M. Garber, *Famous First Bubbles: The Fundamentals of Early Manias* (Cambridge: MIT Press, 2000) with Mike Dash, *Tulipomania* (New York: Crown, 1999).

FOR USEFUL ENTREES to a huge historical literature, see Maxine Berg and Pat Hudson, "Rehabilitating the Industrial Revolution," *Economic History Review* 45 (1992): 24–50; and Joel Mokyr's *The British Industrial Revolution: An Economic Perspective* (Boulder: Westview, 1993), pp. 1–131. For a wider view, see Peter Stearns, *The Industrial Revolution in World History* (Boulder: Westview, 1993). For a history of the concept "industrial revolution" see David S. Landes, "The Fable of the Dead Horse; or, The Industrial Revolution Revisited," in Mokyr, *The British Industrial Revolution*, pp. 132–70; Mikulas Teich and Roy Porter, eds., *The Industrial Revolution in National Context* (Cambridge: Cambridge University Press, 1996), pp. 1–12. Ivy Pinchbeck, *Women Workers and the Industrial Revolution, 1750–1850* (London: Frank Cass, 1930) is an indispensable source for many recent writers.

On London, begin with M. Dorothy George's classic *London Life in the Eighteenth Century* (New York: Capricorn Books, 1965; originally published 1925). Valuable social histories include David R. Green, *From Artisans to Paupers: Economic Change and Poverty in London, 1790–1870* (London: Scolar Press, 1995); L. D. Schwarz, *London in the Age of Industrialisation: Entrepreneurs, Labour Force and Living Conditions, 1700–1850* (Cambridge: Cambridge University Press, 1992); and Roy Porter's entertaining *London: A Social History* (Cambridge: Harvard University Press, 1994). David Barnett, *London: Hub of the Industrial Revolution: A Revisionary History, 1775–1825* (London: Tauris, 1998) analyzes firm-level data.

The literature on Manchester is voluminous, if uneven. I read widely in the following works: Sidney J. Chapman, *The Lancashire Cotton Industry* (Manchester: Manchester University Press, 1904); Martin Hewitt, *The Emergence of Stability in the Industrial City: Manchester, 1832–67* (Aldershot, England: Scolar Press, 1996); Geoffrey Timmons, *The Last Shift* (Manchester: Manchester University Press, 1993; reprint, Clifton, N.J.: A. M. Kelley, 1973); Eric J. Evans, *The Forging of the Modern State: Early Industrial Britain, 1783–1870* (London: Longman, 1983); and Steven

Marcus, *Engels, Manchester, and the Working Class* (New York: Random House, 1974). The best study of the city's business structure is Roger Lloyd-Jones and M. J. Lewis, *Manchester and the Age of the Factory* (London: Croom Helm, 1988).

The Sheffield section draws on David Hey, *The Fiery Blades of Hallamshire: Sheffield and Its Neighbors, 1660–1740* (Leicester: Leicester University Press, 1991); Sidney Pollard, *A History of Labour in Sheffield* (Liverpool: Liverpool University Press, 1959); Geoffrey Tweedale, *Steel City: Entrepreneurship, Strategy and Technology in Sheffield, 1743–1993* (Oxford: Clarendon Press, 1995); Tweedale, *Sheffield Steel and America: A Century of Commercial and Technological Interdependence, 1830–1930* (Cambridge: Cambridge University Press, 1987); and K. C. Barraclough, *Sheffield Steel* (Ashbourne, England: Moorland, 1976).

For entrées to the large literature on "alternative paths" to industrialization, see Charles F. Sabel and Jonathan Zeitlin, eds., *World of Possibilities: Flexibility and Mass Production in Western Industrialization* (Cambridge: Cambridge University Press, 1997); Philip Scranton, *Endless Novelty: Specialty Production and American Industrialization, 1865–1925* (Princeton: Princeton University Press, 1997).

FOR IMPERIALISM and technology, start with Daniel Headrick's several volumes: *The Tools of Empire: Technology and European Imperialism in the Nineteenth Century* (Oxford: Oxford University Press, 1981); *The Tentacles of Progress: Technology Transfer in the Age of Imperialism, 1850–1940* (Oxford: Oxford University Press, 1988); and *The Invisible Weapon: Telecommunications and International Politics, 1851–1945* (Oxford: Oxford University Press, 1991). For the economics of British imperialism see Lance E. Davis and Robert A. Huttenback, *Mammon and the Pursuit of Empire: The Economics of British Imperialism* (Cambridge: Cambridge University Press, 1988) as well as numerous discussion articles in *Past and Present, Economic History Review, The Economic Journal,* and *The Historical Journal.* A recent synthesis is P. J. Cain and A. G. Hopkins, *British Imperialism, 1688–2000,* 2nd edition (Harlow, England: Longman, 2002).

On technology and British imperialism, see Daniel Thorner, *Investment in Empire: British Railway and Steam Shipping Enterprise in India, 1825–1849* (Philadelphia: University of Pennsylvania Press, 1950; Arno reprint 1977); Roy MacLeod and Deepak Kumar, eds., *Technology and the Raj: Western Technology and Technical Transfers to India, 1700–1947* (New Delhi: Sage, 1995); Zaheer Baber, *The Science of Empire: Scientific Knowl-*

edge, Civilization, and Colonial Rule in India (Albany: SUNY Press, 1996); Michael Adas, *Machines as the Measure of Men* (Ithaca: Cornell University Press, 1989); Michael Adas, ed., *Technology and European Overseas Enterprise* (Aldershot, England: Variorum, 1996). There are many valuable case studies in *Indian Economic and Social History Review*. For new research on native Indian industry, see Tirthankar Roy, *Artisans and Industrialisation: Indian Weaving in the Twentieth Century* (New Delhi: Oxford University Press, 1993); and Roy, *Traditional Industry in the Economy of Colonial India* (Cambridge: Cambridge University Press, 1999).

On "railway imperialism," see the indispensable Clarence B. Davis and Kenneth E. Wilburn, Jr., eds., *Railway Imperialism* (Boulder: Greenwood, 1991) and John H. Coatsworth, *Growth Against Development: The Economic Impact of Railroads in Porfirian Mexico* (De Kalb: Northern Illinois Press, 1981). Detail on the Indian railways can be found in Ian J. Kerr, *Building the Railways of the Raj, 1850–1900* (New Delhi: Oxford University Press, 1995) and his anthology *Railways in Modern India* (New Delhi: Oxford University Press, 2001).

THE SECTION on German chemistry draws on John Joseph Beer, *The Emergence of the German Dye Industry* (Urbana: University of Illinois Press, 1959); Jeffrey Allan Johnson, *The Kaiser's Chemists: Science and Modernization in Imperial Germany* (Chapel Hill: University of North Carolina Press, 1990); Henk van den Belt and Arie Rip, "The Nelson-Winter-Dosi Model and Synthetic Dye Chemistry," in W. E. Bijker et al., eds., *The Social Construction of Technological Systems* (Cambridge: MIT Press, 1987), pp. 135–58; J. B. Morrell, "The Chemist Breeders: The Research Schools of Liebig and Thomas Thomson," *Ambix* 19, no. 1 (March 1972): 1–46; R. Steven Turner, "Justus Liebig versus Prussian Chemistry: Reflections on Early Institute-building in Germany," *HSPS* (*Historical Studies in the Physical Sciences*) 13, no. 1 (1982): 129–62; Anthony S. Travis, *The Rainbow Makers: The Origins of the Synthetic Dyestuffs Industry in Western Europe* (Bethlehem: Lehigh University Press, 1993); L. F. Haber, *The Poisonous Cloud: Chemical Warfare in the First World War* (Oxford: Clarendon Press, 1986); and Peter Hayes, *Industry and Ideology: IG Farben in the Nazi Era* (Cambridge: Cambridge University Press, 1987).

Paul Israel's *Edison: An Inventive Life* (New York: Wiley, 1998) supersedes earlier popular biographies by Josephson, Conot, Clark, and Harris. Additional sources include Reese V. Jenkins et al., eds., *The Papers of Thomas A. Edison* (Baltimore: Johns Hopkins University Press, 1989 et seq.); Paul Israel, *From Machine Shop to Industrial Laboratory: Telegraphy and the Changing Context of American Invention, 1830–1920* (Baltimore: Johns Hopkins University Press, 1992); Thomas P. Hughes, *Networks of Power: Electrification in Western Society, 1880–1930* (Baltimore: Johns Hopkins University Press, 1983); Thomas P. Hughes, *American Genesis* (New York: Viking, 1989); and Robert Friedel and Paul Israel, *Edison's Electric Light: Biography of an Invention* (New Brunswick: Rutgers University Press, 1986).

On corporate restructuring in the United States, see Louis Galambos and Joseph Pratt, *The Rise of the Corporate Commonwealth* (New York: Basic Books, 1988); W. Bernard Carlson, *Innovation as a Social Process: Elihu Thomson and the Rise of General Electric, 1870–1900* (Cambridge: Cambridge University Press, 1991); Alfred D. Chandler, Jr., *The Visible Hand: The Managerial Revolution in American Business* (Cambridge, Mass.: Belknap Press, 1977); and Naomi R. Lamoreaux, *The Great Merger Movement in American Business, 1895–1904* (Cambridge: Cambridge University Press, 1985).

Terry Smith, *Making the Modern: Industry, Art, and Design in America* (Chicago: University of Chicago Press, 1993); and Thomas P. Hughes, *American Genesis* (New York: Viking, 1989) are indispensable sources for modernism in art, architecture, and technology. The essential source on Italian Futurism is Pontus Hulten, ed., *Futurismo e Futurismi* (Milan: Bompiani, 1986), a brilliant exhibition catalogue that includes an invaluable 200-page "dictionary" with biographies and manifestos. For the section on de Stijl, I have drawn on Carsten-Peter Warncke, *De Stijl, 1917–31* (Cologne: Taschen, 1998) and Joost Baljeu, *Theo van Doesburg* (London: Studio Vista, 1974). On the Bauhaus I used Reyner Banham, *Theory and Design in the First Machine Age* (Cambridge: MIT Press, 1980) and John Willett, *The New Sobriety, 1917–1933: Art and Politics in the Weimar Period* (London: Thames & Hudson, 1978).

Sources on household rationalization and the Frankfurt building program include Peter Noever, ed., *Margarete Schütte-Lihotzky: Soziale Architektur Zeitzeugin eines Jahrhunderts* (Vienna: Böhlau Verlag, 1996); Peter Noever, ed., *Die Frankfurter Küche* (Berlin: Ernst & Sohn, n.d.); as well as relevant chapters in Mary Nolan, *Visions of Modernity: American Business and the Modernization of Germany* (New York: Oxford University Press, 1994); Angela Oedekoven-Gerischer et al., eds., *Frauen im Design: Berufsbilder und Lebenswege seit 1900*, vol. 1 (Stuttgart: Design Center Stuttgart, 1989); Susan Strasser, Charles McGovern, and Matthias Judt, eds., *Getting and Spending: European and American Consumer Societies in the Twentieth Century* (Cambridge: Cambridge University Press, 1998); and Debra Coleman, Elizabeth Danze, and Carol Henderson, eds., *Architecture and Feminism* (New York: Princeton Architectural Press, 1996).

On "modern" materials, see Pearce Davis, *The Development of the American Glass Industry* (Cambridge: Harvard University Press, 1949; reprint, New York: Russell & Russell, 1970; Harvard Economic Studies, vol. 86); Richard John O'Connor, "Cinderheads and Iron Lungs: Windowglass Craftsmen and the Transformation of Workers' Control, 1880–1905" (Ph.D. diss., University of Pittsburgh, 1991); and Thomas J. Misa, *A Nation of Steel* (Baltimore: Johns Hopkins University Press, 1995).

ON THE GERMAN MILITARY, see William Carr, *Arms, Autarky and Aggression: A Study in German Foreign Policy, 1933–1939* (New York: W. W. Norton, 1972); Alan Milward, *War, Economy, and Society, 1939–1945* (Berkeley: University of California Press, 1977); Mark Walker, *German National Socialism and the Quest for Nuclear Power, 1939–1949* (Cambridge: Cambridge University Press, 1989).

Of the many works on the Manhattan Project see Richard G. Hewlett and Oscar E. Anderson, Jr., *The New World, 1939–1946*, vol. 1 of *A History of the United States Atomic Energy Commission* (University Park: Pennsylvania State University Press, 1962); Richard Rhodes, *The Making of the Atomic Bomb* (New York: Simon & Schuster, 1986); and Barton C. Hacker, *The Dragon's Tail: Radiation Safety in the Manhattan Project, 1942–1946* (Berkeley: University of California Press, 1987).

On postwar computing and electronics, see Stuart W. Leslie, *The Cold War and American Science* (New York: Columbia University Press, 1993); Kenneth Flamm, *Creating the Computer* (Washington, D.C.: Brookings Institution, 1988); Thomas J. Misa, "Military Needs, Commercial Realities, and the Development of the Transistor, 1948–1958," in *Military Enterprise and Technological Change*, ed. Merritt Roe Smith (Cambridge: MIT Press, 1985), pp. 253–87; Arthur L. Norberg and Judy E. O'Neill, *Transforming Computer Technology: Information Processing for the Pentagon, 1962–1986* (Baltimore: Johns Hopkins University Press, 1996); Kent C. Raymond and Thomas M. Smith, *Project Whirlwind: The History of a Pioneer Computer* (Bedford, Mass.: Digital Press, 1980); Paul N. Edwards, *The Closed World: Computers and the Politics of Discourse in Cold War America* (Cambridge: MIT Press, 1996).

WRITINGS ON GLOBALIZATION vary widely in perspective and quality. Contrast Thomas L. Friedman's upbeat view in *The Lexus and the Olive Tree* (New York: Farrar, Straus & Giroux, 1999) with the selections in Jerry Mander and Edward Goldsmith, eds., *The Case Against the Global Economy* (San Francisco: Sierra Club Books, 1996). The best wide-ranging collection I have found is *Globalization and the Challenges of a New Century,* edited by Patrick O'Meara et al. (Bloomington: Indiana University Press, 2000). For a Marxist analysis, see Michael Hardt and Antonio Negri, *Empire* (Cambridge: Harvard University Press, 2000). The *Economist* magazine's pro-globalization perspective, with an international outlook, finds its way into John Micklethwait and Adrian Wooldridge, *A Future Perfect: The Essentials of Globalization* (New York: Crown Business, 2000).

Views on the antiglobalization protests in Seattle and beyond are sharply divided: for the view from the "inside" see Jeffrey J. Schott, ed., *The WTO After Seattle* (Washington: Institute for International Economics, 2000); contrast that with the view from the "streets" in Alexander Cockburn and Jeffrey St. Clair, *Five Days that Shook the World: Seattle and Beyond* (London: Verso, 2000); Janet Thomas, *The Battle in Seattle: The Story Behind and Beyond the WTO Demonstrations* (Golden, Colo.: Fulcrum Publishing, 2000). Compare Susan A. Aaronson, *Taking Trade to the Streets: The Lost History of Public Efforts to Shape Globalization* (Ann Arbor: University of Michigan Press, 2001).

The section on fax machines draws on the work of Jonathan Coopersmith, who is writing a global history of facsimile transmission. See his "The Failure of Fax: When Vision Is Not Enough," *Business and Economic History* 23, no. 1 (1994): 272–82; and "Facsimile's False Starts," *IEEE Spectrum* 30 (February 1993): 46–49. I have also relied on Susanne K. Schmidt and Raymund Werle, *Coordinating Technology: Studies in the International Standardization of Telecommunications* (Cambridge: MIT Press, 1998).

Sources on McDonald's include John F. Love, *McDonald's: Behind the Arches* (New York: Bantam, 1995; originally published 1986); James L. Watson, ed., *Golden Arches East: McDonald's in East Asia* (Stanford: Stanford University Press, 1997); Robin Leidner, *Fast Food, Fast Talk: Service Work and the Routinization of Everyday Life* (Berkeley: University of California Press, 1993); and John Vidal, *McLibel: Burger Culture on Trial* (New York: New Press, 1997). For insider accounts see Ray Kroc, *Grinding It Out: The Making of McDonald's* (Chicago: Henry Regnery, 1977); and George Cohon's lively *To Russia with Fries* (Toronto: McClelland & Stewart, 1999).

On the Internet and World Wide Web, compare Katie Hafner and

Matthew Lyon's journalistic *Where Wizards Stay Up Late: The Origins of the Internet* (New York: Simon & Schuster, 1996) with Janet Abbate's scholarly *Inventing the Internet* (Cambridge: MIT Press, 1999). On computer networks see Urs von Burg, *The Triumph of the Ethernet: Technological Communities and the Battle for the LAN Standard* (Stanford: Stanford University Press, 2001). On Internet privatization, see Jay P. Kesan and Rajiv C. Shah, "Fool Us Once Shame on You—Fool Us Twice Shame on Us: What We Can Learn from the Privatizations of the Internet Backbone Network and the Domain Name System," *Washington University Law Quarterly* 79 (2001): 89–220. For the World Wide Web one can start with Tim Berners-Lee's popular book *Weaving the Web* (New York: HarperBusiness, 1999), but I found the WWW documents cited in my endnotes more revealing of his design philosophy and the Web's technical details.

INDEX